Lecture Notes in Computer Science 10914

Commenced Publication in 1973
Founding and Former Series Editors:
Gerhard Goos, Juris Hartmanis, and Jan van Leeuwen

More information about this series at http://www.springer.com/series/7409

Gabriele Meiselwitz (Ed.)

Social Computing and Social Media

Technologies and Analytics

10th International Conference, SCSM 2018
Held as Part of HCI International 2018
Las Vegas, NV, USA, July 15–20, 2018
Proceedings, Part II

 Springer

Editor
Gabriele Meiselwitz
Department of Computer and Information
  Sciences
Towson University
Towson, MD
USA

ISSN 0302-9743   ISSN 1611-3349 (electronic)
Lecture Notes in Computer Science
ISBN 978-3-319-91484-8   ISBN 978-3-319-91485-5 (eBook)
https://doi.org/10.1007/978-3-319-91485-5

Library of Congress Control Number: 2018944281

LNCS Sublibrary: SL3 – Information Systems and Applications, incl. Internet/Web, and HCI

Printed on acid-free paper

This Springer imprint is published by the registered company Springer International Publishing AG
part of Springer Nature
The registered company address is: Gewerbestrasse 11, 6330 Cham, Switzerland

Foreword

The 20th International Conference on Human-Computer Interaction, HCI International 2018, was held in Las Vegas, NV, USA, during July 15–20, 2018. The event incorporated the 14 conferences/thematic areas listed on the following page.

A total of 4,373 individuals from academia, research institutes, industry, and governmental agencies from 76 countries submitted contributions, and 1,170 papers and 195 posters have been included in the proceedings. These contributions address the latest research and development efforts and highlight the human aspects of design and use of computing systems. The contributions thoroughly cover the entire field of human-computer interaction, addressing major advances in knowledge and effective use of computers in a variety of application areas. The volumes constituting the full set of the conference proceedings are listed in the following pages.

I would like to thank the program board chairs and the members of the program boards of all thematic areas and affiliated conferences for their contribution to the highest scientific quality and the overall success of the HCI International 2018 conference.

This conference would not have been possible without the continuous and unwavering support and advice of the founder, Conference General Chair Emeritus and Conference Scientific Advisor Prof. Gavriel Salvendy. For his outstanding efforts, I would like to express my appreciation to the communications chair and editor of *HCI International News*, Dr. Abbas Moallem.

July 2018

Constantine Stephanidis

HCI International 2018 Thematic Areas
and Affiliated Conferences

Thematic areas:

- Human-Computer Interaction (HCI 2018)
- Human Interface and the Management of Information (HIMI 2018)

Affiliated conferences:

- 15th International Conference on Engineering Psychology and Cognitive Ergonomics (EPCE 2018)
- 12th International Conference on Universal Access in Human-Computer Interaction (UAHCI 2018)
- 10th International Conference on Virtual, Augmented, and Mixed Reality (VAMR 2018)
- 10th International Conference on Cross-Cultural Design (CCD 2018)
- 10th International Conference on Social Computing and Social Media (SCSM 2018)
- 12th International Conference on Augmented Cognition (AC 2018)
- 9th International Conference on Digital Human Modeling and Applications in Health, Safety, Ergonomics, and Risk Management (DHM 2018)
- 7th International Conference on Design, User Experience, and Usability (DUXU 2018)
- 6th International Conference on Distributed, Ambient, and Pervasive Interactions (DAPI 2018)
- 5th International Conference on HCI in Business, Government, and Organizations (HCIBGO)
- 5th International Conference on Learning and Collaboration Technologies (LCT 2018)
- 4th International Conference on Human Aspects of IT for the Aged Population (ITAP 2018)

Conference Proceedings Volumes Full List

1. LNCS 10901, Human-Computer Interaction: Theories, Methods, and Human Issues (Part I), edited by Masaaki Kurosu
2. LNCS 10902, Human-Computer Interaction: Interaction in Context (Part II), edited by Masaaki Kurosu
3. LNCS 10903, Human-Computer Interaction: Interaction Technologies (Part III), edited by Masaaki Kurosu
4. LNCS 10904, Human Interface and the Management of Information: Interaction, Visualization, and Analytics (Part I), edited by Sakae Yamamoto and Hirohiko Mori
5. LNCS 10905, Human Interface and the Management of Information: Information in Applications and Services (Part II), edited by Sakae Yamamoto and Hirohiko Mori
6. LNAI 10906, Engineering Psychology and Cognitive Ergonomics, edited by Don Harris
7. LNCS 10907, Universal Access in Human-Computer Interaction: Methods, Technologies, and Users (Part I), edited by Margherita Antona and Constantine Stephanidis
8. LNCS 10908, Universal Access in Human-Computer Interaction: Virtual, Augmented, and Intelligent Environments (Part II), edited by Margherita Antona and Constantine Stephanidis
9. LNCS 10909, Virtual, Augmented and Mixed Reality: Interaction, Navigation, Visualization, Embodiment, and Simulation (Part I), edited by Jessie Y. C. Chen and Gino Fragomeni
10. LNCS 10910, Virtual, Augmented and Mixed Reality: Applications in Health, Cultural Heritage, and Industry (Part II), edited by Jessie Y. C. Chen and Gino Fragomeni
11. LNCS 10911, Cross-Cultural Design: Methods, Tools, and Users (Part I), edited by Pei-Luen Patrick Rau
12. LNCS 10912, Cross-Cultural Design: Applications in Cultural Heritage, Creativity, and Social Development (Part II), edited by Pei-Luen Patrick Rau
13. LNCS 10913, Social Computing and Social Media: User Experience and Behavior (Part I), edited by Gabriele Meiselwitz
14. LNCS 10914, Social Computing and Social Media: Technologies and Analytics (Part II), edited by Gabriele Meiselwitz
15. LNAI 10915, Augmented Cognition: Intelligent Technologies (Part I), edited by Dylan D. Schmorrow and Cali M. Fidopiastis
16. LNAI 10916, Augmented Cognition: Users and Contexts (Part II), edited by Dylan D. Schmorrow and Cali M. Fidopiastis
17. LNCS 10917, Digital Human Modeling and Applications in Health, Safety, Ergonomics, and Risk Management, edited by Vincent G. Duffy
18. LNCS 10918, Design, User Experience, and Usability: Theory and Practice (Part I), edited by Aaron Marcus and Wentao Wang

http://2018.hci.international/proceedings

10th International Conference on Social Computing and Social Media

Program Board Chair(s): **Gabriele Meiselwitz**, *USA*

- James Braman, USA
- Cristóbal Fernández Robin, Chile
- Nick V. Flor, USA
- Panagiotis Germanakos, Germany
- Sara Anne Hook, USA
- Rushed Kanawati, France
- Carsten Kleiner, Germany
- Niki Lambropoulos, UK
- Marilia Mendes, Brazil
- Hoang Nguyen, Singapore
- Anthony Norcio, USA
- Michiko Ohkura, Japan
- Cristian Rusu, Chile
- Christian Scheiner, Germany
- Shubhi Shrivastava, USA
- Abraham Van der Vyver, South Africa
- Giovanni Vincenti, USA
- Jose Viterbo, Brazil
- Yuanqiong (Kathy) Wang, USA
- June Wei, USA
- Brian Wentz, USA

The full list with the Program Board Chairs and the members of the Program Boards of all thematic areas and affiliated conferences is available online at:

http://www.hci.international/board-members-2018.php

HCI International 2019

The 21st International Conference on Human-Computer Interaction, HCI International 2019, will be held jointly with the affiliated conferences in Orlando, FL, USA, at Walt Disney World Swan and Dolphin Resort, July 26–31, 2019. It will cover a broad spectrum of themes related to Human-Computer Interaction, including theoretical issues, methods, tools, processes, and case studies in HCI design, as well as novel interaction techniques, interfaces, and applications. The proceedings will be published by Springer. More information will be available on the conference website: http://2019.hci.international/.

General Chair
Prof. Constantine Stephanidis
University of Crete and ICS-FORTH
Heraklion, Crete, Greece
E-mail: general_chair@hcii2019.org

http://2019.hci.international/

Contents – Part II

Social Network Analysis

Agents, Models and Algorithms in Social Media

Contents – Part I

Individual and Social Behavior in Social Media

Privacy and Ethical Issues in Social Media

Motivation and Gamification in Social Media

Exploring the Use of Social Media in Education from Learners Perspective

Yara A. AlHaidari[1,2(✉)]

[1] Princess Nourah University, Riyadh, Saudi Arabia
[2] Takamol Holding Company, Riyadh, Saudi Arabia
Y.alhaidari@Takamol.com.sa

Abstract. Social media has affected different fields of our lives, students are using and contributing to social media tools formally and informally, to benefit from networking, research, engagement and motivation. In this research, we explored the use of social media tools by students in term of formal and informal learning, inside and outside the learning management system (LMS). We designed an online survey distributed to higher education students in Saudi Arabia. The number of participants were (102), and (88) of them were female student, the rest were male students. The results show that social networking tools are the top consumed tools with (81%) of responses, on the other hand, document sharing tools are the top tool the participants contribute to with (58%) responses. There is a significant difference in the use of social media based on scope of use, inside and outside LMS, this difference is due to the nature of the tools, academic level and scope of interactions.

Keywords: Social media · Online learning · Learning management system

1 Introduction

1.1 Social Media in Education

Social media have effected different fields of our lives, education is one of the major fields that benefit from social media. The ubiquity of social media is no more apparent than at the higher education (Tess 2013). Based on previous literature, research has shown that students use social media tools for general purpose more than personal or social purpose, and developing connections and professional networks are the primary usage of these tools. Also, research have shown that social media has a positive effect on academic engagement and satisfaction for both student and instructor (Han et al. 2016).

Lewis (2014) has conceptualize social online tools into three modes: absorbing information, sharing information and engaging with others. He concludes that social media is an essential mode of communication that can not be ignored. Also, he defined the roles of social media to include, but not limited to, a source of news, a platform for global conversation, a vehicle for promoting or disseminating content, and a hub for professional networking.

© Springer International Publishing AG, part of Springer Nature 2018
G. Meiselwitz (Ed.): SCSM 2018, LNCS 10914, pp. 3–11, 2018.
https://doi.org/10.1007/978-3-319-91485-5_1

1.2   Challenges

In term of exploring the use of social media in education, we have to take into consideration the challenges; in order to address them or even document them for further investigation. One of the challenges in exploring the role of social media in education is the recognizing user identity, and also, recognizing user identity amongst different platforms or applications. Absar et al. (2016) found that one of the critical issues in their research is identifying and tracking the same individual across multiple platforms.

Also, one of the challenges is the need of a learning environment that helps students to adapt to the massive development of knowledge in the 21st century. Educational institutions must provide students with dynamic learning mediums such as social media to meet the continuous change of people and environment, to be able to adapt and embrace this paradigm shift in education (Hirsch and Ng 2011).

1.3   Benefits for Learners

In addition to the benefits of online learning for students, which are primly in the flexibility of time or place, online learning also enables learners to complete their learning or training with variant learning mediums when they are employed (Anderson 2011).

Based on the constructivist theory, learner is the center of the learning process, and the instructor role is to facilitate learning process more than teaching. Knowledge has been defined as what learner process by senses, it is not something to receive from the outside. Anderson (2011) concludes that learning activities that help learner to contextualize knowledge have to be through online instruction.

We have to differentiate between different online categories; social media, user generated content and web 2.0. One of the classification of social media groups applications currently subsumed under the generalized term into more specific categories by characteristic: collaborative projects, blogs, content communities, social networking sites, virtual game worlds, and virtual social worlds (Kaplan and Haenlein 2010).

Faiz et al. (2013) have found that social media is an effective way to enhance students' engagements as it enables shy, intimidated or bored students to share ideas and to express their opinions in a more comfortable way.

1.4   Instructor Role

Jung and Lee (2016) showed an empirical evidence that social media tools can help instructors to enhance connection with the students, and this will contribute to improving student's academic performance. Social media tools show new ways of delivering knowledge to students. Students prefer to use social media as a channel to acquire new information, and instructors have new essential role to support their students in social media (Dneprovskaya et al. 2014).

Successful integration of social media in learning and teaching requires a thoughtful planning based on instructional design standards and best practices. There are different theoretical approaches that help on the integration of social media in education, Associative Learning theory, Cognitivist learning theory and Situative learning theory (Thota 2015). Educators use for social media have been counted as an application of connectivist learning theory (Friesen and Lowe 2012).

1.5 Formal and Informal Learning

Absar et al. (2016) have shown that learners are using various forms of social media to bridge the gap between in school and out of school learning, in both activity types, instructors led or learner initiated.

1.6 Academic and Research

Gandhi (2014) discussed four different social networking platform for researchers, ResearchGate[1] as an example of networking site, academia.edu as a network platform for academicians, and Graduate Junction is a networking resource for early stage researcher. Social media platforms also have supported research in term of bringing together researchers from different disciplines, Method Space is an example of a social network service for that.

Also, they discussed different tools for social bookmarking services (e.g. Delicious, Diigo[2] and BibSonomy[3]), social citation tools (e.g. CiteULike[4] and Zotero[5]), and blogging tools, e.g. Blogger[6], WordPress[7] and Twitter[8] for microblogging.

1.7 LMS and Social Media

Learning management system (LMS) have been counted as an institutional safe environment for learners that provide learning content through an interactive learning experience. Learners and instructors need to enrich their practices by using social media in education, Vickers et al. (2014) concluded on their experience that the existing Higher Education Learning Management Systems (LMS) or Virtual Learning Environment (VLE) are not fit for the socially networked, always connected today's learners. LMs provides effective tools for the management and administration of learning, but offer no enhancement to learning experience or the attainment of learning outcomes.

---

[1] https://www.researchgate.net/.

[2] https://www.diigo.com/.

[3] https://www.bibsonomy.org/.

[4] http://www.citeulike.org/.

[5] https://www.zotero.org/.

[6] https://www.blogger.com/.

[7] https://wordpress.com/.

[8] https://twitter.com/.

1.8 Cultural Influence

Jones et al. (2014) have conducted a cross cultural comparison in higher education institutions between the United States and Singapore, and they found that organizational use of social media is accompanied by internal changes in structure and processes. They found a significant evidence that different cultures have different influence or interpersonal, or relationship oriented work-related outcomes, such as negotiation, and trust.

1.9 Future Research

Piotrowski (2015) concludes that future research needs to focus on (a) optimizing the academic potential of modern interactive technology, (b) addressing solutions regarding drawbacks and limitations of interactive learning, and (c) engaging business school administrators to address these emerging educational challenges. Gandhi (2014) discussed the task of social media in research and academic, they found that social media can be used not only for the promotion of research but also for research development.

Researchers have presented social media tools into three categories, the first category is communication network, such as blogging, microblogging, location, social networking and aggregation. Second category is collaboration network tools, such a conferencing, wikis, social bookmarking, social bibliography, social news, asocial documents and project managements. Third category is multimedia networking tools, such as, photographs, video, presentation sharing, and virtual worlds.

2 Methodology

2.1 Research Questions

This research aims to investigate student's perspective about the main research questions (RQ):

- RQ1: What are the social media tools that student consume and contribute to?
- RQ2: What are the social media tools that student use inside and outside the learning management system (LMS)?

2.2 Research Design

The survey contains of demographic questions (gender, age, major), followed by (4) core questions to answer research main questions. The online survey was created by Google Forms[9], and distributed to higher education graduate and post graduate students in Saudi Arabia. The survey takes about (5) minutes to complete.

---

[9] https://www.google.com/forms/about/.

2.3   Data Collection

The survey was distributed to a sample of (102) higher education graduate and under-graduate students in Saudi Arabia. The total number of participants were (102) partici-pants, (88) of them were female, and (14) of them were male. The participants' age distribution is as the following: (58) of participants were 20 to 25 years old, (18) were more than 25 to 30 years old, (16) were more than 30 to 35 years old, (9) of them were more than 35 to 40 years old, and (1) of them was more than 40 years old.

3   Results

The First Research Question Was: What are the social media tools that student consume and contribute to? To answer this question, the participants have been requested to answer the following two sub questions:

- What are the tools you consume in education?
- What is the tools you contribute to?

The responses to the first sub question are represented in Table 1, the table gives a list of the response rate for each social media tool. It shows the most consumed social media tool in education was social networking with (81%) response rate. The second social media tool was document sharing with (66%) response rate. The third social media tool was presentation sharing with (46%) response rate. The least social media tool consumed was asynchronous discussion with (6) response rate.

Table 1.   What is the tools you consume in education?

Social media tool	Responses (%)
Social networking	81%
Document sharing	66%
Synchronous discussion	14%
Microblogging	20%
Academic social networking	30%
Blogs	32%
Asynchronous discussion	6%
Multimedia repositories	23%
Presentation sharing	46%
Wikis	20%
Academic bookmarking	15%
Virtual worlds	8%

The responses to the second sub question are represented in Table 2, the table gives a list of response rate for each social media tool. The most consuming social media tool in education was social networking with (81%) of responses. The second social media tool was document sharing with (66%) of responses. The third social media tool was presentation sharing with (46%) of responses. The least social media tool consuming was asynchronous discussion with (6%) of responses.

Table 2. What is the tools you contribute to education?

Social media tool	Responses (%)
Social networking	42%
Document sharing	58%
Synchronous discussion	23%
Microblogging	17%
Academic social networking	20%
Blogs	6%
Asynchronous discussion	7%
Multimedia repositories	18%
Presentation sharing	30%
Wikis	10%
Academic bookmarking	6%
Virtual worlds	6%

The Second Research Question Was: What are the social media tools that student use inside and outside the learning management system (LMS)? Table 3 shows the percentage of responses for the usage of each tool inside and outside LMS (Fig. 1).

Table 3. What are the social media tools that student use inside and outside the learning management system (LMS)?

Social media tool	Inside LMS	Outside LMS
Social networking	22%	77%
Document sharing	45%	23%
Synchronous discussion	22%	8%
Microblogging	12%	12%
Academic social networking	12%	12%
Blogs	12%	18%
Asynchronous discussion	10%	8%
Multimedia repositories	16%	14%
Presentation sharing	23%	12%
Wikis	10%	13%
Academic bookmarking	3%	6%
Virtual worlds	9%	4%

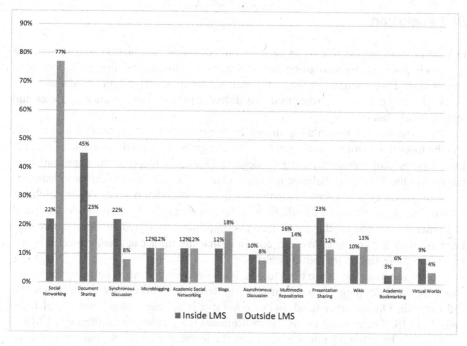

Fig. 1. The use of social media tools inside and outside LMS

4 Discussion

Research results have shown different uses of social media in education. Also, results emphasized on the importance of social media tools categories. In term of tools consuming by learners, the top consumed tools' category is social networking; this due to the importance of establishing and maintaining academic professional networking. The order of the rest categories is: document sharing, presentation sharing, blogs, academic social networking, microblogging, wikis, academic bookmarking, synchronous discussion, virtual words and asynchronous discussion. In term of tools learners contribute to, the top tools' category that learners contribute to is document sharing; due to the ease of use and the collaborative features provided by these tools.

The results also have shown a significant difference between the use of social media tools inside and outside learning management system. The most social media tools that have been used outside of LMS is social networking tools; due to the nature of this tool that depends on professionals and peers inside and outside of the academic institution. Learners mostly use document sharing tools inside LMS; due to the high level of privacy and institutional support. Also, LMS is considered as a safe learning environment with high privacy and security levels.

5 Conclusion

In term of studying the role of social media in education, we have to consider the use from two different aspects: using the tool as a consumer, and using the tool as a contributor. This research explored the integration of social media in education by investigating the Learner's role from two different perspectives, learner as a user and learner as a consumer.

This research has shown a significant difference between the tools that learner use, and the tools that learner contributes to. The difference is due to the purpose of use, the tool features and learner's academic degree. Learner's tendency to contribution is relevant to the learner's academic maturity. One of the primary indicators to learner's academic maturity is learner's academic degree (e.g. bachelor degree).The research has shown that the tendency to contribution is developed gradually with the development of learner's academic level. Learner tend to contribute in postgraduate levels more than learners in under graduate. We can conclude that the higher academic level a leaner reach, the more he/she tended to contribute.

One of the main issues we have to take into consideration when studying the use of social media is defining the scope of use. The basic definition of the scope is internal and external. On the other hand, the external use is the use of tools outside of institution's LMS. The internal use is the institutional use, inside the institution's LMS.

Also, we have to take into consideration the learning context, we have to be aware of the implementation context of the external use. It is important to differentiate between formal and informal uses. The formal use of social media tools is planned and monitored by the instructor and aligned with specific learning object/s. On the other hand, the informal use will be initiated and managed by the learner but still contribute to the achievement of learning objectives.

In term of the scope, this paper has shown benefits of the external use of social media tools. The external use provides students with wider interaction opportunities. Also, enrich learner's engagement with a global range of mentors and peers. The external use in more sustainable and open, since the internal user is limited to the academic study period.

This paper results related to the scope of use, provide an indicator to differentiate and study the formal and informal uses of social media, which the formal use means the planned use of the tool by the instructor as an essential or supporting element of the learning experience. The informal use means the student personal use which is not requested by the instructor as a part of curriculum.

References

Absar, R., Gruzd, A., Haythornthwaite, C., Paulin, D.: Linking online identities and content in connectivist MOOCs across multiple social media platforms. In: The 25th International Conference Companion on World Wide Web, pp. 483–488. Montréal, Québec (2016)

Anderson, T.: The Theory and Practice of Online Learning, 2nd edn. AU Press, Athabasca University (2011)

Dneprovskaya, N., Koretskaya, I., Dik, V., Tiukhmenova, K.: Study of social media implementation for transfer of knowledge within educational milieu. Nat. Min. Univ. **4**, 146–151 (2014)

Faizi, R., El Afia, A., Chiheb, R.: Exploring the potential benefits of using social media in education. Int. J. Eng. Pedagogy **3**(4), 50–53 (2013)

Friesen, N., Lowe, S.: The questionable promise of social media for education: connective learning and commercial imperative. J. Comput. Assist. Learn. **28**(3), 183–194 (2012)

Gandhi, S.: Incorporating social media in research and academics. Int. J. Inf. Sci. Comput. **8**(2) (2014)

Han, K., Volkova, S., Corley, C.: Understanding roles of social media in academic engagement and satisfaction for graduate students. In: The 2016 CHI Conference Extended Abstracts of Human Factors in Computing Systems, pp. 1215–1221. ACM, New York (2016)

Hirsch, B., Ng, J.W.P.: Education beyond the cloud: anytime-anywhere learning in a smart campus environment. In: The International Conference for Internet Technology and Secured Transactions. IEEE, Abu Dhabi (2011)

Jones, S., Lee, C., Welch, E., Eason, J.: Social media use in higher educational organizations: a comparison of the us and Singapore. In: The 5th ACM International Conference on Collaboration Across Boundaries: Culture, Distance and Technology, pp. 103–106. ACM, New York (2014)

Jung, S., Lee, S.: Developing a model for continuous user engagement in social media (2016)

Kaplan, M., Haenlein, M.: Users of the world, unite! The challenges and opportunities of Social Media. Bus. Horiz. **53**(1), 59–68 (2010)

Lewis, T.: Tweeting @ work: the use of social media in professional communication. Inf. Serv. Use **34**, 89–90 (2014)

Piotrowski, C.: Pedagogical Applications of social media in business education: student and faculty perspective. J. Educ. Technol. Syst. **43**(3), 257–265 (2015)

Tess, P.: The role of social media in higher education classes (real and virtual) – A literature review. Comput. Hum. Behav. **29**(5), A60–A68 (2013)

Thota, N.: Connectivism and the use of technology/media in collaborative teaching and learning. New Dir. Teach. Learn. **2015**(142), 81–96 (2015)

Vickers, R., Cooper, G., Field, J., Thayne, M., Adams, R., Lochrie, M.: Social media and collaborative learning: hello Scholar: In: The 18th International Academic Mind Trek Conference: Media Business, Management, Content and Services, New York, USA, pp. 103–109 (2014)

Development of Methods to Enhance Staff Members' Chats in Refresh Areas in Workplaces for Encouraging Their Knowledge Sharing

Hidenori Fujino[1][✉], Motoki Urayama[2], Takayoshi Kitamura[3],
Hirotake Ishii[2], Hiroshi Shimoda[2], Kyoko Izuka[1], Ryo Shimano[1],
Misato Tanemoto[1], Misaki Maeda[1], Manabu Goto[4],
and Masaki Kanayama[4]

[1] Fukui Prefectural University,
4-1-1 Matsuoka Kenjo-Jima, Eiheiji-Cho, Yoshida-Gun, Fukui 910-1195, Japan
fujino@fpu.ac.jp
[2] Kyoto University, Yoshida Honmach, Sakyo-Ku, Kyoto 606-8501, Japan
[3] Ritsumeikan University, 1-1-1 Noji-Higashi, Kusatsu, Shiga 525-8577, Japan
[4] Institute of Nuclear Safety System,
Inc. 64 Sata, Mihama-Cho, Mikata-Gun, Fukui 919-1205, Japan

Abstract. The purpose of this study is to develop the method to encourage organizational members to chat with each other in a refresh area in workplace for fostering their knowledge sharing. First, authors focused on their chat-topics and developed the method to induce their chats on expected topics with applying a big-sized display. The result of the experiment to examine the effectiveness of proposed method showed that the proposed method would be effective on members familiar with each other, on the other hand not effective on members unfamiliar with each other. Next, authors looked the problem that unfamiliar members hardly talk to each other in a refresh room, and developed the method to encourage members to talk with unfamiliar members each other with applying smile recognition and a display. The pilot experiment to examine the effectiveness of that method was conducted and the expected effects were observed in some extent. Furthermore, some improvable points of the experimental procedures were found.

Keywords: Knowledge sharing · Informal communication · Refresh areas

1 Introduction

It is believe as a common sense on business administration sciences that informal communication among staff members is important for knowledge sharing in the organization. Therefore, on many organizations, it is often practiced to improve environments of refresh areas so that more members can get together, be relaxed and talk with each other more openly in order to encourage their knowledge sharing.

© Springer International Publishing AG, part of Springer Nature 2018
G. Meiselwitz (Ed.): SCSM 2018, LNCS 10914, pp. 12–27, 2018.
https://doi.org/10.1007/978-3-319-91485-5_2

However, there are following limits of those practiced methods:

1. Members do not always talk about topics related to their jobs and connected to enhancing knowledge sharing among them.
2. In most case, refresh-areas-chats are only held among members who have a friendship or good relationship with each other. That is, it is rare to occur members' chats across the boundary of their daily communities.

Therefore, the purpose of this study is to develop the method to solve these problems and evaluate the effectiveness of the developed method.

2   Development of a Method for Promoting Staff Members' Job-Related Chats

2.1   Hypothesis

Process of Emerging a Chat. The general process of emerging a chat among members can be hypothesized as following; first, any information that can be a seed of the chat-topic come into the scene and be recognized by person "A". After that, anything associated to that information in A's memory will be recalled on A's mind. Then, if A has any motivation to utter, that recalled thing will be voiced. At that time, A's this utterance is just A's monologue. But, if A's monologue will be heard by anyone (called "B" here) in that place, it will lead B's recalling anything associated with A's monologue in his memory. Further if B also have any motivation to utter, B will voice his/her recalled things as a response to A's utterance. After that, A and B will hear what each voices, and further voice what be recalled in each's mind to each other (Fig. 1).

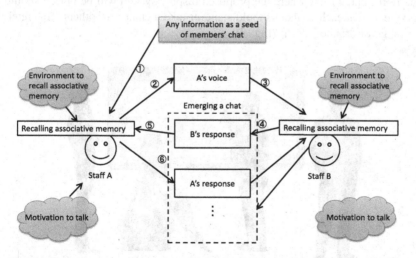

Fig. 1. The hypothesized process of emerging a chat on a topic between members.

Key Requirements for Promoting Their Job-Related Chats. According to this model, satisfying the 3 conditions as following will be required in order to promote member's chat with a specific topic:

- Seed information should be related to the topic which is expected to be talked about among members.
- Environment around members should be related to the topic which is expected to be talked.
- Members should have a motivation to voice.

Especially respecting the 2nd condition, recalled memory is believed associated semantically to what is activated in his/her mind at that time, according to a spreading activation theory [1]. If they will be in the environment where a lot of things be considered related to the expected topic, the activation level of what he/she see or hear will be heightened up unintentionally and the tendency of recalling the memory related to the expected topic will be increased. Therefore, the 2nd condition is listed up here.

2.2 Proposed Methods

Based on these conditions, the proposed method is constituted of the followings:

- A display as large as that can be seen by anyone and from anywhere in the area should be placed in a refresh area, and a lot of information related to the expected chat-topic (e.g., headlines of news articles related to the topic) should be presented on that display one by one like a slideshow, the method that could attract members' attention to the display.
- A type of the used display should be a touch panel display. If anyone touch on a presented information, the popup window will be appeared and the more detail information (e.g., the body of the news article) be presented on that display.
- The refresh area itself where the proposed display system will be placed should be as wide as that each of them should stand or take a chair near others (concretely in the range of around 1.5 m) (Fig. 2).

Fig. 2. The image of appearance of the proposed display system.

The first point mainly covers both the first and second requirement stated above. By showing a lot of information on the display, it is expected that some of those information will become a seed for a following chat. Furthermore, even though those information will not be a seed of a chat topic directly, members' memories associated to displayed information will be likely to be recalled in their mind as their memories will be in state to be activated easily by seeing the display regardless of whether or not intentionally. Because information is presented with moving like a slideshow, members who do not have what to do at that time will see the display even if they do not have such a strong interest to the information.

The third requirement is covered by the second and third points and the display size referred in the first point. The second point and the display size in the first point aim to prevent decreasing their motivation to talk to others. In workplace, some of members say that it is not so easy to speak to others because they cannot know what the others are interested in at that time and cannot find the topic about which they can enjoy talking with the others. This can be a reason that their motivation to talk to others be decreased. By the proposed method, they can share the situation that they see the same display since the display is as large as all of them can see at the same time. Therefore, if they speak to others about information presented on the display as a topic, it is likely that the others will come to talk to them. Furthermore, the function showing the popup window stated above can make the interested things of members who have touched on the display visible. That is, if a member touches on any information on the display and let the popup window shown, that information must be interesting for him/her and be likely to be a topic to enjoy a talk with him/her. Therefore, this function can be expected to prevent members' decreasing motivation to speak to others. On the other hand, the third point aims to increase members' motivation to speak to others. According to the theory of Proxemics, interpersonal distance affects human communication behavior [2], and if anyone enter one's personal space, he/she feel psychological pressure to talk to the person because entering this pace means that they have an intimate friendship with each other [2, 3]. Therefore, if it is possible to make members enter their personal area each other by setting a narrow area as a refresh area as stated in the third point, members will be motivated to talk to each other as a response to such a pressure.

2.3 Experiment

In order to examine the effectiveness of the proposal method, participant experiment was conducted in the laboratory.

Methods of the Experiment. The outline of this experiment is as following; first, the environment that can simulate the situation that members have a break time in a refresh area in workplace would be constructed in the laboratory. And then the proposed display system would be placed in that environment. In order to examine the proposed method, contents of members' chats in both of the conditions of operating the display and of not operating the display would be captured and compared to each other.

In order to simulate a situation of an actual refresh time in workplace in the laboratory, the following method was adopted: The dummy task would be prepared.

About that dummy task, participants would be directed that what theirs to do in this experiment would be to perform that task continually with taking a break time sometimes. It must be secreted for participants that the presented task be just a dummy. Also, about the purpose of this experiment, a dummy purpose was explained to them that the purpose of this experiment would be to evaluate their cognitive performance by the presented task. A 15 min break time would be placed in every 30 min task time, and a refresh room would be prepared separately from a task room too. Participants would be instructed that they can spend freely a break time in the refresh room.

Environments. The appearance of developed display system is shown in Fig. 3. Used hardware was 60 inches multi touch Full HD display named "PN-L603B" produced by Sharp corp. There were 8 tiles on the display and the information was described on each tile. Information was gathered from the internet by searching news articles with some keywords related to participants' jobs. If participants would touch on the tile, the popup window with the body of news article would appear on the display.

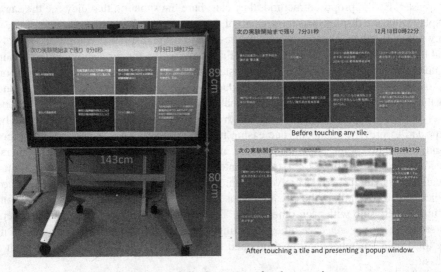

Before touching any tile.

After touching a tile and presenting a popup window.

Fig. 3. Developed display system for the experiment

The constructed refresh room is shown in Fig. 4. Participants were ordered to stay in this room in a break time. On the other hand, in a task time they moved to the other room for performing a prepared task. For fulfilling the third point of proposed method, a table of which the width was around 1.5 m by 1.2 m and two chair for participants were placed on the front of the display like Fig. 4. For letting participants having the expected chairs in a break time, the experimenters sit the chairs as shown in Fig. 4 at the beginning of experiment and made participants anchor it in their mind that the chairs not been taken by experimenters should be their places.

Time Table. The time table of this experiment is shown in Table 1. While there were 5 times of a break time and a task time in Table 1, only 1st to 4th break times were targeted in this study. 0th break time was placed in order to let participants be accustomed to the

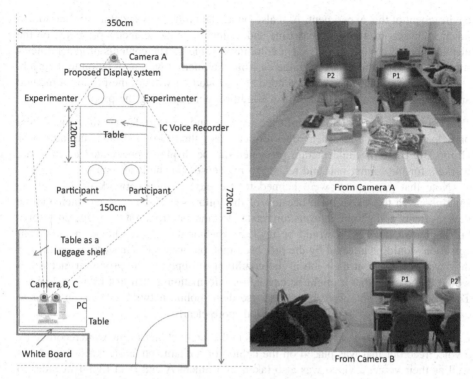

Fig. 4. The constructed refresh room for the experiment in 1st study.

Table 1. Time table of the experiment in 1st study

Time	Action	Title presented on display
10:10–10:30	Prior explanation and questionnaire	General news
10:30–11:20	Task instruction and practice	
11:20–11:35	0th Break time	General news
11:35–12:05	1st task time	
12:05–12:40	Lunch Time	General news
12:40–12:45	Flicker measurement	
12:45–13:00	**1st Break time**	No title presented
13:00–13:30	2nd task time	
13:30–13:45	**2nd Break time**	Participants' job related news
13:45–14:15	3rd task time	
14:15–14:30	**3rd Break time**	No title presented
14:30–15:00	4th task time	
15:00–15:15	**4th Break time**	Participants' job related news
15:15–15:45	5th task time	
15:45–16:15	Post explanation, questionnaire and interview	Participants' job related news

environment of this experiment. And also on a lunch time, they spent a time and took a lunch in this room. In each break time and a lunch time, the contents presented on the display were also shown in the right column of Table 1. The contents of general news were the hot topics at that time gathered from the internet with no relation to their job.

1st and 3rd break time were intended for a condition without the proposed method and 2nd and 4th break time were for a condition with applying the proposed method.

Participants. A pair of female childcare workers and a pair of female masseuses were participated in this experiment (the former was called Group A and the latter was Group B). Therefore, information presented on the display were related to childcare workers for the former pair and related to masseuses for the latter pair.

Note that participants were banned to use their mobile network devices like a smartphone in this experiment because of the following: If they can use their mobile devices freely and get the information related to their job from not the prepared display but their own devices, it would be unclear whether their topic choice in their conversation would be affected by the proposed method or not. Other word, Even though they would talk about their job in the condition of applying the proposed method, it would be possible that it might be caused on information gotten just by their devices. Therefore, participants were banned to use their mobile network devices in order to examine the effect of the proposed method more clearly.

Captured Data. Whole of their chat voices in the targeted break time were recorded by a voice recorder that was placed on the behind of the tabletop as shown in Fig. 4. As well as their voice, a video was also taken by Camera A and B in Fig. 4 in order to observe their behavior in a break time[1]. While it was kept a secret for all of participants to record their voices and take a video until finishing the experiment in order to let them talk naturally in a break, the fact that their voices and a video had been recorded was disclosed to them in the post-experiment explanation and their allowance to use and analyze them as the experiment data was confirmed. All the participants allowed it.

Result. Recorded voices of each condition[2] were classified by the relatedness to participants' jobs by three persons who were not experimenters separately. If there would be a voice which the three's classifications were differed each other, they would discuss and decide to its class cooperatively. After classification, both the voices related to their jobs and not related to their jobs were countered and calculated the rate of each voice to the total number of voices in each condition in each pair.

The calculated rate of the job-related and the not-job-related in each condition in each pair were shown in Fig. 5. By Chi-square test, it was shown that there was a significant difference between the rate of the job-related and not-job-related voices in the condition of applying the proposed method (Group A), while the difference between the ratio of ones in the condition of not applying the method (Group B) was not significant.

---

[1] Camera C was placed for camouflaging that all cameras were just put on without operating as an equipment of the laboratory.
[2] The recorded data of two break times in each condition in each pair were packed to one data in the analysis.

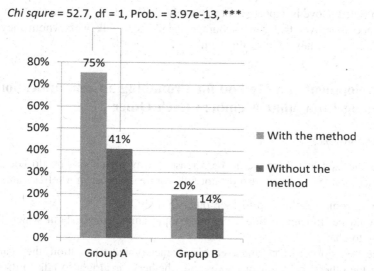

Fig. 5. The rate of participants' job-related and not job-related voices to all of their voices in the conditions with the proposed method and without the proposed method in each pair.

Respecting the differences between the results of Group A and B, it was found from the interview to them that participants in Group A met and talked with each other in daily because they worked in the same office, while participants in Group B did not meet and talk with each other so often because each of them worked usually by herself. Furthermore, from the observation of participants' behavior in a break time, participants of Group A had been talked continually without any sense of hesitation. On the other hand, participants of Group B had not been talked so frequently and easily. Both of participants Group B had looked feeling some nerves and it looked that there had been a polite atmosphere between them. It is considerable that participants of Group B had not talked so continually and naturally on any topics in the experiment because their daily relationship would not be so matured and they would feel nerves to talk easily to each other. Therefore, the reason why there was not significant difference between each condition in Group B is considered that they had not talked with each other enough to appear the significant difference in the first place because of their immature relationship.

2.4   Discussion

As a result of the experiment, the followings were found: If members in the area often talk with each other in a daily life, their chat topic would become related to information shown on the display. On the other hand, if members do not talk with each other so often and do not have a mature relationship with each other, the proposed method would not always have an effect to encourage them to talk about the expected topic.

From the first result, the first problem described on the top of this paper is expected to be solved by our proposal method. However, from the latter result, the second problem stated above is expected not to be solved. Therefore, it is necessary to develop the method moreover that can encourage members to talk with whom they do not usually talk as the next step of this study.

3 Development of a Method for Promoting a Communication Among Unfamiliar Members Each Other

3.1 Hypothesis

We hypothesized that there would be 3 reasons why members who do not talk with each other so often in usual also do not talk with each other in a refresh area:

1. They have no relationship to each other in daily life.
2. It would be difficult to find a common topic among them about which they can enjoy to chat.
3. Since they do not know about each's characters and disposition, they cannot see from the others' behavior or expression whether it is alright to talk to them.

3.2 Proposed Method

As a solution of this problem, the method composed of two elements is proposed; preparing smile recognition snack box called "Okashi-Bako[3]" and applying the display system developed in the first study stated above.

Smile Recognition Snack Box – "Okashi-Bako". This box contains snacks and paper cups for taking coffee or other beverages, which many of users of a refresh area usually want to get. This box is locked normally, and in order to open it, it is need that more than two members must show their smile face to the display and let the computer recognize their smile for some seconds long. If the computer can recognize their smiles for the set seconds continuously, the compute unlock the box and open the top of the box, and members can get snack or paper cups.

There are two expected effect of this box: The first is to show their smile each other autonomously through trying opening the box, and the second is to build the fact that they will have cooperated for one purpose.

Showing their smiles to other, according to Inoue [4] and Yoshikawa et al. [5], is believe to let others feel more friendly and attractive to them in spite that they are unfamiliar with each other. Thus after they show their smile each other, it is expected they feel more easy to talk to each other. Here, how to let them show their smile each other is a problem to be solved. As a solution to this, it is conceptualized that what many persons usually want to do or have to do in a refresh area like having a cup of coffee, eating snacks, sitting a sofa, smoking, or even entering a refresh area, should be connected to showing their smiles. Based on this concept, the smile recognition snack box illustrated above is proposed in this study.

---

[3] While Japanese characters are different on each meaning, the pronunciation of "Okashi" has two meanings in Japanese; "funny" and "snack". "Bako" means "a box" in Japanese.

This method is also expected a countermeasure to the problem referred at first above that no-relationship among them would cause their not chatting. Through showing their smiles each other in order to open the box and get snacks, the cooperative relationship among them will be constructed. Therefore, after this operation, it is expected that they feel any relations among them and more easy to talk to each other than before.

Display System. On the other hand, the countermeasure to the second problem stated above that they cannot find a suitable topic to talk to each other is to apply the display system proposed in the first study. As already having described in the above section, the display show many information with an attractive movement. Therefore, they will see the display regardless of whether intentionally or not and sometimes they could find a topic from information on the display.

3.3 Experiment

In order to examine the effectiveness of the proposal method, participant experiment was conducted in the laboratory.

Methods of Experiment. The basic procedure of this experiment was modeled on the experiment in the first study. The environment that is similar to the situation of having a break time in workplace was constructed and the dummy task was prepared like the first experiment. Participants would be explained the dummy purpose of the experiment instead of the real purpose. However, there are some deferent points between the experiments: The design of this experiment was a between-participants one while the former experiment was a within-participants one. Furthermore, while participants continually performed the task and had a break time over around 6 h of experimental period in the former experiment, participant would perform the task and have a break time only one time actually.

Environment. The sketch map of the environment for the experiment is shown in Fig. 6. Used display in this experiment was 27 inches multi touch Full HD display called "ProLite T2735MSC" produced by iiyama corp. The width was 67 cm and the height was 41.5 cm. The blue top box on the table in the example picture from camera A in Fig. 6 was the smile recognition snack box. To open the box, participants should show their smile to the display on the steel shelf. The PC for processing smile recognition was placed behind the display. Raspberry pi 3 model B+ and servo motor were equipped on the box to open the top of the box. If the PC could succeed recognizing participants smile for 3 s, the PC would send the message to Raspberry Pi to open the top of box through Wi-Fi, and then Raspberry pi would actuate the servo motor to open the top of the box.

The contents of the display designed for this experiment is shown in Fig. 7. There were four tiles and the contents were news about pops music, information of ramem restaurants and café around the institute in which the laboratory is, news related to sports and health, and news about Japanese show business. These contents themes were selected on accordance with the result of the questionnaire survey conducted before this experiment, the purpose of which was to grasp the topic about which persons can talk

Fig. 6. The concept image of operating smile recognition snack box

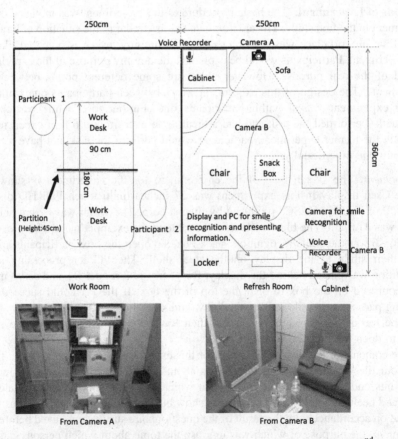

Fig. 7. The constructed work room and refresh room for the experiment in 2nd study.

to others who are not so familiar with them. All the information were gathered from the internet by authors from the view point of interestingness for authors themselves.

At the beginning of the experiment in the condition applying the proposed method, the screen of the display was for smile recognition. If the smile recognition would be completed, the scree would change to the one to present the information.

Time Table. The time table of the experiment was shown in Table 2. Respecting after the break time in this table, the schedule in the left part was presented to participants at the beginning of the experiment. However, this was just a dummy. Actually, the experiment would be finished and the post explanation, questionnaire and interview were conducted immediately after the 15 min break time (Fig. 8).

Table 2. Time table of the experiment in 2nd study.

Time	Action	
10 min.	Instruction for task and prior questionnaire	
20 min.	1st task time	
15 min.	Break time	
10 min.	Instruction for 2nd task	Post explanation, questionnaire and group interview
20 min.	2nd task time	
15 min.	Post questionnaire and interview	

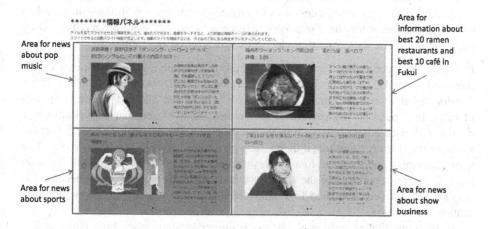

(Red frames were not shown on actually.)

Fig. 8. Contents and appearance of the display in 2nd experiment.

Participants. Participants of this study were 13 pairs of university students. In each pair, participants were not familiar with each other. 5 of 13 pairs were in the condition that both of the snack box and the display were placed. 2 pairs were in the condition of only placing the display and other 2 pairs were in the condition of only placing the snack box. Remained 4 pairs were in the condition neither the snack box nor the display was placed. In the conditions of not placing the smile recognition snack box, the simple snack box in which the same amount of snack and paper cups were put was placed on the table instead of the smile recognition snack box.

Here, what has to be paid special attention in this study is the influence of mobile network devices like a smartphone. If they would be allowed to use such a device, participants would gaze only their devices for a period of the experiment in order to avoid the psychological pressure to talk to each other. As a result, the proposed method would not show any effect to help them to talk with each other. Therefore, in order to examine the effect of the proposal method more clearly, such a device should be banned to be used in this experiment. On the other hand, in the actual situation of a break time, they can use any devices freely; therefore to ban them to use such a device would be so unnatural that it is possible that the result under such a situation would not be valid to be generalized. Thus, in this experiment either situation of allowing a device or one of banning a device was set as a pilot study.

Captured Data. Whole of pairs' behavior including their voices in a break time were taken a vide with keeping it a secret to them until finishing the experiment. In the post explanation, that fact was disclosed and the allowance to use the data were confirmed to each participants' group. All the participants allowed it.

Result. The recorded videos were analyzed as following processes: First, three members of authors separately observed those videos with taking a note about participants' communication and other behavior driven by their concerning about each other. After that, the effectiveness was discussed qualitatively with those notes taken before among the same three members. If necessary, videos were observed again by them together. The data from the interviews was referred in discussion.

Behavior. Respecting the result of the condition of applying both elements of the proposed method, 3 of 5 pairs had been performed the smile task to open the box. Other 2 pairs had not been performed. Those 2 pairs had been allowed to use their own mobiles. In 3 pairs having performed the smile task, it looked that they could enjoy talking with each other easily with using the proposed display well. That is, the scenes were observed many times that if their chat with a topic had been reached end, they had seen the display and they had found a new topic and stated to talk again. On the other hand, 2 pairs not performing the smile task had not used the display too because the display had not been switched from the screen for the smile task to the screen presenting information.

In the condition of only applying the smile recognition snack box, one pair had performed the smile task, while another pair had not performed. Not-performed pair had allowed using their mobiles. Performed pair had talked with each other continually until around the half of a break time after performing the smile task, but in the latter half, they gradually could not have continued talking with each other. In the last part of a break time, each of them had seen other than the partner or fingered things around the room nervously. On the other hand, not-performed pair had spent a time with reading a comic or watching her own smartphone individually and without talking to each other.

In the condition of only applying the display, one of two pair, which had not been allowed to use a mobile device, had talked with each other continually during a break time without watching the display almost at all. Another pair, which had been allowed to use a mobile device, also had talked continually during a break time. However, the experiment of this latter pair had been failure because the system had been troubled around 6 min after the experiment had been started and the display had not shown any information since then. Until that trouble, this pair had seen the display though the display had not given them any chat topic. Note that they had almost not handed their smartphone during a break time in spite that they had been allowed to use it.

Respecting the condition of not applying either the snack box or the display, 3 of 4 pair had been banned to use their mobile devices and remained one pair been allowed to. One of 3 banned-pairs had started to talk with each other immediately after a break had been started and their talks had been continued with various topics until the end of a break time. The other one banned-pair also had talked during the experiment. However unlike the pair described above, their talk had been stopped every about 3 or 4 min. During such a time of their not talking, they had eaten snacks on the table, drunken beverages, read those labels or seen around the room nervously. The remained one banned pair' behavior had been similar to the second banned pair until around the half of a break time. However during the latter half of a break time, gradually they had come to talk with each other freely and naturally. On the other hand, respecting the allowed-pair, they had not talked at all during a break time. They had spent a time with watching a smartphone, taking snacks and beverages, stretching their bodies or seeing around the room nervously.

Interviews. Here, some of participants' voices, which should be remarked, are described. Related to the effectiveness of the smile recognition snack box, 4 of 7 pairs who had been in the conditions of applying this box said that the smile task had let them become easy in spite that they had been unfamiliar with each other. Two of the other three pairs referred the contents of the snack box, related to whether they had been motivated to the smile task. One of them said that they had performed the smile recognition in order to get paper cups though the smile recognition box itself had not been interesting for them. The other pair, who had not performed the smile task, answered that they had not performed because they had thought that there would have been only paper cups in the box and they had not want paper cups at that time. The last one pair who also had not performed the smile task said that it had been difficult for them to ask unfamiliar persons to perform the smile task together.

Respecting the effectiveness of the display, 2 of 5 pairs who had been in the conditions of applying the display[4] said that they had often looked the display when their talk had been stopped. Both of them were also applied the smile recognition snack box. One of them, furthermore, said that they would be able to continue to talk with

---

[4] Actually 6 pairs had been set in the condition of applying the display. However, one pairs of them, who had been applied the smile recognition snack box too, had not performed the smile task. Therefore, the display had not change from the screen for the smile task to the screen to present information. Thus, this pair was omitted from pairs who had been in the condition of applying the display here.

getting a topic from the display, if the display would be placed. On the other hand, one of the other three pairs said that they had not looked the display intentionally because they consider it would be impolite for the partners. This pair was in the condition of only the display. The other one of that three pairs said, which was also in the condition of only the display, that it was no need for this pair to see the display because they had been able to continue their chats during a break with introducing themselves and taking other topics about themselves. This pair added that they might see the display if their chats were stopped.

3.4   Discussion

As a result of the experiment, it was appeared that the effect of showing their smiles each other, which is aimed to let them feel more friendly each other and develop a good relationship, was only suggested on their interviews and not appeared as the deference of their behavior among the experiment conditions. That is expected because even many pairs in the conditions of not applying the proposed method could have started talking by introducing themselves or taking other topics rather than had been embarrassed toward the situation of staying with an unfamiliar person and had hesitated to talk.

To connect taking snacks and paper cups, which is thought of what usually to be wanted to do in a refresh room, to showing their smiles was suggested effective in some extent to promote them to show their smiles each other naturally. Because there was a pair said in the interview that they had performed that in order only to get a paper cup.

Respecting the display, it was appeared effective because it was observed that many pairs had found a topic for their chat from the display in the condition applied it, while many pairs had looked nervous and embarrassed in the condition not applied it. Furthermore, it was founded as an unexpected side effect that the proposed display would have an effect to gather participants' eyes when they did not have anything to talk. At that time participants had looked not embarrassed to that situation even though they had kept silent. This effect have 2 meanings that they can avoid to feel the stress from the situation to staying with unfamiliar person and that they can know the other person's state that he/she watch the display at that time. Resulting from the second meaning especially, they can expect that information presented on the display would be a suitable chat-topic and can talk to each other about that information. That is, the proposed display will have this ecological effect as well as the expected effect connecting associative human memories and the effect as a machine of giving a suitable topic stated in Sect. 2.

The experiment of this study was thought of insufficient to evaluate the effectiveness of the proposed method strictly. Therefore, the further experiment has to be conducted in order to examine the effectiveness of the proposal method more precisely, referring the result of this experiment.

4 Conclusion

This study consisted of 2 studies aimed at developing the method to solve the two problems related to informal communication in a refresh area in workplace; how to encourage members to talk about their jobs and how to encourage members to talk to unfamiliar members each other. As a fruit of this study, the methods applying a big-sized touch panel display and smile recognition technology were developed.

While the experiments were conducted to examine the effectiveness of developed method, it should be just a basic study only in the very controlled environment of the laboratory. Therefore, as a next work, the further experiment should be conducted under the more various conditions simulated more precisely to the real situation in a refresh room in workplace to evaluate the effectiveness more strictly. Furthermore, more practical experiment that those methods be applied to an actual refresh area in any real workplace should be done as well.

References

1. Collins, A.M., Loftus, E.F.: A spreading-activation theory of semantic processing. Psychol. Rev. **82**(6), 407–428 (1975)
2. Hall, E.T.: The Hidden Dimension. Doubleday & Company, New York (1966)
3. Nishide, K.: Architectural planning on human psychology and ecology: the distance between persons. Archit. Pract. **8**(11), 95–99 (1985). (in Japanese)
4. Inoue, K.: First time impressions given by facial expressions. Bull. Living Sci. **36**, 183–194 (2014). (in Japanese)
5. Yoshikawa, S., Nakamura, M.: Facial information and the regulation of speech behaviors. Trans. Inst. Electr. Inf. Commun. Eng. **A 00080**(00008), 1324–1331 (1997)

Rewarding Fitness Tracking—The Communication and Promotion of Health Insurers' Bonus Programs and the Use of Self-tracking Data

Maria Henkel[1]([⊠]), Tamara Heck[2], and Julia Göretz[1]

[1] Heinrich Heine University, Universitätsstr. 1, 40225 Düsseldorf, Germany
maria.henkel@hhu.de
[2] University of Southern Queensland, West Street, Toowoomba 4350, Australia

Abstract. This paper analyzes German and Australian health insurer programs that offer self-tracking options for customers. We considered aspects of program promotion, program goals, and data privacy issues. Results are based on scanning current information available online via insurer websites. Seven Australian and six German insurers apply self-tracking. Programs in both countries vary, whereas most Australian insurers build their programs on third-party providers, and German insurers offer direct financial rewards. Those differences may be reasonable due to diverse health systems in both countries. Commonalities regarding the programs' intentions are obvious. Furthermore, concerns about data policies arise across countries. The reward systems and intended program goals vary. The outcomes give insights into the status quo of self-tracking health insurer programs and contribute to a better understanding of the use of self-tracking data by providers. Moreover, further questions arise about the benefits of those programs and the protection of sensitive self-tracking data.

Keywords: Fitness tracker · Activity tracking · Wearables · Health insurance
Self-tracking · Data privacy

1 Introduction

Self-tracking one's health and fitness status have become very popular. A GfK study from 2016 says that 33% of online users (across 16 countries) are monitoring or tracking their health or fitness either via an application (online or mobile) or via wearables (fitness bands, clips, smartwatches). The market for wearables that track a person's activities is still growing and new vendors that sell luxury and fashion devices are entering the market [1, 2].

The popularity and the ability of those technologies have attracted health insurers as well. About three years ago, the first health and life insurers, mainly in the United States, but in other countries as Australia and Germany as well, started to offer discounts for customers that track their fitness and reach a determined fitness goal. Life insurer AIA started its program in Australia in 2014 and cooperates with health insurers [3]. In 2015,

G. Meiselwitz (Ed.): SCSM 2018, LNCS 10914, pp. 28–49, 2018.
https://doi.org/10.1007/978-3-319-91485-5_3

a local affiliation of the AOK was the first health insurer in Germany subsidizing the Apple Watch [4].

Since then, discussions emerged on the use and misuse of self-monitored fitness data [5]. Moreover, research on fitness and activity trackers showed the technologies' inaccuracies [6] and low long-term effects on people's health behavior [7, 8]. However, health insurers seem to get ready to establish new programs that include self-monitoring and tracking of health data. Tedesco et al. [9] summarize current reports that forecast insurers will invest in new technologies like wearables and partner with digital technology providers to engage customers and offer new services.

The following paper aims at giving an overview of the current status quo of health insurances that investigate self-tracking opportunities and possible rewards for customers that share their fitness and health activities. We are interested in how insurers promote their health and well-being programs (intended program goals) and motivate customers to live healthier (incentives). We introduce research in progress while firstly focusing on the countries Germany and Australia. We discuss the current situation of health insurance clients' data use, data security issues as well as long-term health benefits regarding those programs based on recent research on self-tracking activities. The research questions are:

1. Which health insurers offer options for clients to self-track health and fitness data?
2. How do insurers communicate about the programs?
3. How do those insurers communicate about data security?

2 Methods and Limitations

To find current health insurance programs that offer self-tracking, we decided that this could be done best while reviewing health insurance websites as well as media reports (news articles, blogs, and insurance comparison services). We did this first review with regard to scope the landscape of current programs to be able to investigate future research based on this status quo (compare Gough et al. [10] and Arksey and O'Malley on the aims of a scoping review [11]). Health insurance programs are all activities health insurers offer to their customers and that relate to aspects concerned with the customers' health and well-being. Some insurers explicitly name their programs, i.e. they establish specific programs that include diverse services related to fitness, well-being, and lifestyle (accessed for example via an application). Others list their activities as part of their overall services. We aimed at finding all activities related to self-tracking.

As the German and Australian health system differs, we have to distinguish between insurance types. Australia has a public health insurance (Medicare) funded by tax levy. Medicare does not offer any additional health program services. Besides, there are currently 38 private health insurances, whereof 25 are open to all employee groups and available in all eight states[1]. Private health insurance is additional health cover to

---

[1] http://www.privatehealth.gov.au/dynamic/insurer.

that provided under Medicare, to reimburse all or part of the cost of hospital and/or ancillary services incurred by an individual. In 2014–15 there were 10.1 million adult Australians with private health insurance (57.1% of all people 18 years and over). This was the same rate as in 2011–12, but an increase from 2007–08 (52.7%) [12]. Australia has 24.5 million citizens (Mar 2017) [13]. We considered all 38 Australian insurers officially listed by the government.

Germany has a law enforced statutory health insurance (or public health insurance). 72.26 million (about 88%) of the population are covered by statutory health insurance (SHI) while private health insurance (PHI) covers 8.77 million people (about 11%) [14]. There are currently 110 health insurers [15] listed of which we identified the 30 biggest SHIs and 10 biggest PHIs, defined by the number of insured persons [14]. As we lack an official resource, we based our number on diverse sources to define the 40 SHI and PHI with the highest number of customers [14, 16].

The following paper discusses health programs of 38 Australian and 40 German health insurers. A full list of all insurers examined is published online [17]. We reviewed the programs based on website information by the insurers in December 2017, and searched for updated information in February 2018 as many insurers change their programs every year.

The described review method is not free of limitations. First, result rely on information published by health insurers. During our research, we realized that finding specific information about health programs is hard. Some insurers, specifically Australian providers, base their services on personal customer relations, therefore. In Germany, there are many health insurances and no central service that stores all necessary information consistently and exhaustively. Especially information on private health insurance is scarce. While collecting information for our analysis, we did so to the best of our ability, but cannot guarantee absolute exactness. Furthermore, as we did not reach out to health insurers to ask them personally, this first review study might miss relevant details on programs. Second, some of the apps that provide main access to services are only accessible for recent customers. Regarding the apps' functions, we relied on the descriptions by insurers and at application store sites whenever there was no (demo) version open to the public.

The article is structured as follows. We will introduce current literature and reports on self-tracking and eHealth in Sect. 3, before we show results on the German and Australian health insurance programs (Sect. 4). Afterwards, we will discuss the findings (Sect. 5) and conclude with arising questions that need to be investigated in future (Sect. 6).

3   Related Literature

Research on digital health technologies discusses effects and impact on people and their health behavior, including fitness tracking devices, mobile applications, and online services, but as well health data usage by health providers, insurers and research. Moreover, studies investigate new forms of communication between public and health professionals and aspects of health information communication that can be summarized under the term *health promotion* [18, 19]. The following study is interested in two issues: Offered self-tracking options by insures - via using new technologies like tracking devices and mobile apps - as well as information disseminated about this offer.

Self-tracking, self-monitoring or 'quantified self' refer to all aspects concerned with personal data, its analytics and sharing with others. It is about "the practice of gathering data about oneself on a regular basis and then recording and analyzing the data to produce statistics and other data (such as images) relating to regular habits, behaviors and feelings [20]." The 'self' hereby emphasizes the fact that oneself is the actor of collecting this data and is aware of being tracked [21]. Another relevant aspect is the focus of improving oneself, i.e. "'live by numbers' [...] to quantify and then optimize areas of one's life [22]."

Today, digital developments like apps and devices support this process. Smart devices instantly track activities, store data and automatically analyze them to provide insights into the quantified self. The website 'Quantified Self'[2] gives an overview of technologies and application. Regarding the numbers, health and fitness support seem to be most popular. Many digital services hereby provide automatic self-tracking, where the device – a smartphone, smartwatch, or trackers and life-loggers specifically designed to track personal actions – registers one's activities. This paper focuses on both, self-tracking activities that are traced either with the help of devices or registered by oneself.

Various studies regarding fitness and health concerns focus on user motivation. They discuss users' ability to understand data and fitness goals. Asimakopoulos et al. [23] worked with Fitbit and Jawbone users and applied self-determination theory elements. They remark that users have difficulties to interpret device data and gain motivation from it, due to a critical user experience regarding the fitness tracker apps. For example, users found some graphical visualizations useful, other means of data presentation were found less useful. Users were not aware of being able to edit their goals (like steps per day or sleeping hours) either. Moreover, the authors stress that goals are kind of pre-defined by the fact that the device has predefined tracking options. They conclude that "motivational relevancy of content [...] should support a users' immediate as well as overall intrinsic goals" [23]. Donnachie et al. [24] conducted a 12-week study with 28 obese men, analyzing factors of self-determination. They found out that intrinsic motivation leads to better long-term effects, i.e. participants continued their personal activity and were more satisfied in doing so. In contrast, participants with extrinsic motivations adopted a more rigid approach to exercise behavior but failed in long-term activity behavior changes. The last group reported pedometers as "undermining" or "controlling" rather than motivating. In addition, for these participants, group motivation seemed more important and once the activity program they participated ended, they stopped their activities. Rowe-Robertset et al. [25] conducted a 7-month study with 212 employees. Results showed that activity trackers can improve a person's fitness and health risks. Moreover, participants with high health risks were more active and more engaged, i.e. they had more steps/day on average and used the tracker for a longer period than participants classified as low- or medium risk. However, further survey data revealed that the devices themselves might not be the main factor for behavior change. Participants that were not active reported a general

---

[2] http://quantifiedself.com/guide/.

lack of motivation or lack of time, and suggested a need for additional motivation, like fitness games for example. Other studies showed that game mechanics [26], remote coaching [27] and social networking [28] are the driving motivators to change activity behavior. Glance et al. [29] showed that participants in teams are significantly more active than individuals (average daily steps). If health insurers' tracking programs intend to support their clients to get healthier and more active, insurers need to consider factors of communicating and engaging in those programs. In the following study, we are interested in the way insurers promote their programs (intended program goals) and try to increase user motivation (incentives).

Related to the above aspects are social, economic and political issues of digitized and tracked health data. Lupton [19] raises concerns about "socio-political implications of digitizing health promotions." Digital health data containing sensitive information on people's health and well-being are already part of a large network of a complex economy, where providers of self-tracking devices, health promoters and health insurers participate. Currently most of the health promoting programs aim at changing people's behavior towards a healthier lifestyle with a rather prescriptive "top-down approach", while not considering individual circumstances and the limitations of the meaning of quantified health data [19, 30]. Purpura et al. [30] discuss the boundaries between persuasion and coercion. Health programs and fitness devices often apply generalized standards and an ideological assumption by designer that then tell users best ways to get healthier and fitter: "[…] the key distinguishing feature that concerns us with the persuasive computing literature is that users do not get to choose their own viewpoints" [30]. Those points are critical when examining self-tracking programs by health insurers. Which goals do insurers have with their programs? How do those programs communicate health and fitness aspects to users? Which economic and social concerns do those programs implicate?

Another issue arising with the advent of digital health technology for personal activity and fitness tracking is the handling and protection of personal sensitive data. While retailers recognize many opportunities regarding individualized data mining and personalized product recommendations [31], the fear of data breaches, inference, identity theft or other risks connected to creating and transmitting health-related data grows. Indeed, among current "personal IoT devices, fitness trackers are those that have the most number of sensors and capable of collecting the most sensitive information" [32]. In the context of sharing such data with health insurers, the fear of "inference attacks" might be the most pressing one:

> Though third parties are authorized to access and process the personal information of users [there is the risk that] these trusted third parties can breach privacy statements and infer sensitive information that was not shared by the user. These attacks are typically known as "inference attacks" where private information can be derived by exploiting the available data provided by the user to the third party [32].

Besides self-tracking program promotion and user motivation increase, this paper as well reviews data privacy statements of self-tracking programs. In the following, we will report on the results of the scoping of current health insurer self-tracking programs in Germany and Australia, before we discuss the introduced issues around digital health technologies.

4   Results

In the following, we show results of our review on the status of health insurance programs that relates to self-tracking activities, and refer to our three research questions. We distinguish between two different types of self-tracking. Self-tracking via wearables (fitness tracking devices) that allows for automatic synchronization of data, and self-tracking via manual user input, where users themselves share information, for example about their daily steps, heart rate, weight loss etc.

4.1   Do Insurers Offer Self-tracking Programs?

Seven Australian and six German health insurers offer self-tracking related programs (Table 1). We will summarize the findings in the following.

Table 1. Insurers offering self-tracking options as part of their health programs.

Health insurer	Country	Measurement tool	Synchronizing via wearables	Activity measured	Rewards
AOK Nordost	GER	FitMit AOK App	YES	Steps, heart rate, caloric deficit	Monetary and gift rewards, donations
AOK Plus	GER	AOK Bonus App	YES (synchronize with Google Fit or Apple Health)	Caloric deficit, heart rate, steps	Monetary reward
BARMER	GER	Fit2Go App	YES (and manually)	Active minutes	NO
IKK Südwest	GER	Familienabenteuer mit Fred App	YES (and manually)	Steps	NO
Signal Iduna	GER	Sijox AppLife	NO	Steps	Cashback (up to 42%)
Techniker Krankenkasse	GER	TK App	YES (synchronize with Google Fit or Apple Health)	Steps	Monetary reward or "health dividend" (credit for additional treatments)
GMHBA Limited	AUS	AIA website and app	YES	Steps, activity, kj, etc.	Rewards via AIA Vitality points
HBF	AUS	My Pocket Health App	NO	Manual input of data	NO
HCF	AUS	My Health Guardian website/app by Healthways	NO	Steps, heart rate, weight etc.	NO
Medibank Private Limited	AUS	Flybuys account (website/app)	YES	Fitbit features like steps, km	Rewards via Flybuys points

(*continued*)

Table 1. (*continued*)

Health insurer	Country	Measurement tool	Synchronizing via wearables	Activity measured	Rewards
myOwn Health[a]	AUS	AIA account (website/app)	YES	Steps, activity, kj, etc.	Rewards via AIA Vitality points
NIB Health Funds Ltd. (Qantas Assure[b])	AUS	Qantas Assure (website, app)	YES (synchronize via Google Fit, Fitbit, Strava)	Steps, cycle, swimming, manual activities	Rewards via Qantas Frequent Flyer points, premium, upgrade possible
St. Lukes Health	AUS	My Health Guardian website/app by Healthways	NO	Steps, heart rate, weight etc.	NO

[a]Cofounded by GMHBA and AIA Vitality (life insurer)
[b]Qantas Assure is an additional new health insurance service by NIB, arranged by Qantas Airways Limited

German Health Insurance. For 25 out of the 40 insurer's websites, we can say that we found all needed information regarding bonus programs, activity tracking (if applicable) and data handling. All other websites had restrictions, i.e. you had to log in or contact the insurance company to get more information. This is especially the case with all 10 PHIs we analyzed. For the SHIs, there were only 5 out of 30 cases where not all information was publicly accessible, and only 2 of those did not have sufficient information available to confidently determine whether an activity tracking program or activity tracker subsidy was offered.

We identified six German insurers, five SHIs and one PHI, who support activity tracking in their programs and/or via apps provided (see Table 1). We also found ten insurers who at least subsidize the purchase of an activity tracker or even apple watch in any way.

"AOK Bonus-App" (AOK Plus), "FitMit AOK" (AOK Nordost) and "TK-App" (Techniker Krankenkasse) are applications that offer automatic activity tracking to their customers. Regarding the content and the reward system, "AOK Bonus-App"and "FitMit AOK" have many similarities: Customers can link to a tracking application (Google Fit, Apple Health) to convert steps, caloric deficits, and heart rate into points for the insurance's bonus program. In the apps, collecting points is also possible by sending proof of treatments or preventive check-ups (e.g., blood donations or cancer screening). The point system relates to a level system (Fig. 1). Depending on the level of the user (e.g. beginner, expert, champion) different bonuses and discounts are available for local cooperation partners from the sports and health sector. Furthermore, customers can get a cash bonus reward through the app at the end of each year.

Techniker Krankenkasse offers a much more comprehensive application. It as well can synchronize with Google Fit or Apple Health to transfer steps into the insurer's system. 600.000 steps per week are necessary to get 500 points for the health insurance bonus program. These, in turn, can be paid out as cash bonus. In addition to track one's activities, customers get an overview of medication and prescriptions. Furthermore, they can send sick reports and receive letters from the health insurer. "Fit2Go"

(BARMER) and "Family Adventure with Fred" (IKK Südwest) offer self-tracking, but provide no rewards in return. While "Fit2Go" supports the usual step counting functionalities, "Family Adventure with Fred" is meant to engage whole families in hiking and walking adventures (Fig. 1).

Fig. 1. From left to right: AOK Bonus-App with level and point system as well as cooperation discounts; Family Adventure with Fred in a virtual environment; Sijox AppLife with step-counter and cashback visualization.

"Sijox AppLife" (SIGNAL IDUNA) is only available with the purchase of one specific insurance package. The synchronization with Apple Health or Google Fit allows users to convert steps into a cashback of up to 42% for SIGNAL IDUNA's disability insurance. Performance graphs show user activity. The application has a timer that measures the duration of an activity. Customers can see their cashback amount clearly at the top center of the dashboard (Fig. 1).

While we could only identify six German insurers, who support activity tracking right now, we also found ten insurers who at least subsidize the purchase of an activity tracker: Techniker Krankenkasse, DAK Gesundheit, AOK Plus, AOK Nordost, KKH Kaufmännische Krnakenkasse, BKK Mobil Oil, IKK Südwest, Bahn-BKK, hkk Krankenkasse, and BIG direkt gesund. These insurers might develop activity tracking programs in the future.

Australian Health Insurance. In Australia, only the insurer HBF offers its own app (My Pocket Health, Fig. 2). Customers can manually input their fitness activities, store health information (e.g. x-rays or test documents), and manage a calendar (e.g. with doctor appointment, vaccinations and medication times). Automatic self-tracking is not possible, there is no reward system related those activities. All other Australian insurers use services from external providers, i.e. they offer customers the opportunity to use those services. Two insurers use the service of a health management provider, Healthways[3] (Fig. 2). Healthways is a B2B-provider that offers personalized health programs like assessments, care management and coaching. Neither Healthways nor their insurance partners HCF and St. Lukes Health offer any rewards participating in programs.

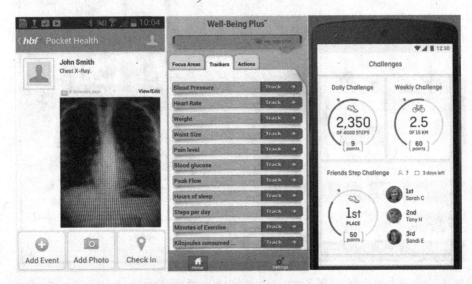

Fig. 2. From left to right: HBF app 'My Pocket Health'; Well-Being Plus app by Healthways, used by health insurers HCF and St. Lukes Health; Qantas Assure app, offered by NIB Health.

Four insurers use external providers and as well the reward system offered by those. Customers can collect points for their activities. Life insurer AIA[4], who entered the Australian market in 2014, cooperates with two health insurers (GMHBA Limited and myOwn Health, cofounded by GMHBA and AIA) and offers its Vitality program. Costumers collect points, get a higher 'AIA Vitality Status' and receive discounts on services and products of third-party providers. Similarly, Medibank Private and Flybuys cooperate. Medibank customers are able to get extra Flybuys points (10 for 10,000 steps a day) when they synchronize their fitness activities. Manually uploaded steps are not considered. Flybuys[5] is a large Australian loyalty program, where

---
[3] http://www.healthwaysaustralia.com.au/.

[4] https://www.aiavitality.com.au/.

[5] https://www.flybuys.com.au/.

customers collect point when shopping with participating retailers. Medibank customers are able to receive extra points when they link their Flybuys and Medibank membership. For example, they get extra points for buying fresh fruit at Coles supermarket. To get 10 points for achieving 10,000 steps per day, customers have to link a Fitbit device to their Flybuys account.

NIB Health Funds has a cooperation with Qantas Airways. Together they established a new health insurance service, Qantas Assure that offers rewards for fitness activities Fig. 2). Customers can get 25 points for achieving 10,000 steps a day. The Qantas app measures activities with a phone or via a wearable device. NIB as well offers discounts on premium insurances (via Qantas Assure) for customers that have collected adequate fitness points. GMHBA established a similar offer. Their extra insurance packages are cheaper for customers that participate in the AIA program and receive a high AIA Vitality status.

All programs that offer rewards require automatic synchronization via a phone or wearable fitness tracker. The relevant activity that counts for rewards points is daily steps, referring to the current general recommendation of steps per day [33]. AIA offers diverse options to gather points, e.g. through tracking calories, speed and heart rate data – here, specific conditions the number of points that a user can collect apply.

It is striking that German insurers seem to rely on their own applications for managing their self-tracking reward programs. Only one Australian insurer built its own application (without the option to automatically self-track). However, applications that try to facilitate customers services are established in Australia as well. Those apps allow customers to make claims, to upload health-related documents, or to contact insurers.

4.2   How Do Insurers Communicate About Programs?

German Health Insurance. German insurers do not prominently advertise activity tracking on the websites. In most cases, the established bonus programs include activity tracking. While all SHIs we analyzed offer a bonus program (2017 and 2018), many still use paper sheets, which members and/or their practitioners can fill out to record any preventive treatment or check-up. The insurances that offer activity tracking do not actively advertise this feature, but describe the bonus programs and activities more generally on the related websites and/or in digitized flyers and booklets. We found that information by searching for "bonus program" or similar terms via the sites' own search box or by looking for the appropriate category in the website menu. Most insurers state that their bonus programs exist to prevent illness and promote health, but do not describe details regarding for example particular health benefits. They use more general motivational phrases, as "take your health seriously"[6], "show what you got", or "don't lose time"[7]. The reward, which in the cases mentioned is money (cashback or credit),

---

[6] https://plus.aok.de/inhalt/bonusprogramm/.

[7] https://www.tk.de/techniker/service/gesundheit-und-medizin/praevention-und-frueherkennung/tk-bonusprogramm/programm-2010356.

is mentioned in phrases like "a healthy lifestyle pays for you"[8], "good for your health and for your pocket" (see footnote 6), or "get money back with every step"[9]. Almost all insurers develop and use their own app and underline the ease of use and convenience ("available 24 h", "fast", "easy"). AOK Nordost particularly emphasizes the novelty of their "digital program" and describes it as "completely mobile, fast and save and "a new way of communication" (see footnote 8). PHI Signal Iduna promotes its Sijox AppLife on its own website under its own domain. Here, users are encouraged to "protect [themselves] against occupational disability" and warned that "every 4th German becomes unfit for work during their working life" (see footnote 9).

As BARMER and IKK do not offer any rewards for their activity tracking apps, they focus on the health and motivational aspects of tracking. While BARMER describes the "Fit2Go App" as "your perfect partner on your way to more activity" and its use as "easier than you think"[10], IKK concentrates on family well-being and values, e.g. "so that you and your family stay healthy", "promote health literacy among your children", "let your children develop a healthy lifestyle".[11]

Australian Health Insurance. All programs are considered as additional benefits to the general health insurance services. Terms often used are fitness, well-being, wellness, and lifestyle (Fig. 3), whereas programs that relate to popular rewarding points like the cooperation between Medibank and Flybuys emphasize the rewards. In contrast, HBF emphasizes on the easy management of health and fitness related information, data, and documents. Users are able to store any relevant information in the app and are as well able to share this information via My Pocket Health app. GMHBA (Fig. 3) promotes its new insurance packages (AIA Vitality V Plus) with the 'know your health, improve your health, enjoy the rewards'[12]. There is a focus on the aspect of improving one's fitness and at the same time enjoying a better lifestyle, i.e. being able to afford enjoyable products on discount (like spa and wellness treatments, travel offers). The HCF website offers an online brochure that gives very precise information about My Health Guardian. It emphasizes that users can personalize fitness and eating plans to choose what fits best for them "into their daily routine". Qantas Assure describes its self-tracking program as "wellness rewards". It points out that any activity gets tracked within their program. Members who own an appropriate device can track their running, walking, cycling, swimming. They as well promote family membership and the opportunity for kids to earn points through activities. The information website of St. Lukes (My Health Guardian by third-party provider Healthways) emphasizes that the program offered is for "preventative health" care and for "members who are generally healthy [...] to help you look after your health" (Fig. 3[13]). Moreover, the website uses terms that focus on the self-management of improving one's health. St. Lukes as well hints to daily well-being challenges to keep users motivated.

---

[8] https://nordost.aok.de/inhalt/neues-digitales-praemienprogramm-der-aok-nordost/.

[9] https://www.app-life.de/.

[10] https://www.barmer.de/gesundheitscampus/apps/fit-to-go-8536.

[11] https://www.muuvit.com/ikk.

[12] https://www.gmhba.com.au/aia-vitality#Products.

[13] https://www.gmhba.com.au/aia-vitality; http://www.stlukes.com.au/MyHealthGuardian.aspx.

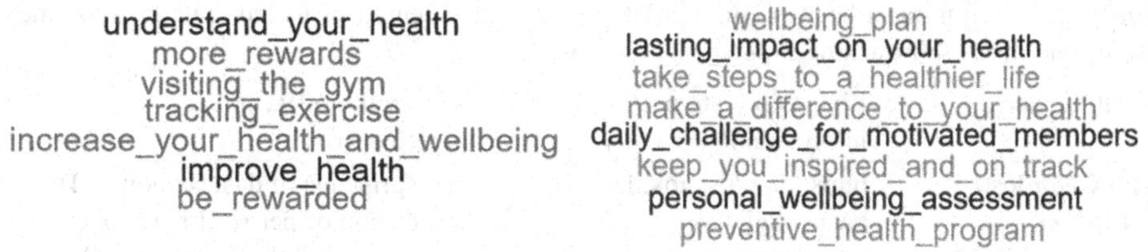

understand_your_health
more_rewards
visiting_the_gym
tracking_exercise
increase_your_health_and_wellbeing
improve_health
be_rewarded

wellbeing_plan
lasting_impact_on_your_health
take_steps_to_a_healthier_life
make_a_difference_to_your_health
daily_challenge_for_motivated_members
keep_you_inspired_and_on_track
personal_wellbeing_assessment
preventive_health_program

Fig. 3. Terms used by GMHBA (left) and St. Lukes to describe their offered health programs, both using third-party providers (AIA and Healthways, respectively).

Health insurers clearly distinguish their tracking and reward programs from their general health-related services. One main reason for this is legislation to which health insurers are bound, and that regulates payments of services offered by insurers [34]. In contrast, all German health insurers offer reward programs as kind of 'bonus' for customers. Customers that behave accordingly (e.g., doing preventive medical care), are able to get financial or item rewards directly from the health insurance.

Some insurers did inform about self-tracking fitness devices, for example on their blogs. In most cases, they give neutral information. Medibank has a health and fitness blog and talks about tracking and devices: "Tracking your activity can be an excellent way to motivate yourself and keep you reminded of your goals."[14] However, they do not mention their program they started with Flybuys in 2016. Other insurers clearly state that they are not interested in offering self-tracking reward options, for example due to data security reasons [35].

4.3 Data Privacy

When analyzing health insurer websites, we as well focused on data handling. Do insurers use the data to create statistics? Is there a possibility of inference? We have not found any evidence for that so far. Although data is handled differently in the two countries, most providers state that actual activity data is not saved by the insurance. However, please note that we are not interested in proving data privacy statements according to national standards and legal requirements. We are mainly interested in how insurers communicate about data privacy issues – specifically when other providers are involved – and if they provide information on personal health data sharing.

German Health Insurance. German Health Insurances put a strong emphasis on data security when communicating about their tracking services and apps. AOK Nordost, for instance, describes the protection of personal data as a "central concern" (Table 2). Regarding activity data they clearly state that only the bonus points for the processing of the bonus program and no actual fitness data is collected and stored. This is also true for AOK Plus, BARMER and Techniker Krankenkasse (Table 2). It becomes clear that these insurers want their customers to trust them and their technology – not only

---

[14] https://www.medibank.com.au/livebetter/be-magazine/exercise/13-ways-to-be-more-active-in-the-workplace/.

Table 2. Excerpts from health insurers' data privacy statements that include reference on the use of data from self-tracking programs.

Health insurer	Link to private policy information	Excerpts of data privacy statements[a]
AOK Nordost, GER	https://nordost.aok.de/inhalt/ neues-digitales-praemienprogramm-der-aok-nordost/	The principle of data economy: The protection of personal insurance data is a central concern of AOK Nordost in this innovative program. The data protection concept was therefore developed under scientific advice of the data protection expert Prof. Dr. med. Dirk Heckmann and agreed with the responsible data protection officer of the state of Brandenburg. According to the principle of data minimization, FitMit AOK only collects the data which is absolutely necessary for the execution of the rewards program. For example, only the bonus points collected by sports, but not the details of the activities, are forwarded to the AOK. Fitness data such as distance and speed or sensitive health data such as the results of a check-up are neither collected nor stored
AOK Plus, GER	https://plus.aok.de/inhalt/ bonusprogramm/	No fitness or vital data will be stored in the AOK Bonus App and/or transferred to AOK PLUS. Only the result of the physical activity test and the activity date will be saved by the AOK PLUS
BARMER, GER	https://www.barmer.de/ serviceapp/fit2go/fit2go-datenschutz	We want you to feel confident in using our app. The protection of your personal information is very important to us. We tell you when we save what data and what we use it for. Personal data is being recorded in the technically necessary amount only. In no case will the data collected be sold or otherwise passed on to third parties without your consent. BARMER adheres strictly to the regulations on data protection.

(*continued*)

Table 2. (*continued*)

Health insurer	Link to private policy information	Excerpts of data privacy statements[a]
		The moving minutes and activities stored in the FIT2GO App will not be communicated to BARMER or third parties. They are just for your own account of how much you have already moved
IKK Südwest, GER	https://www.muuvit.com/ikk/ data_privacy/574	The protection of your personal data is very important to us. Personal data are individual details about personal or factual circumstances of a specific or identifiable natural person. This includes information such as the civil name and the e-mail address. Since these data enjoy special protection, they are only collected by us to the extent technically necessary. Below we describe what information we collect during your visit to our website and how it is used. Our data protection practice is in accordance with the provisions of the Telemedia Act (TMG) and the Social Code (SGB). We will use your personal data exclusively for the performance of the IKK family adventure. We expressly exclude any disclosure of your data to third parties for advertising purposes
Signal Iduna, GER	https://www.app-life.de/	I agree that the data given by me will be electronically recorded and stored. My data is only used strictly for this purpose and used limited to processing my request. I can revoke this consent at any time with effect for the future. The data will also be deleted without my revocation if they are no longer required for the processing of the business transaction, taking into account existing retention periods

(continued)

Table 2. (*continued*)

Health insurer	Link to private policy information	Excerpts of data privacy statements[a]
Techniker Krankenkasse, GER	https://www.tk.de/techniker/ unternehmensseiten/ unternehmen/die-tk-app/tk-app-fitnessprogramm-2023654	The completed steps are only transferred from the connected data source to the TK-App for the purpose of mapping your progress in the program. The steps are not permanently saved in the TK-App. The Insurance therefore has no access to your actual step or other data
HCF, AUS, cooperation with Healthways	https://www.hcf.com.au/ members/manage-your-health/health-action-plans	Healthways will not disclose personal information to HCF other than your name, postcode, date of birth, gender, date of participation with *My Health Guardian*, member number and customer number in order to allow HCF to evaluate the service. With your permission Healthways may disclose your personal information to your regular doctor or other healthcare provider and may also collect your health information from HCF
myOwn Health, AUS, cooperation with AIA Vitality	https://www.myown.com.au/ en/privacy-policy.html	MO Health will also collect your personal information from AIA Australia Limited (AIAA) which provides and administers the AIA Vitality program. This includes viewing the activities you have undertaken as part of the AIA Vitality program and the Vitality points you have earned, and collecting your Vitality status from AIAA to determine whether you are eligible for a premium discount
NIB Health Funds Ltd. (Qantas Assure), AUS	https://support.qantasassure. com/hc/en-us/articles/ 217853218	Data that we access from HealthKit framework [note: used to measure fitness data within app] will not be used by Qantas, or shared with third parties, for the purpose of serving advertising

(*continued*)

Table 2. (*continued*)

Health insurer	Link to private policy information	Excerpts of data privacy statements[a]
St. Lukes Health, AUS	http://www.stlukes.com.au/MyHealthGuardian.aspx	Any personal or medical information you provide through the program remains confidential and is not released to St. LukesHealth
	https://myhealthguardian.com.au/Login/Login.aspx	Self-tracking provider Healthways: We may disclose your personal information to persons or organisations such as […] the entity that funds your participation in our programs (such as your private health insurer […])"

[a]German texts translated

through the high level of transparency and comprehensibility of these information, but also through formulations such as "we want you to feel confident in using our app" and "the protection of your personal information is very important to us."

In contrast, other insurers like Signal Iduna ("Sijox AppLife") and IKK ("Family Adventure with Fred" developed by Muuvit[15]) do not seem to be concerned to communicate data privacy issues clearly. Data handling and protection information on their websites is somewhat hidden and the language used rather vague. This is comparable to what we found on other websites, Moreover, it stays unclear, which data they store.

Australian Health Insurance. Most health insurers clearly state that they do not have access to any personal health and fitness data self-tracked by customers. However, regarding the external provider party apps, insurers refer to the provider's data privacy statements and their conditions. In Table 2 we collected excerpts of those data privacy statements regarding the handling of self-tracked data.

Insurers' third-party providers need to share information to offer rewards to eligible customers. MyOwn Health for example states they need to know customers activities to check their eligibility. They are very clear about the data they receive. In other cases, it remains ambiguous what personal information means for the insurers. Insurers that do not offer rewards, like St. Lukes, explicitly state that they do not have access to data collected by Healthways. However, Healthways does state to disclose personal information to insurers. Unlike examples in German statements, Australian insurers do not mention if they store personal data and for how long.

---

[15] https://www.muuvit.com/.

5 Discussion

Besides the number of health insurers – 38 in Australia and over 100 in Germany – self-tracking fitness data options are still an exception. However, the number of insurers offering self-tracking options has been increasing since the last three years, with currently seven Australia and at least six German insurers investigating self-tracking (Feb 2018). Medibank started its cooperation with Flybuys in 2014, whereas the Qantas Assure program was launched in 2016. Moreover, new players like life insurer AIA or health program provider Healthways cooperate with insurers. Healthways and AIA currently work together with two Australian insurers (offering two diverse funds each). Life insurer AIA started its program in Australia in 2014 [3].

It is striking that the business models and ways of offering self-monitoring services differ remarkably. Tracking options for clients varies. All Australian insurers except one (HBF) cooperate with third-party providers that establish health assessments, well-being programs and user interfaces (online and app) to make those programs accessible. Those partners as well offer rewards for fitness activities. In Germany, six health insurers promote self-tracking programs. In contrast to Australian insurers, most German insurers are able to reward their customers. For example, clients of the Techniker Krankenkasse receive cash bonuses for tracking their fitness with the mobile application.[16] Furthermore, most build their own mobile applications to allow self-tracking.

In the following, we will discuss the program in relation to outcomes of self-tracking research and data privacy concerns, and we will finish with open questions that arise from the status of self-tracking health insurance programs.

5.1 Effects of Insurer Programs

Most current programs rely on either a reward system that offers direct financial benefits (Germany) or indirect benefits via third-party products (Australia, e.g. Flybuys or Qantas flyer points). Another motivational factor to increase people's activity rates seems social engagement. Studies reveal that users want to engage and interact with other users (family members or online users), like having support, challenges and competitions [18, 30]. A study by Zhang et al. [36] showed that the competitive factor is even more relevant than the support factor. Participants that had access to rankings of peers and their activities (no matter if they got individual or team incentives), were more motivated to join exercise classes than those that were not in a team or those, who only got team support, but no access to any comparative peer rankings. Rooksby et al. [22] speak of *social tracking* and emphasize that play, competition, but also friendship and peer support are important for people. We see that current programs try to implement challenge features (like AIA, Qantas and My Health Guardian). Australian health insurer (HBF) enables customers to share their activities with friends, a feature that they emphasize in their program promotion. In Germany, we could not find

---

[16] https://www.tk.de/techniker/service/leistungen-und-mitgliedschaft/leistungen/praevention/tk-bonusprogramm/vorteile-tk-bonusprogramm-2006040.

anything similar – this may be due to the high value both insurers and users put on privacy and data protection.

Another factor to consider is the generalization of program intended goals. Exercise and Sports Science Australia (ESSA) notes[17] that the use of activity tracking devices and the general goals related a healthy lifestyle, like 10,000 steps per day, might not be desirable and beneficial by everyone. People with chronic conditions might need to achieve different fitness behavior. Aiming for a more beneficial and long-term effect on personal fitness behavioral, insurance programs need to be able to adapt to personal conditions and circumstances. AIA Vitality already combines its self-tracking options with a health assessment test that users do before they start tracking their activities. Activity levels refer to the assessment results. As well, HCF talks of a personalized health program (My Health Guardian by Healthways) that supports customers to "progress towards [their] chosen health goal" and is able to schedule "exercise activities into [their] daily routine"[18].

However, the question is if self-tracking programs consider diverse ranges of customers in a way that is beneficial for customers. Lupton [19] raises this concern as well and criticizes that current health promotion acts against people who already suffer from socioeconomic disadvantages. Social inequities for example exist regarding people's health and digital literacies. Those groups are not able to take advantage of digital health programs, or as well do not understand the information their personal data is able to reveal about themselves. Moreover, current health insurer programs show another concern. Financial unprivileged people might not be able to afford access to current self-tracking health programs. For example, programs offered by GMHBA in cooperation with AIA or the Sijox AppLife program by Signal Iduna are only available for customers who purchase specific insurance packages. These social aspects relate to the following concern, the use and interpretation of sensitive personal data.

5.2 Data Privacy Concerns

Until now, there are not many user studies, which give insights on people's opinion, perception and behavior regarding health and activity data privacy. Yoon, Shin, and Kim report two directions regarding privacy concerns: "unnecessary anxiety" and "vague fear". Lehto and Lehto [37] conducted qualitative interviews about the sensitivity of health data. They found that "information collected with wearable devices is not perceived as sensitive or private" by most users. The authors state that handling of tracked data "needs to be described clearly and transparently to mitigate any privacy concerns from the individuals." In another attempt to understand the privacy concerns of fitness tracker users, Lidynia et al. [38] conducted an online survey (n = 82). Here, participants preferred to keep logged data to themselves.

When it comes to the legal situation, however, many researchers agree, that "[i]n most countries, laws that govern the collection, storage, analysis, processing, reuse, and

---

[17] https://www.essa.org.au/media_release/do-you-wear-a-step-counter-on-your-wrist-be-aware-this-may-not-be-the-answer-to-long-term-health-benefits/.

[18] http://www.hcf.com.au/HealthGuardianVT/index.html.

sharing of data (…) fail to adequately address the privacy challenges associated with human tagging technologies" because they were "enacted decades ago" [39]. Although those legal issues are being worked on, the current situation is one of uncertainty.

Till [40] reports that although companies state that they do not sell identifiable information, studies found proof that they share user data. There are numerous reports about company activities exchanging data, including personal sensitive data like names and email addresses, activity and diets information [41]. Data privacy policies of health and fitness apps are not always clear about sharing and personal user data [42]. Some Australian health insurers refer to data privacies of the cooperating companies that arrange self-tracking options. Many do not. The definition of personal and sensitive data is ambiguous and inconsistent. Another critical issue is the relationships between health insurers and external health providers in Australia. Those providers rely on successful programs and active customers to be profitable. Interest in the commercial value of personalized data is obvious.

Moreover, recent research shows that activity tracker data like steps, heart rate, sleep, and location, might infer latent sensitive information, like drunkenness, fever or smoking [43–45]. Thus, tracker data becomes more powerful to reveal details about personal health and lifestyles. Users and experts alike fear that sharing tracked data with health insurance might mean that disadvantaged people will end up paying higher insurance premiums in the future. Health insurers offering those programs need to establish mechanisms that allow for credibility and user trust. One aspect might be transparency regarding data privacy policies. The question is, if those policies – regardless of them being applied according to effective law – need to be changed to increase trust and credibility.

6 Conclusion

We gave an overview of the status of German and Australian health insurance programs that allow for self-tracking by their customers. We found 13 insurers and examined their programs, as well as the communication about the programs' intentions and data privacy policies.

German insurers offer their own programs and apps, whereas most Australian insurers rely on third-party providers. Communication about programs is similar in both countries, but twofold. Insurers that do not offer rewards try to persuade customers to join the programs with emphasizing support in managing personal health and fitness plans. Insurers that offer rewards are emphasizing the benefit of the rewards, like cashback, discounts for products or services. Some products or services relate to health and fitness aspects, many are not. Although participation in those programs is voluntary, questions regarding disadvantages of unprivileged people arise. Major concerns arise with data privacy statements and ambiguous information on data sharing practices between companies. This is specifically the case, when third-party providers are involved. Statements do not always give a clear definition of personal information. Data sharing is required to prove if customers are eligible for rewards, as insurers and third-party providers state. German insurers seem more concerned about data privacy. They as well communicate about aspects of sensitive data storage. Future research needs to investigate if insurer programs with self-tracking options are contributing to the improvement of people's individual health and well-being, if they cause any discrimination on unprivileged groups, and if we need to reconsider privacy policies to prevent misuse driven by economic and political ambitions.

References

1. Van der Meulen, R., Forni, A.A.: Gartner says worldwide wearable device sales to grow 17 percent in 2017 (2017). https://www.gartner.com/newsroom/id/3790965. Accessed 15 Dec 2017

2. Ubrani, J., Llamas, R., Shirer, M.: Wearables aren't dead, they're just shifting focus as the marketgGrows 16.9% in the fourth quarter, according to IDC (2017). https://www.idc.com/getdoc.jsp?containerId=prUS42342317. Accessed 15 Dec 2017

3. Liew, R., Binsted, T.: Your insurer wants to know everything about you (2015). http://www.smh.com.au/business/retail/your-insurer-wants-to-know-everything-about-you-20151201-gld5t1.html. Accessed 15 Dec 2017

4. Mihm, A.: Erste Krankenkasse zahlt für Apple Watch (2015). http://www.faz.net/aktuell/wirtschaft/unternehmen/aok-erste-krankenkasse-zahlt-fuer-apple-watch-13736118.html. Accessed 15 Dec 2017

5. Boyd, A.: Could your Fitbit data be used to deny you health insurance? (2017). https://theconversation.com/could-your-fitbit-data-be-used-to-deny-you-health-insurance-72565. Accessed 15 Dec 2017

6. Yang, R., Shin, E., Newman, M.W., et al.: When fitness trackers don't 'fit'. In: Proceedings of the 2015 ACM International Joint Conference on Pervasive and Ubiquitous Computing (UbiComp 2015), pp. 623–634. ACM, New York (2015). https://doi.org/10.1145/2750858.2804269

7. Shih, P.C., Han, K., Poole, E.S., et al.: Use and adoption challenges of wearable activity trackers. In: Proceedings of iConference. iSchools (2015). http://hdl.handle.net/2142/73649. Accessed 15 Dec 2017

8. McMurdo, M.E.T., Sugden, J., Argo, I., et al.: Do pedometers increase physical activity in sedentary older women? A randomized controlled trial. J. Am. Geriatr. Soc. **58**(11), 2099–2106 (2010). https://doi.org/10.1111/j.1532-5415.2010.03127.x

9. Tedesco, S., Barton, J., O'Flynn, B.: A review of activity trackers for senior citizens: research perspectives, commercial landscape and the role of the insurance industry. Sensors (Basel) **17**(6) (2017). https://doi.org/10.3390/s17061277

10. Gough, D., Oliver, S., Thomas, J. (eds.): An Introduction to Systematic Reviews, 2nd edn. Sage, London (2017)

11. Arksey, H., O'Malley, L.: Scoping studies: towards a methodological framework. Int. J. Soc. Res. Methodol. **8**(1), 19–32 (2005). https://doi.org/10.1080/1364557032000119616

12. Australian Bureau of Statistics: Health Service Usage and Health Related Actions, Australia, 2014–15: Private Health Insurance (2017). http://www.abs.gov.au/ausstats/abs@.nsf/Lookup/by%20Subject/4364.0.55.002~2014-15~Main%20Features~Private%20health%20insurance~5. Accessed 20 Feb 2018

13. Australian Bureau of Statistics: Australian Demographic Statistics, June 2017 (2017). http://www.abs.gov.au/AUSSTATS/abs@.nsf/mf/3101.0. Accessed 20 Feb 2018

14. Bundesministerium für Gesundheit: Gesetzliche Krankenversicherung - Mitglieder, mitversicherte Angehörige und Krankenstand (2018). https://www.bundesgesundheitsministerium. de/fileadmin/Dateien/3_Downloads/Statistiken/GKV/Mitglieder_Versicherte/KM1_Januar_ 2018.pdf. Accessed 20 Feb 2018

15. GKV-Spitzenverband: Anzahl der Krankenkassen im Zeitverlauf (2018). https://www.gkv-spitzenverband.de/media/grafiken/krankenkassen/Grafik_Krankenkassenanzahl_Konzentrati onsprozess_300dpi_2018-01-03.jpg. Accessed 20 Feb 2018

16. Wikipedia: Private Krankenversicherung (2018). https://de.wikipedia.org/wiki/Private_ Krankenversicherung. Accessed 20 Feb 2018

17. Henkel, M., Heck, T., Göretz, J.: Dataset of 'Rewarding fitness tracking – the communication and promotion of health insurers' bonus programs and the usage of self-tracking data' (2018). https://doi.org/10.5281/zenodo.1183635

18. Lupton, D.: Beyond techno-utopia: critical approaches to digital health technologies. Societies 4(4), 706–711 (2014). https://doi.org/10.3390/soc4040706

19. Lupton, D.: Health promotion in the digital era: a critical commentary. Health Promot. Int. 30(1), 174–183 (2015). https://doi.org/10.1093/heapro/dau091

20. Lupton, D.: Self-tracking cultures. In: Robertson, T. (ed.) Designing Futures: The Future of Design. Proceedings of the 26th Australian Computer-Human Interaction Conference (OzCHI 2014), pp. 77–86. ACM, New York (2014). https://doi.org/10.1145/2686612. 2686623

21. Lupton, D.: The Quantified Self. Polity Press, Cambridge, Malden (2016)

22. Rooksby, J., Rost, M., Morrison, A., et al.: Personal tracking as lived informatics. In: Proceedings of the SIGCHI Conference on Human Factors in Computing Systems, pp. 1163–1172. ACM, New York (2014). https://doi.org/10.1145/2556288.2557039

23. Asimakopoulos, S., Asimakopoulos, G., Spillers, F.: Motivation and user engagement in fitness tracking: heuristics for mobile healthcare wearables. Informatics 4(1), 5 (2017). https://doi.org/10.3390/informatics4010005

24. Donnachie, C., Wyke, S., Mutrie, N., et al.: 'It's like a personal motivator that you carried around wi' you': utilising self-determination theory to understand men's experiences of using pedometers to increase physical activity in a weight management programme. Int. J. Behav. Nutr. Phys. Act. 14(1), 61 (2017). https://doi.org/10.1186/s12966-017-0505-z

25. Rowe-Roberts, D., Cercos, R., Mueller, F.: Preliminary results from a study of the impact of digital activity trackers on health risk status. Stud. Health Technol. Inf. 204, 143–148 (2014). https://doi.org/10.3233/978-1-61499-427-5-143

26. Lin, J.J., Mamykina, L., Lindtner, S., Delajoux, G., Strub, H.B.: Fish'n'Steps: encouraging physical activity with an interactive computer game. In: Dourish, P., Friday, A. (eds.) UbiComp 2006. LNCS, vol. 4206, pp. 261–278. Springer, Heidelberg (2006). https://doi. org/10.1007/11853565_16

27. Brodin, N., Eurenius, E., Jensen, I., et al.: Coaching patients with early rheumatoid arthritis to healthy physical activity: a multicenter, randomized, controlled study. Arthritis Rheum. 59(3), 325–331 (2008). https://doi.org/10.1002/art.23327

28. Toscos, T., Faber, A., An, S., et al.: Chick clique. In: Extended Abstracts on Human Factors in Computing Systems (CHI06), pp. 1873–1878. ACM, New York (2006). https://doi.org/ 10.1145/1125451.1125805

29. Glance, D.G., Ooi, E., Berman, Y., et al.: Impact of a digital activity tracker-based workplace activity program on health and wellbeing. In: Kostkova, P., Grasso, F., Castillo, C. (eds.) Proceedings of the 2016 Digital Health Conference (DH 2016), pp. 37–41. ACM, New York (2016). https://doi.org/10.1145/2896338.2896345

30. Purpura, S., Schwanda, V., Williams, K., et al.: Fit4life. In: Proceedings of the SIGCHI Conference on Human Factors in Computing Systems, pp. 423–432. ACM, New York (2011). https://doi.org/10.1145/1978942.1979003

31. Rosenbaum, M.S., Ramírez, G.C., Edwards, K., et al.: The digitization of health care retailing. J. Res. Interact. Mark. **11**(4), 432–446 (2017). https://doi.org/10.1108/jrim-07-2017-0058

32. Torre, I., Sanchez, O.R., Koceva, F., et al.: Supporting users to take informed decisions on privacy settings of personal devices. Pers. Ubiquit. Comput. **12**(2), 1 (2017). https://doi.org/10.1007/s00779-017-1068-3

33. Tudor-Locke, C., Bassett, D.R.: How many steps/day are enough? Sports Med. **34**(1), 1–8 (2004). https://doi.org/10.2165/00007256-200434010-00001

34. Australian Government: Private Health Insurance Act 2007 (2007). https://www.legislation.gov.au/Details/C2016C00911. Accessed 20 Feb 2018

35. IKK Südwest: Digitale Diät: Alle reden von Digital Health – und kaum jemand nutzt es (2017). https://www.ikk-suedwest.de/2017/06/digitale-diaet-alle-reden-von-digital-health-und-kaum-jemand-nutzt-es/. Accessed 20 Feb 2018

36. Zhang, J., Brackbill, D., Yang, S., et al.: Support or competition? How online social networks increase physical activity: a randomized controlled trial. Prev. Med. Rep. **4**, 453–458 (2016). https://doi.org/10.1016/j.pmedr.2016.08.008

37. Lehto, M., Lehto, M.: Health information privacy of activity trackers. In: Proceedings of the European Conference on Cyber Warfare and Security (ECCWS), pp. 243–251 (2017). https://www.jyu.fi/it/fi/tutkimus/julkaisut/tekes-raportteja/health-information-privacy-of-activity-trackers.pdf. Accessed 20 Feb 2018

38. Lidynia, C., Brauner, P., Ziefle, M.: A step in the right direction – understanding privacy concerns and perceived sensitivity of fitness trackers. In: Ahram, T., Falcão, C. (eds.) AHFE 2017. AISC, vol. 608, pp. 42–53. Springer, Cham (2018). https://doi.org/10.1007/978-3-319-60639-2_5

39. Voas, J., Kshetri, N.: Human tagging. Computer **50**(10), 78–85 (2017). https://doi.org/10.1109/mc.2017.3641646

40. Till, C.: Exercise as labour: quantified self and the transformation of exercise into labour. Societies **4**(4), 446–462 (2014). https://doi.org/10.3390/soc4030446

41. Kaye, K.: FTC: Fitness Apps Can Help You Shed Calories – and Privacy (2014). http://adage.com/article/privacy-and-regulation/ftc-signals-focus-health-fitness-dataprivacy/293080/. Accessed 20 Feb 2018

42. Peppet, S.R.: Regulating the Internet of Things: first steps toward managing discrimination, privacy, security and consent. Tex. Law Rev. 78 p. (2014). https://ssrn.com/abstract=2409074. Accessed 20 Feb 2018

43. Kawamoto, K., Tanaka, T., Kuriyama, H.: Your activity tracker knows when you quit smoking. In: Proceedings of the 2014 ACM International Symposium on Wearable Computers (ISWC), pp. 107–110. ACM, New York (2014). https://doi.org/10.1145/2634317.2634327

44. Ertin, E., Stohs, N., Kumar, S., et al.: AutoSense: unobtrusively wearable sensor suite for inferring the onset, causality, and consequences of stress in the field. In: Proceedings of the 9th ACM Conference on Embedded Networked Sensor Systems (SenSys), pp. 274–287. ACM, New York (2011). https://doi.org/10.1145/2070942.2070970

45. Yan, T., Lu, Y., Zhang, N.: Privacy disclosure from wearable devices. In: Proceedings of the 2015 Workshop on Privacy-Aware Mobile Computing (PAMCO), pp. 13–18. ACM, New York (2015). https://doi.org/10.1145/2757302.2757306

Motivations to Join Fitness Communities on Facebook: Which Gratifications Are Sought and Obtained?

Aylin Ilhan(✉)

Department of Information Science, Heinrich Heine University Düsseldorf,
Düsseldorf, Germany
aylin.ilhan@hhu.de

Abstract. Activity trackers are providing their users data on health and fitness. They measure, for instance, heart rates, record exercises and sleeping quality, and display burned calories. On Facebook, there are many activity tracker- and fitness-related groups. Why are users of activity trackers joining and consequently using such groups? In order to answer this basic question two theoretical approaches are adapted. Firstly, the *Uses and Gratifications Theory (U&GT)* identified gratifications, which are sought and obtained – in our case within those Facebook groups. Secondly, the *Self-Determination Theory (SDT)* is used to understand if the activities of users are caused by extrinsic or intrinsic motivations. For the purpose of this study an online survey was developed and distributed in 20 activity tracker- or fitness-related Facebook groups. All in all, data from 445 participants, who all are group members and are using an activity tracker, were evaluated.

Keywords: Activity tracker · Motivation · Gratification · Social media
Facebook · Uses and gratifications theory · Self-determination theory
Facebook groups

1 Introduction

In recent years activity trackers attract more and more the attention of researchers, especially within the Human Computer Interaction (HCI) community. They are not only focusing on technical improvements such as the enhancement of the measurement quality and the collection and visualization of the data [1–3], they are doing studies on user-based research related to the use and non-use of activity trackers as well [4–7]. Communities or rather the social online setting is less investigated related to activity trackers. According to Lee et al. [1] "products and services that promote health-related behavior, such as activity trackers, have increased dramatically in the market, little attention has been given to their social influences, such as social reinforcement from mediators." Also Rooksby et al. [8] describe activity trackers as social tracking devices and not only as health devices collecting data. Lee et al. [1] show that "in social media, the participants tried to make ideal presentations of themselves and gain emotional support, such as attention and reputation, from their social media friends." The social

© Springer International Publishing AG, part of Springer Nature 2018
G. Meiselwitz (Ed.): SCSM 2018, LNCS 10914, pp. 50–67, 2018.
https://doi.org/10.1007/978-3-319-91485-5_4

environment supports not only participant's improvement of health behavior, it enables the emotional support (relief and motivation), too [1]. Users of activity trackers regularly have the possibility to upload their activity records to Facebook. Fig. 1 shows that posts within Facebook groups can be diverse. The left part in Fig. 1 shows a discussion starting from a question. A Facebook user is searching after activity tracker-related information and received information by other users. The right part shows an overview (Fitbit dashboard) of succeed goals (steps, miles, active minutes and burned calories). User6 is self-presenting her-/himself by posting the succeed aims; and User6 got positive feedback from another Facebook user (User7).

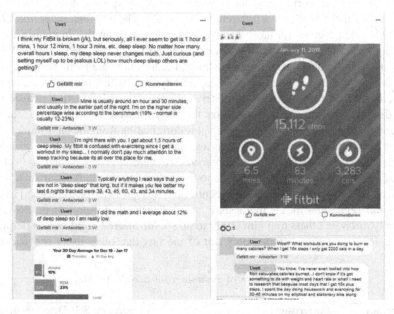

Fig. 1. Posts of a Facebook group (anonymized); left: User1 needs information; right: User6 realizes her-/himself (screenshot of a Fitbit dashboard).

Within Social Networking Services (SNSs), here, Facebook, we are able to identify numerous different fitness and health groups. Why do activity tracker- or rather fitness-orientated users cooperate with such Facebook groups if activity trackers provide functionalities that enable the improvement of health and fitness? Do they need the social reinforcement, competitions, information, entertainment, self-presentation or the motivation for the perseverance of fitness aims? To answer those questions the contribution is based on the *Uses and Gratifications Theory (U&GT)* and on the *Self-Determination Theory (SDT)*. The latter one "has increasingly become a basis for interventions in the areas of health promotion and physical activity" [9]. Ang, Talib, Tan, Tan, and Yaacob point out that *U&GT* is not sufficient to be able to comprehend why humans use and seek and obtain gratifications. Therefore, they used a mixed approach model (*U&GT* and *SDT*) for the analysis of online friendships [10]. We agree

that for a deeper understanding of motivational reasons and needs the uses and gratifications theory supports the comprehension but is not per se sufficient or the only
approach to understand completely the media use of individuals [10–12]. Therefore,
our study combines the two theoretical frameworks (*U&GT* and *SDT*). The purpose of
this study concentrates on the needs and motivational forces, why members of activity
tracker- or fitness-related Facebook groups are using this SNS.

2 Theoretical Background: SDT and U&GT

Humans all over the world carry out activities caused by specific needs. The motivations to satisfy those needs have different backgrounds. The *Self-Determination Theory*
(SDT) [9, 13–15] focuses on those backgrounds and point out that humans are doing
something based upon intrinsic or extrinsic motivation. The former is limited to the
activity itself. Individuals are doing something, because they are interested in it. There
is no exterior influence or pressure. It is the activity itself which motivates individuals.
The decision to do something is completely self-determined [13, 14]. Extrinsic motivation describes the situational condition that activities are done, because they are
expedient or an instrument to reach some values originating from the environment [13].
Extrinsic motivation has four subcategories, namely *external regulation, introjected
regulation, identified regulation* and *integrated regulation*. These subtypes (Table 1)
are built related to the strength of autonomy (self-determination) from own values
recognized in the environment to fully controlled through exterior influence [15].

External regulation means that a user joins a Facebook activity- or fitness-related
group only, because others told him or her to do so. *Introjected regulation* is defined as
the behavior to use those groups only out of the fact that other users and friends of

Table 1. Subtypes of extrinsic regulation [9, 14, 16].

Subtypes	Characteristics	Degree on Self-Determination
External	• Punishment, • Controlled rewards, • Compulsion.	Fully Controlled ◉ ○ ○ ○ Self-Determined
Introjected	• Predetermined consequences, • Worth conscience, • Partial internalization.	Fully Controlled ○ ◉ ○ ○ Self-Determined
Identified	• Identification with external values.	Fully Controlled ○ ○ ◉ ○ Self-Determined
Integrated	• Own values are coherent with exterior values, • Self-Endorsement.	Fully Controlled ○ ○ ○ ◉ Self-Determined

activity trackers are using those groups, too. Otherwise, if they do not join and use them they get a worth conscience, because it seems as not supporting other participants. This kind of extrinsic regulation is "a partial internalization in which regulations are in the person but have not really become part of the integrated set of motivations, cognitions, and affects that constitute the self" [13]. Even if *identified regulation* is more self-determined than *introjected* and *external regulation* it is still the activity itself which is instrumentalized to gain something. "[I]f people identified with the importance of exercising regularly for their own health and well-being, they would exercise more volitionally" [13]. Here, *identified regulation* is defined as the importance to support and help other users of activity trackers, for example, to reach their aims by forcing the social solidarity or to answer questions. The strongest autonomous regulation related to the extrinsic subtypes is described as *integrated regulation*. It is still a kind of extrinsic motivation, because of the fact, that individuals are doing something "to attain separable outcomes rather than for their inherent enjoyment" [16]. Activities or adapted values conditioned by integrated regulation "have been evaluated and brought into congruence with one's other values and needs" [16]. Besides individuals who are doing something out of intrinsic or extrinsic motivation, there are individuals who are not willing to do something.

Ryan, Williams, Patrick, and Deci describe three different reasons why people are feeling *unmotivated*: (1) lack of skills or knowledge to do an activity, (2) missing coherence between activity and desired results, and (3) missing interest [9].

The *Uses and Gratifications Theory (U&GT)* is first used for traditional media channels and examines why people decide to use a medium and which "needs" should be satisfied [17–21]. This theory leads on Katz et al. [18,19]. *U&GT* is also applied to SNSs and other social media channels [22–24]. Hsu et al. [25] point out that social media and their popularity triggers a lot of research attention, especially for the better understanding why people decide to use it [25, 26]. Hsu et al. [25] showed that the motives why people use social media are divided in two categories: Firstly, psychological needs and gratifications [27–30] and secondly, social interaction [29, 31]. According to Quan-Haase and Young [32], the *U&GT* is one "of the more successful theoretical frameworks from which to examine questions of 'how' and 'why' individuals use media to satisfy particular needs."

Therefore, *U&GT* stressed out "that users of media are active and goal-oriented" and that the selected medium depends on the satisfaction of those gratifications which satisfied the needs [25, 33]. The process of user's media use begins with a social and psychological need [18]. Such human needs lead to user's choice of a medium (e.g., Facebook) based on the expectation that the use of this medium can gratify the social and psychological need. Gratification is described as the behavior of seeking satisfaction of certain needs [34]. The satisfaction of certain needs, and therefore the motivation of using a specific medium, here Facebook activity tracker- or fitness-related groups, is based on the categorization related to McQuail [35], namely *information seeking, self-presentation, socialization,* and *entertainment.* Hsu et al. [25] show that a lot of researchers worked with the four categories in the context of social media [36–40]. Besides seeking of gratification, gratification can be obtained as well [41–43]. Palmgreen et al. point out that sought gratification and the obtained gratification are not always the same [43]. If an individual is searching for information he or she can obtain other aspects, too. The information content itself can be assumed entertaining as well. Additionally, the need of information can cause to keep in touch with other individuals to get specific information, too. Therefore, by searching information an individual can obtain social contacts (socializing) as well. Klenk et al. did research about fitness applications based on the theoretical framework of *U&GT* and *SDT* [44]. They found out that a combination of fitness applications with social media supports social gratifications [44]. "Sharing the results of physical activities via Facebook can provide social support through friends' encouraging comments or their own status information, allowing the comparison of one's own results with others" [44]. Furthermore, Park, Kee, and Valenzuela did already research Facebook groups' user's gratifications and found out that all needs, *U&GT* defines (seeking information, self-presentation, socialization, and entertainment) play an important role [30].

There is research on activity trackers, also related to motivational aspects, usefulness, ease of use, and gamification [4, 6, 7, 45–48]. But if one monitors the data of his/her activity tracker itself (e.g. steps, burned calories, heart rates), why does she/he participate in fitness-related Facebook groups? Our research idea includes four dimensions (D1–D4) (Fig. 2). The first dimension shows our target group, here, users of activity trackers. Based on the theory of *SDT* individuals are doing something out of intrinsic or extrinsic motivation (D2). "Something" is in this study defined as using Facebook related to activity tracker- or fitness-related groups (D3). To understand the needs of individuals who joins and uses those Facebook groups we applied *U&GT* and its four categories (D4).

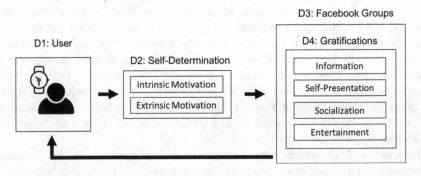

Fig. 2. Research model.

Based on our research model (Fig. 2), the study is going to answer the following research questions (RQs):

RQ1: Which gratifications are sought and which are obtained?
RQ2: Is there a correlation between sought and obtained gratifications?
RQ3: Are users more intrinsically or more extrinsically motivated?
RQ4: Do sought gratifications cohere with extrinsic or intrinsic motivation?
RQ5: Do obtained gratifications support the use of activity trackers?

3   Methods

This section describes the investigation's study design. It consists of the collection of quantitative data as outcomes of an online survey and the analysis of the quantitative data in order to answer the five research questions.

An online survey, with the help of eSurvey Creator[1], was created with all in all 26 items. Some of those scale items could not adopt from previous studies as they do not apply *U&GT* and *SDT* to comprehend the general use of Facebook groups around health, fitness, activity trackers, weight loss, nutrition and similar topics. The items are formulated by having regard to the core characteristics of the mentioned two theories. Our online survey was divided into three sections. The first section covers demographic information (gender, age, country), activity tracker related information ('Do you have an activity tracker?', 'I have been using an activity tracker since: …' and 'Without the Facebook group I would stop to use my activity tracker'), the type of user (producer, consumer, participant), testing item ('I'm currently a member of the following Facebook group: Name/Link') and a general free field for further comments. If participants answered the testing item with 'no' the survey was finished. The testing item was necessary to confirm that the participants are really a member of those Facebook groups.

The second section (see Appendix 1) examines the needs why users of Facebook groups use those activity- or fitness-related groups based on the *U&GT*. The theoretical framework considers both, gratifications sought and gratifications obtained. Participants having an activity tracker got the items (Appendix 1: #9–12), too. All items of the second section are equipped with a seven-point Likert scale, from 1 'It is absolutely not true', to 7 'It is absolutely true'. Participants got the possibility to choose "No Answer", too. The motives of the *U&GT* are completed by examples for the participants to support the easier understanding of each motive. The third section (see Appendix 2) deals with the *SDT*. Besides the intrinsic motivation (Appendix 2: #1), this section includes extrinsic motivation items (Appendix 2: #2–5) as well. The third section is equipped with the same seven-point Likert scale and the category "No Answer," too.

Before distributing the questionnaire, a pretest with five test persons was conducted to clarify discrepancies and vague descriptions. For German Facebook groups, the questionnaire was translated into German otherwise the questionnaire was in English.

The target group for this investigation is restricted. Only Facebook users who joined and use Facebook groups related to investigation's constrained topics come into consideration. Therefore, for each of the 20 analyzed Facebook groups the survey was duplicated with the only adaption of the testing question. Facebook groups such as 'Fitbit Charge 2 Group', 'Garmin vivosmart hr', 'Apple Watch', 'Freeletics Cologne', 'Fitbit For Women', 'Fitbit Weight Watchers Addict', 'Women-Fitness and Nutrition'[2] are examples where the survey was distributed. If one looks to some description of those Facebook groups one can find statements such as "let's post our accomplishments and met other fitbit users and change FITBIT ID'S. Let's form friendships […] and motivate each other to move!!!", "Feel free to share recipes, ideas, photos of your

---

[1] https://www.esurveycreator.com.

[2] Translated from German: Frauen-Fitness und Ernährung.

walks, celebrate stepping milestones or whatever else you want to discuss with the group [...]", "This group is for women [...] here you can ask questions, post recipes, fitness successes, and so on", "A group to discuss the Apple 1, 2 and 3 Series Watches! Post your questions, comments and pictures here!". The frequency of members varies from around 300 to 34,000 members ($\emptyset \sim 5,300$). On Facebook there are much more activity tracker- and fitness-related groups but only those are considered if their admins approved the distribution. A lot of Facebook groups did not allow the distribution of the questionnaire. There was no compensation for participants.

All in all, after data preparation 445 of 452 questionnaires, where participants affirmed the use of those groups, the using of activity trackers and completing the survey, are evaluable. The demographic data of our participants is shown in Table 2. As the data are not normally distributed, we worked with the Spearman-Rho correlation for identifying interrelationships between variables. The interpretation of the effect sizes are based on Cohen [49].

Table 2. Demographics of respondents.

Item	Answer	Frequency	Percent
Gender	Male	67	15%
	Female	375	84%
	I prefer not say	3	1%
Age	Silent Generation (1925–1945)	2	1%
	Baby Boomers (1946–1960)	26	6%
	Generation X (1961–1980)	184	41%
	Generation Y (1981–1998)	227	51%
	Generation Z (>1998)	6	1%
Country	Europe	368	84%
	North America	63	14%
	South America	1	0%
	Asia	7	2%
	Australia	6	1%
Operating time	N.A.	3	1%
	<2013	22	5%
	2014	30	7%
	2015	66	15%
	2016	105	24%
	2017	185	42%
	2018	34	8%
Without the group I would stop to use my activity tracker	Yes	3	1%
	No	442	99%

4 Results

4.1 Gratifications Obtained and Sought Within Activity Tracker- and Fitness-Related Facebook Groups (RQ1)

Based on the description of some Facebook groups, participants have the possibility not only to search for information on activity tracker products, wristbands, exercises and recipes, they are also able to post images of weight loss before and after a diet, to post achievements, to search for friends and to be motivated through other users. Which of those aspects are the most preferred seeking ones? To determine the most sought and obtained motive we calculated statistical values such as median, mean and the standard derivation (the last two values only for additional information as the data is not normally distributed) (Table 3). The key motive why participants are using those Facebook groups is explained by the motive "Information" (Median: 6). At least half of our participants confirm that it is true that they use this group to receive information.

Table 3. Gratifications sought and obtained within activity tracker- and fitness-related Facebook groups.

Gratifications	Statistical Values	
	Sought	Obtained
Information	Median = 6 (IQR = 2) Mean = 5.86 (±1.38) N = 436	Median = 6 (IQR = 2) Mean = 5.62 (±1.44) N = 436
Self-Presentation	Median = 3 (IQR = 4) Mean = 3.17 (±1.98) N = 371	Median = 4 (IQR = 3) Mean = 3.69 (±2.09) N = 371
Socialization	Median = 4 (IQR = 4) Mean = 4.11 (±2.01) N = 399	Median = 4 (IQR = 3) Mean = 4.28 (±2.00) N = 399
Entertainment	Median = 4 (IQR = 3) Mean = 4.39 (±1.71) N = 418	Median = 5 (IQR = 2) Mean = 4.65 (±1.69) N = 418

Scale: 1 (It is absolutely not true) – 7 (It is absolutely true); ± (Standard Deviation); IQR (Interquartile Range).

The possibility to realize oneself is the less preferred reason why users are using those groups (Median: 3). Participants reveal that it is not true that they are looking for the possibility to realize themselves. The receiving of achievements, to recognize one's successes (weight loss, stepping milestones) is one aspect, the need to sharing those successes with other, another. The participants do not deny that they occasionally use (Median: 4) those groups out of the fact that they seeking entertainment and socialization. Individuals who are searching for information can assume, based on the results, that they will receive needed information. They confirm that it is true that the use of the Facebook groups enables them to receive information (Median: 6). Conspicuously, while participants do not prospect the possibility to realize themselves more than half of participants state that they nevertheless posted successes occasionally (Median: 4). The Facebook groups try to be in general an informative platform, where users can satisfy their information needs related to activity tracker- or fitness-related topics. Participants agree (Median: 5) that it is rarely true that the use of those Facebook groups enables to feel entertained and to have fun. All in all, comparing the median values between gratifications sought and obtained, in two cases the experience (obtaining a gratification) is higher than the expectation (seeking a gratification), namely for self-presentation and for entertainment. Users do not explicitly seek for self-presentation and for entertainment; however, they get it.

4.2   Correlations Between Sought and Obtained Gratifications (RQ2)

Are there correlations between gratifications sought and obtained? Here, the correlations have to be interpreted always bidirectional. Table 4 shows the significance levels as well as the effect size. The effect size $r = .10$ is characterized as low, $r = .30$ as medium and $r = .50$ as strong. Based on the results there are nearly in all cases significant correlations between gratifications sought and obtained.

Participants obtained the gratification they sought (and – of course – vice versa). When a user seeks for information, she/he obtains information (.669***); looking for self-presentation, the participant gets it (.649***); hoping of socialization, it indeed happens (.686***); and finally, seeking entertainment is correlated with obtaining entertainment (.698***). All are strong and statistically highly significant correlations.

Table 4. Bivariate rank correlation (Spearman's rho) between gratifications sought and obtained.

		Gratifications Obtained			
		Information	Self-Presentation	Socialization	Entertainment
Gratifications Sought	Information	.669*** (N = 436)	.050 (N = 370)	.099 (N = 395)	.179*** (N = 414)
	Self-Presentation	.035 (N = 432)	.649*** (N = 371)	.359*** (N = 394)	.165*** (N = 412)
	Socialization	.189*** (N = 436)	.460*** (N = 374)	.686*** (N = 399)	.328*** (N = 416)
	Entertainment	.216*** (N = 437)	.272*** (N = 374)	.445*** (N = 399)	.698*** (N = 418)

Significance values: * p < .05, ** p < .01, *** p < .001

The seeking for a certain gratification results in many cases in obtaining different additional gratifications as well. Seeking for information correlates lowly with entertainment, but not with self-presentation and socialization. Seeking for self-presentation correlates with socialization with a medium effect and with entertainment (however, only on a low level). Looking for socialization, it correlates with all other gratifications obtained, namely information (low effect size), entertainment (medium effect size) and self-presentation (medium effect size). If one seeks for entertainment, she/he obtains except of entertainment itself information (low effect size), self-presentation (low effect size) and socialization (medium effect size).

4.3   Users' Intrinsic and Extrinsic Motivations (RQ3)

In fact, participants of activity tracker- or fitness-related Facebook groups are using them based mainly on intrinsic motivation (Median: 6) (Table 5). More than half of the participants confirmed that they like it to join those groups and that they do not have any external expectations by joining and doing something within this group. Nobody is compelling the users to join and use those groups (external regulation: Median: 1; introjected regulation: Median: 1).

Two extrinsic motivational factors have some influence, but both are tending to be self-determined. Participants tell that it occasionally happens that they use those groups because they identify with the values and behavior of those groups (identified regulation; Median: 4). Participants confirm that those values are occasionally coherent with the individual's own values and behavior (integrated regulation; Median: 4).

Table 5.   Use of Facebook groups caused by intrinsic and extrinsic motivation.

Item	Statistical Values
Motives, why I use this Facebook group:	
Intrinsic Motivation	Median = 6 (IQR = 3) Mean = 5.31 (±1.54) N = 439
External Regulation	Median = 1 (IQR = 0) Mean = 1.23 (±1.05) N = 439
Introjected Regulation	Median = 1 (IQR = 0) Mean = 1.26 (±1.05) N = 438
Identified Regulation	Median = 4 (IQR = 4) Mean = 3.89 (±2.02) N = 424
Integrated Regulation	Median = 4 (IQR = 3) Mean = 3.54 (±1.85) N = 383

Scale: 1 (It is absolutely not true) – 7 (It is absolutely true); ± (Standard Deviation); IQR (Interquartile Range).

4.4   Motivational Background of Sought Gratifications (RQ 4)

What is the motivational background of sought gratification (Table 6)? We would like to know which kind of motivation (intrinsic or extrinsic) is prevalent with the four defined gratification categories (information, self-presentation, socialization, and entertainment). The negative correlation delivers that the interpretation of the correlation of two data is contrary. If people are seeking for information within those Facebook groups, it does not correlate positively with external (−.149**) and introjected regulation (−.137**). If people would be compelled to use those groups or they would have a bad conscience it is not founded by the need to seek information.

Table 6. Correlations between intrinsic and extrinsic motivations and sought gratifications.

| | | Intrinsic | Subtypes of Extrinsic Regulation | | | |
			External	Introjected	Identified	Integrated
Gratifications Sought	Information	.250*** (N = 434)	−.149** (N = 433)	−.137** (N = 432)	.105* (N = 420)	.090 (N = 380)
	Self-Presentation	.032 (N = 430)	.186*** (N = 430)	.233*** (N = 429)	.326*** (N = 418)	.319*** (N = 378)
	Socialization	.123* (N = 434)	.084 (N = 434)	.122* (N = 433)	.476*** (N = 421)	.435*** (N = 381)
	Entertainment	.263*** (N = 433)	.030 (N = 433)	.078 (N = 432)	.303*** (N = 420)	.259*** (N = 381)

Significance values: *p < .05, ** p < .01, *** p < .001

The more people using those groups, because they are looking for the possibility to receive information the more it is intrinsically (.250***) motivated. Searching for self-presentation does not significantly correlate with the intrinsic motivation, but with extrinsic subtypes (external: .186***; introjected: .233***; identified: .326***; integrated: .319***). If the decision to use a Facebook group is controlled through looking for the possibility to socialize it exists a significant strong correlation between socialization and identified regulation (.476***) as well as with integrated regulation (.435***). The more the intrinsic motivation or the introjected and identified regulations (extrinsic motivation) is predominant the more participants are seeking for entertainment.

4.5   Gratifications Supporting the Use of Activity Tracker (RQ5)

The study shows that participants do not really need the Facebook groups to continue the use of their activity trackers. 99% of our participants mentioned that they would not stop to use the activity tracker without the use of those groups. Based on the possibility to obtain the chance to receive information, to socialize and to be entertained participants agree that it is indifferent that it supports the use of their activity trackers

(Median: 4) (Table 7). Self-presentation via a Facebook group does not support the use of an activity tracker (Median: 3).

Table 7. Gratifications and the use of activity trackers.

Item	Statistical Values
The use of your activity tracker is supported by…	
Information	Median = 4 (IQR = 4) Mean = 4.11 (±2.02) N = 427
Self-Presentation	Median = 3 (IQR = 4) Mean = 3.35 (±2.07) N = 368
Socialization	Median = 4 (IQR = 3) Mean = 3.78 (±2.06) N = 391
Entertainment	Median = 4 (IQR = 3) Mean = 4.01 (±1.91) N = 399

Scale: 1 (It is absolutely not true) – 7 (It is absolutely true); ± (Standard Deviation); IQR (Interquartile Range).

5   Discussion

The objective of this study was to find out why users of activity trackers are additionally applying activity tracker- or fitness-related Facebook groups. To answer this basic question, the study was based on two approaches: the *U&GT* and the *SDT*. A survey, designed following the basic principles of these two theories, was developed and distributed within different Facebook groups. One main point is to work out the sought and obtained gratifications (information, self-presentation, socialization, entertainment) and the satisfaction related to obtained gratifications. The second main point is to comprehend why Facebook users are tending to do something. Are their activities or their needs to look for gratifications caused by intrinsic or extrinsic motivation?

The demographic values show that participants have not had their activity tracker for too long. Half of 445 participants own their activity trackers since 2017. To the starting time of using a new device (here activity trackers), normally, users need information, for example, related to the ease of use and meaning of measured values (active minutes, sleeping phases). The study shows that an exchange of information is ensured. Within those Facebook groups the information need can be satisfied.

Functionalities of an activity tracker such as step milestones, calories burned, active minutes, sleeping phases are working without social support. But, challenges with other users of activity trackers, obviously, assume that users need other users. Individuals are not equal. Some individuals do not need social reinforcement; they are their own support and motivation. However, there are also individuals who need the support, the feedback, the emotional reinforcement to keep motivated. Facebook groups, which include descriptions like "Let's form friendships" focus on those values and are suited for those individuals.

The participants do not have the need to share their successes in Facebook groups. They are not looking for the possibility to realize themselves by posting and sharing achievements, step milestones, and so on. Indeed, it is assumed that the behavior of other users within those Facebook groups (posting of success), the group description, for example, "Feel free to share recipes, ideas, photos of your walks, celebrate stepping milestones" or invocations of admins to share successes are contagious. Likewise, participants are feeling more entertained in those Facebook groups as they expected. To sum up, based on the *U&GT* the study confirmed that participants are getting exactly or more than expected what they are seeking for in those groups.

Besides the fact, that participants obtained their sought gratifications, the correlations show that looking for the possibility to satisfy ones' need, for example, receiving information, enables also to feel entertained. Participants who are using those groups for seeking self-presentation obtained the chance to socialize, too. This is accounted by the fact that sharing or posting successes is an activity itself. But, the support through positive feedback and emotional support or compliments does not work without other users. The possibility to get to know other users, to feel motivated through the general social reinforcement can reduce participants fears to feel ashamed and strengthens the user's self-confidence. Generally, the analysis of evaluated data states that not always the sought gratification enables the obtaining of other gratifications and vice versa. Participants who look for self-presentation do not obtain information.

Those Facebook groups are not necessary for the continued use of activity trackers. 99% of participants argue that they would not stop to use their activity tracker without those groups. Considered the fact that an activity tracker is a device that enables the self-quantification by showing up the measured aspects, there is no Facebook group needed to get to know how many steps a user did. But, it is recognizable that the possibility to receive information, to socialize and to feel entertained sometimes support the use of the activity tracker. If someone would like to know how he or she can track an exercise the Facebook group can provide an answer and therefore support the use of the activity tracker. Individuals' social environment does not necessarily have an activity tracker. But without other users there is no possibility to start challenges. Mentioned Facebook groups enable to find other friends or, here, competitors for challenges; this supports the continued use of the activity trackers furthermore. Besides receiving information, feeling entertained and to socialize, self-presentation within those groups is rarely a factor that leads to the support of using activity trackers.

To sum up, the continued use of an activity tracker itself is depending on its own functionalities. Other studies show that the impact and ease of use of activity trackers have an influence on the use as well. However, sometimes the support of the Facebook groups in different ways should be considered as supporting aspect, too, but not as a necessary aspect.

Individuals are doing activities or decisions out of intrinsic or extrinsic motivation. Participants of this study provide the fact that seeking information within those groups is caused through one's own intrinsic motivation without any extrinsic influences.

Here, the need to seek information is not induced by external or introjected regulation. Descriptions of Facebook groups "Post and share your experiences, recipes, pictures and so on" encourage users of the group to post something. Self-presentation is caused through subtypes of the extrinsic regulation. Users who share their successes out of identified regulations agree with the values of the group, here to support each other, to share successes and to motivate other persons. Participants who are looking for socialization identify with the rules of the groups. It is important to support other users, to exchange experiences, to meet new users and to give feedback. Participants are not compelled to meet other people or to support each other. The more participants are seeking for entertainment the more this need is caused by their own motivation and the activity itself (to have fun). Beyond that, to seek for entertainment can be caused by identified and integrated regulation (extrinsic motivation) as well. The feeling of being entertained is triggered regularly through different conditions, for example, other users who try to create a funny atmosphere while sharing and chatting with other users. Especially admins try to convey a space of respect but with facility and humor. Based on the evaluated data, it is not possible to indicate for each sought gratification a type of motivation. Instead it is either intrinsic as extrinsic, too.

Ultimately, both theories *U&GT* and *SDT* enable an answer and an understanding of gratifications and motivations why some people are using activity tracker- and fitness-related Facebook groups. The understanding of the chosen shared successes or the kind of postings assume a deeper study with, for example, a content analysis. Which kind of successes users are posting? Health metrics such as calories burned, step milestones, average heart rates, etc. or pictures before and after a diet?

This study has some weak points. Over the entire analysis it should be at the back of one's mind that the groups are not consistent over all. The criteria which Facebook group could be used in this study was depended on the permission of admins and the topic. However, for a first comprehensive analysis of the role of those Facebook groups beside the devices themselves the study enables significant insights.

Appendix 1: Questions Concerning Uses and Gratifications Theory

#		Item
	U&GT: Sought Gratifications	I use this Facebook group, because I'm looking for the possibility …
1		… to receive information
2		… to realize myself (e.g., to show my success, aims and obtained achievements)
3		… to socialize (e.g., for being motivated, for challenges, and emotional reinforcement)
4		… to be entertained (e.g., to have fun)
5	U&GT: Obtained Gratifications	The use of the Facebook group actually enables me … … to receive information

(*continued*)

(continued)

#		Item
6		… to realize myself (e.g., to show my success, aims and obtained achievements)
7		… to socialize (e.g., for being motivated, for challenges, and emotional reinforcement)
8		… to be entertained (e.g., to have fun)
	Support of Activity Tracker	The use of your activity tracker is supported by …
9		… receiving information within this Facebook group
10		… self-realization (e.g., to show my success, aims and obtained achievements) within this Facebook group
11		… socialization (e.g., for being motivated, for challenges, and emotional reinforcement) within this Facebook group
12		… entertainment (e.g., to have fun) within this Facebook group

Appendix 2: Questions Concerning Self-Determination Theory

#		Item
		Motives, why I use Facebook groups:
1	Intrinsic Motivation	I like to use Facebook groups like this one. I don't have any external expectations.
2	Extrinsic Motivation: External Regulation	I was required (compelled) to use this group. I had no choice.
3	Extrinsic Motivation: Introjected Regulation	I use this group, because otherwise I would have a bad conscience as my circle of friends and acquaintances use groups like this one.
4	Extrinsic Motivation: Identified Regulation	I identify with the aims and behavior of this group and adapt them (e.g., the support of others). I'm agreeing with these practices and values.
5	Extrinsic Motivation: Integrated Regulation	The values and practices of the group are coherent with my own values and practices. I completely adapt them.

References

1. Lee, Y., Kim, M.G., Rho, S., Kim, D., Lim, Y.: Friends in activity trackers: design opportunities and mediator issues in health products and services. In: Proceedings of IASDR, pp. 1206–1219 (2015)
2. Choe, E.K., Lee, N.B., Lee, B., Pratt, W., Kientz, J.A.: Understanding quantified-selfers' practices in collecting and exploring personal data. In: Proceedings of the SIGCHI Conference Human Factors Computing Systems - CHI 2014, pp. 1143–1152 (2014). https://doi.org/10.1145/2556288.2557372

3. Fan, C., Forlizzi, J., Dey, A.: Considerations for technology that support physical activity by older adults. In: Proceedings of 14th International ACM SIGACCESS Conference Computer Accessibility - ASSETS 2012, pp. 33–40 (2012). https://doi.org/10.1145/2384916.2384923
4. Ilhan, A., Henkel, M.: 10,000 steps a day for health? User-based evaluation of wearable activity trackers. In: Proceedings of the 51st Hawaii International Conference on System Sciences, pp. 3376–3385. IEEE Computer Society, Washington, DC (2018)
5. Giddens, L., Leidner, D., Gonzalez, E.: The role of fitbits in corporate wellness programs: Does step count matter? In: Proceedings of the 50th Hawaii International Conference on System Sciences, pp. 3627–3635. IEEE Computer Society, Washington, DC (2017)
6. Gao, Y., Li, H., Luo, Y.: An empirical study of wearable technology acceptance in healthcare. Ind. Manag. Data Syst. 115, 1704–1723 (2015). https://doi.org/10.1108/IMDS-03-2015-0087
7. Fritz, T., Huang, E.M., Murphy, G.C., Zimmermann, T.: Persuasive technology in the real world: a study of long-term use of activity sensing devices for fitness. In: Proc. of the SIGCHI Conference on Human Factors in Computing Systems - CHI 2014, pp. 487–496. ACM, New York (2014)
8. Rooksby, J., Rost, M., Morrison, A., Chalmers, C.: Personal tracking as lived informatics. In: Proceedings of the 32nd Annual ACM conference on Human factors in computing systems - CHI 2014, pp. 1163–1172 (2014). https://doi.org/10.1145/2556288.2557039
9. Ryan, R.M., Williams, G.C., Patrick, H., Deci, E.L.: Self-determination theory and physical activity: the dynamics of motivation in development and wellness. Hell. J. Psychol. 6, 107–124 (2009). https://doi.org/10.1080/17509840701827437
10. Ang, C.S., Abu Talib, M., Tan, K.A., Tan, J.P., Yaacob, S.N.: Understanding computer-mediated communication attributes and life satisfaction from the perspectives of uses and gratifications and self-determination. Comput. Hum. Behav. 49, 20–29 (2015). https://doi.org/10.1016/j.chb.2015.02.037
11. Ko, H., Cho, C.-H., Roberts, M.S.: Internet uses and gratifications: a structural equation model of interactive advertising. J. Advert. 34, 57–70 (2005)
12. Ryan, R.M., Rigby, C.S., Przybylski, A.: The motivational pull of video games: a self-determination theory approach. Motiv. Emot. 30, 347–363 (2006). https://doi.org/10.1007/s11031-006-9051-8
13. Deci, E.L., Ryan, R.M.: The "what" and "why" of goal pursuits: human needs and the self-determination of behavior. Psychol. Inq. 11, 227–268 (2000). https://doi.org/10.1207/S15327965PLI1104_01
14. Ryan, R.M., Deci, E.L.: Intrinsic and extrinsic motivations: classic definitions and new directions. Contemp. Educ. Psychol. 25, 54–67 (2000). https://doi.org/10.1006/ceps.1999.1020
15. Deci, E.L., Ryan, R.M.: Intrinsic Motivation and Self-Determination in Human Behavior. Plenum Press, New York (1985)
16. Ryan, R.M., Deci, E.L.: Self-determination theory and the facilitation of intrinsic motivation, social development, and well-being. Am. Psychol. 55, 68–78 (2000). https://doi.org/10.1037/0003-066X.55.1.68
17. Kippax, S., Murray, J.P.: Using the mass media: need gratification and perceived utility. Communic. Res. 7, 335–359 (1980). https://doi.org/10.1177/009365028000700304
18. Katz, E., Blumler, J.G., Gurevitch, M.: Uses and gratifications research. Public Opin. Quart. 37, 509–523 (1973–1974)
19. Katz, E., Blumler, J.G., Gurevitch, M.: Utilization of mass communication by individual. In: Blumler, J.G., Katz, E. (eds.) The Uses of Mass Comunications: Current Perspective on Gratifications Research, pp. 19–32. Sage, Beverly Hills CA (1974)

20. Palmgreen, P., Rayburn, J.D.: Uses and gratifications and exposure to public television: a Discrepancy Approach. Commun. Res. **6**, 155–180 (1979). https://doi.org/10.1177/009365027900600203
21. Rubin, A.M.: Television uses and gratifications: the interactions of viewing patterns and motivations. J. Broadcast. **27**, 37–51 (1983). https://doi.org/10.1080/08838158309386471
22. Flanagin, A.J.: IM online: instant messaging use among college students. Commun. Res. Reports **22**, 175–187 (2005). https://doi.org/10.1080/00036810500206966
23. Larose, R., Mastro, D., Eastin, M.S.: Understanding Internet usage: a social-cognitive approach to uses and gratifications. Soc. Sci. Comput. Rev. **19**, 395–413 (2001). https://doi.org/10.1177/089443930101900401
24. Leung, L.: College student motives for chatting on ICQ. New Media Soc. **3**, 483–500 (2001)
25. Hsu, M.-H., Chang, C.-M., Lin, H.-C., Lin, Y.-W.: Determinants of continued use of social media: the perspectives of uses and gratifications theory and perceived interactivity. Inf. Res. **20** (2015). http://www.informationr.net/ir/20-2/paper671.html#.Wu4AXSP5wgo
26. Kim, Y., Sohn, D., Choi, S.M.: Cultural difference in motivations for using social network sites: a comparative study of American and Korean college students. Comput. Hum. Behav. **27**, 365–372 (2011). https://doi.org/10.1016/j.chb.2010.08.015
27. Lee, C.S., Ma, L.: News sharing in social media: the effect of gratifications and prior experience. Comput. Hum. Behav. **28**, 331–339 (2012). https://doi.org/10.1016/j.chb.2011.10.002
28. Dunne, Á., Lawlor, M., Rowley, J.: Young people's use of online social networking sites – a uses and gratifications perspective. J. Res. Interact. Mark. **4**, 46–58 (2010). https://doi.org/10.1108/17505931011033551
29. Li, D.C.: Online social network acceptance: a social perspective. Internet Res. **21**, 562–580 (2011). https://doi.org/10.1108/10662241111176371
30. Park, N., Kee, K.F., Valenzuela, S.: Being immersed in social networking environment: facebook groups, uses and gratifications, and social outcomes. Cyberpsychol. Behav. **12**, 729–733 (2009). https://doi.org/10.1089/cpb.2009.0003
31. Pempek, T.A., Yermolayeva, Y.A., Calvert, S.L.: College students' social networking experiences on Facebook. J. Appl. Dev. Psychol. **30**, 227–238 (2009). https://doi.org/10.1016/j.appdev.2008.12.010
32. Quan-Haase, A., Young, A.L.: Uses and gratifications of social media: a comparison of Facebook and instant messaging. Bull. Sci. Technol. Soc. **30**, 350–361 (2010). https://doi.org/10.1177/0270467610380009
33. Sangwan, S.: Virtual community success: a uses and gratifications perspective. In: Proc. 38th Hawaii Int. Conf. Syst. Sci, pp. 193c. IEEE Computer Society, Washington, D.C. (2005)
34. Rosengren, K.E.: Uses and gratifications: a paradigm outlined. In: Blumler, J.G., Katz, E. (eds.) The Uses of Mass Communications: Perspectives on Gratifications Research, pp. 269–286. Sage, Beverly Hills (1974)
35. McQuail, D.: Mass Communication Theory: An Introduction. Sage, London (1983)
36. Shao, G.: Understanding the appeal of user-generated media: a uses and gratification perspective. Internet Res. **19**, 7–25 (2009). https://doi.org/10.1108/10662240910927795
37. Chen, G., Yang, S., Tang, S.: Sense of virtual community and knowledge contribution in a P3 virtual community. Internet Res. **23**, 4–26 (2013). https://doi.org/10.1108/10662241311295755
38. Papacharissi, Z., Mendelson, A.: Toward a new(er) sociability: uses, gratifications and social capital on Facebook. In: Papathanassopoulous, S. (ed.) Media Perspectives for the 21st Century, pp. 212–213. Routledge, New York (2011)

39. Boyle, K., Johnson, T.J.: MySpace is your space? examining self-presentation of MySpace users. Comput. Hum. Behav. **26**, 1392–1399 (2010). https://doi.org/10.1016/j.chb.2010.04.015

40. Ellison, N.B., Steinfield, C., Lampe, C.: The benefits of facebook "friends:" social capital and college students' use of online social network sites. J. Comput. Commun. **12**, 1143–1168 (2007). https://doi.org/10.1111/j.1083-6101.2007.00367.x

41. Greenberg, B.S.: Gratifications of television viewing and their correlates for British children. In: Blumler, J.G., Katz, E. (eds.) The Uses of Mass Communications: Current Perspectives on Gratifications Research, pp. 195–233. Sage, Beverly Hills (1974)

42. Katz, E., Haas, H., Gurevitch, M.: On the use of the mass media for important things. Am. Sociol. Rev. **38**, 164–181 (1973)

43. Palmgreen, P., Wenner, L.A., Rayburn, J.D.: Relations between gratifications sought and obtained: a study of television news. Commun. Res. **7**, 161–192 (1980). https://doi.org/10.1177/009365028000700202

44. Klenk, S., Reifegerste, D., Renatus, R.: Gender differences in gratifications from fitness app use and implications for health interventions. Mob. Media Commun. **5**, 178–193 (2017). https://doi.org/10.1177/2050157917691557

45. Giddens, L., Leidner, D., Gonzalez, E.: The role of Fitbits in corporate wellness programs: does step count matter? In: Proceedings 50th Hawaii International. Conference on System Sciences, pp. 3627–3635. IEEE Computer Society, Washington, D.C. (2017)

46. Clawson, J., Pater, J.A., Miller, A.D., Mynatt, E.D., Mamykina, L.: No longer wearing. In: Proceedings of the 2015 ACM International Joint Conference on Pervasive and Ubiquitous Computing - UbiComp 2015, pp. 647–658 (2015)

47. Shin, G., Cheon, E.J., Jarrahi, M.H.: Understanding quantified-selfers' interplay between intrinsic and extrinsic Motivation in the use of activity-tracking Devices. In: iConference, pp. 45–47 (2015)

48. Ledger, D., McCaffrey, D.: Inside wearables: how the science of human behaviour change offers the secret to long-term engagement (2014). https://blog.endeavour.partners/inside-wearable-how-the-science-of-human-behavior-change-offers-the-secret-to-long-term-engagement-a15b3c7d4cf3

49. Cohen, J.: Statistical Power Analysis. Lawrence Erlbaum, Hillsdale (1988)

Conversation Envisioning to Train Inter-cultural Interactions

Maryam Sadat Mirzaei[1,2]([⊠]) [iD], Qiang Zhang[2], and Toyoaki Nishida[1,2]

[1] Human-AI Communication Group, RIKEN AIP,
Nihonbashi, Tokyo 103-0027, Japan
maryam.mirzaei@riken.jp
[2] Graduate School of Informatics, Kyoto University,
Yoshida-honmachi, Kyoto 606-8501, Japan

Abstract. Conversation is an integral part of human's relationship, which involves a large amount of tacit information to be uncovered. In this paper, we introduce the idea of conversation envisioning to disclose the tacit information beneath our conversation. We employ virtual reality for graphic recording (VRGR) to allow both observers and participants to visualize their thoughts in the conversation and to provide a training tool to acquire inter-cultural interactions using situated conversations. We focus on a bargaining scenario to highlight the tacitness of our conversations and use VRGR to make an in-depth analysis of the scenario. The proposed framework allows for performing a detailed analysis of the conversation and collecting different interpretations to provide timely assistance for realizing smoother cross-cultural conversations.

Keywords: Virtual reality · Graphic recording
Conversation envisioning · Inter-cultural interactions

1 Introduction

Conversation is the essence of human relationship and is far beyond simple message or information transfer. Through communication, we tend to build common ground, grow mutual understanding, learn about cultural patterns, develop empathy and many more. When conversing, verbals and non-verbals are combined seamlessly to convey our intentions [1]. A small piece of conversation can include an immense amount of information, not just bounded to what can be seen on the surface, but also deep beneath the surface structure. In this view, a lot of tacit dimensions can be found when we converse, leaving a vast area to explore and huge knowledge to learn for a better understanding of human-human communication and further improving human-AI communication [2]. Extracting such hidden aspects can also shed light on the dynamic and sophisticated structure of conversation and can be used as a training resource for understanding social interactions and cultural implications.

© Springer International Publishing AG, part of Springer Nature 2018
G. Meiselwitz (Ed.): SCSM 2018, LNCS 10914, pp. 68–80, 2018.
https://doi.org/10.1007/978-3-319-91485-5_5

While we learn math and science at school, we are supposed to learn to converse through osmosis and without specific training even when communicating with and relating to people from diverse backgrounds. However, acquiring appropriate cultural competencies is deemed critical in today's culturally and linguistically diverse society that involves sophisticated forms of conversations [3]. Developing cross-cultural skills can be realized by engaging in conversations with people of different cultures [4]. In this view, pedagogy needs to consider the immense tacitness of cross-cultural conversations in order to address people's assumptions about the exchanged messages and their experiences with other cultures. This paper introduces virtual reality graphic recording (VRGR) as a medium to envision tacit dimensions of the conversation, thus provide a training tool for participants to practice cross-cultural interactions, avoid misunderstanding and realize smoother communication (Fig. 1).

Fig. 1. Cross-cultural conversation in virtual reality augmented by interpretations using graphic recording

1.1 Misunderstanding Due to Cultural Differences

Conversation form varies across a wide range of inter-cultural context. Parties of conversation with different backgrounds often have culturally-conditioned assumptions about the messages exchanged, which may cause misunderstandings, misinterpretations or conflicts [5]. Misunderstandings frequently manifest during inter-cultural conversations due to the lack of common ground, which makes it difficult to decode the messages correctly. Common ground is the collection of knowledge, beliefs, and suppositions shared by the participants of interactions [6] and lack of it can lead to unpleasant or unexpected consequences or losing opportunities. The following conversation is an example of a cross-cultural dialogue that highlights the role of cultural differences and the tacit dimensions of one's assumptions and interpretations.

Customer: You know what? I am about to marry (smiling) and I am
              looking for something very special. What do you (exte-
              nding the arm toward the other person) suggest?
Shopkeeper: I usually recommend the brown ones (pointing to items)
              . They are generally popular.
Customer: I see. Well... these are good, but they are a little
              ...(lower pitch) Do you have anything else?
Shopkeeper: Oh, you can choose from those over there (pointing to
              the items), they're much cheaper (moving hand up-down)
Customer: No, I mean something special. For my WEDDING!(high in-
              tensity)
Shopkeeper: (Raised eyebrows, shrug) Why don't you consider these
              brown ones?
Customer: Humph (giving a snort, shaking head no), thank you
              (leaving the shop).

In this conversation, the customer from a feminine culture tries to socialize and make a relationship with the shopkeeper by sharing a personal story and asking for individualized suggestion prior to the bargaining or purchasing phase. The customer is expecting the shopkeeper to acknowledge her marriage (e.g., by saying "*congratulations*") and to recommend a very special item to her, since she is asking for an exclusive suggestion (extending arms toward the shopkeeper). The shopkeeper, on the other hand, is mainly focusing on the business, which is normally the case in the masculine societies [7]. The shopkeeper's behavior is rather recognized as an ignorance and indifference by the customer ("*I usually recommend...*"), which induces an unpleasant feeling. Moreover, the shopkeeper's general suggestion implies this message to the customer that he is not considering her as a special case, which fortifies her assumption of being ignored.

However, from the shopkeeper's viewpoint, the conversation is going on as a simple question-answer, which is a normal case at a shop. What meant to be a friendly conversation, is now more of a misinterpretation due to cultural differences. The customer, however, chooses to avoid being direct, rooted in feminine and collectivist society and tries to gently reject the suggestion by showing her hesitation.

From a business viewpoint, the focus of such conversation should be centered on the price, or the deal itself (typical for individualist and masculine society). Accordingly, in such a case, the customer's message could be interpreted as an issue with the price. With such assumption, the shopkeeper suggests cheaper products to show his helpfulness. This suggestion unnerves the customer, making her specify why she is not happy with this and the previous recommendation ("*I mean something special...*"). She indirectly informs the shopkeeper that she is not happy for being treated as an ordinary customer and does not like the commonly-used items. The shopkeeper is more confused, wondering why she is not considering the first recommendation, which is the best suggestion he could ever give to the customer. He shows this by shrugging and raising eyebrows.

What made the customer disappointed is not clear to the shopkeeper and he could not see the point that the customer is unhappy mainly because of being ignored rather than disappointed by the suggested items. This misunderstanding occurs due to lack of common ground and awareness between the shopkeeper and the customer and leaves the customer with no choice but to end this conversation and leave the shop. Even in doing so, the customer tries to cover the disappointment by saying "*thank you*" before leaving, although it is understood as a form of sarcasm.

1.2   Training Inter-cultural Interactions

The example that is given above clearly demonstrates the importance of socio-cultural competencies in evolving common ground and avoiding misunderstanding. This highlights the role of inter-cultural training and explains the motivation of this study. Cultural competency training has been the focus of many studies that tried to raise the awareness and develop inter-cultural skills by designing practical experiences [8,9]. The use of technologies such as machine learning techniques has been also playing roles for extracting surface information and using them for smoothing conversation [10]. However, less emphasize was given to abstract messages, high-level interpretations and common ground formation in cross-cultural interactions. This paper focuses on providing simulated experience in a virtual reality environment accompanied by learners'·collaboration and self-explanation, allowing for collecting explanations and presenting interpretations as a source of timely assistance.

Self-explanation is a knowledge-building activity and a process to assist the learners to understand the external input by explaining one's thoughts, which is done in an attempt to raise the conscious awareness of the mental process [11]. With inferencing and elaborating, learners are supposed to disclose and improve their mental representations. In addition, explanation-to-others as another principle also allows the learners to convey meaning and causality relations and learn by collaboration and discussion. This would allow us to make a corpus of interpretations and collective utterances derived from participants' collaborations, which can be further used by AI-agents or other participants. The utterances collected from self-explanation may be fragmented or incomplete whereas the utterances from the explanation to others are often more consistent and coherent [12]. Both would provide us with the rich knowledge to augment the conversation. Moreover, such collaboration in a situated environment would allow the participants to actively engage in role-plays and experience situated learning, where learning takes place as a result of social co-participation in a situation in which it occurs [13]. Finally, learners can practice inter-cultural interactions, while benefiting from observers' or other participants' interpretations derived from self or other-explanations. Empowered by the given interpretations, participants are supposed to train and improve their inter-cultural competencies and converse more smoothly. In this view envisioning and augmenting the conversation is anticipated to minimize misunderstandings, prevent conflict, reduce risks and lead to building rapport and good relations.

2   Envisioning as High-Level Conversation Analysis

To envision the tacit mental states of the participants in a particular situation, the conversation is required to be augmented to explicitly express the inner dynamics of the thoughts. To facilitate this augmentation, the participants are assumed to *"do things that they think will get them what they want"* [14]. Theory of mind specifies the reasoning behind actions to a set of belief-desire states. This type of analysis entails a sophisticated interconnecting graph of interpretative, predictive and explanatory resources that can be plentiful in the context of shared cultures or groups. More sophisticated interactions, such as inter-cultural interaction, however, have limited shared resources and raise new questions regarding the level of awareness of the parties, and their mental strategies manifested in their intentions and actions. Therefore, each action (e.g., utterance or gesture) in a given context can be augmented using a rubric shown in Fig. 2.

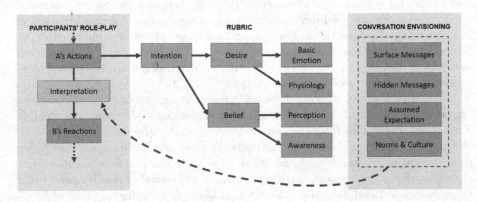

Fig. 2. Envisioning rubric of VRGR to make an in-depth analysis of the conversation: Intention, desire, and belief that induced participant (A)'s action is analyzed by meta-participants and used to identify the surface message and to interpret the intended or hidden message. In addition, participants' expectation of the forthcoming reaction is analyzed and compared with participant (B)'s actual reaction. In all of these analyses, norms and cultural context form the basis of envisioning.

In this model, a party perceives the incoming signals of the interaction such as linguistic structures, prosodic information, gestures, eye gaze, etc. and forms a belief about the current state of the interaction. This belief can range from an educated guess to a cultural expectation, a probable supposition from the experience, or an unjustified suspicion. This belief depends on the knowledge about the other party, and whether he/she is aware of the mentioned cultures, norms or rules [14].

On the other hand, basic emotions and physiological states affect the desires. The desire derives one to perform the action, whether it is based on wishes or hopes, or it is out of obligation [15]. Temporal and imposed emotions (e.g., anger, fear) may mask the long-term wishes and ideals that are the innate desires of a person, while physiology can shift the desires temporarily (e.g. hunger, pain).

According to Wellman [14] When reasoning about the actions, beliefs are especially useful because they are distinctly directed at the world (although they are hidden and implicit) and they stem in perceptual and evidential experiences of a person. Nevertheless, the desire can be indirectly inferred from verbal and non-verbal cues, e.g., facial expression signaling sort of anger, or body motions indicating pain.

Beliefs and desires form the one's intention to perform the action and to achieve his/her desired goal. In this view, a multitude of possible actions may come to the mind which fulfills the intention differently. When evaluating different actions, one is expecting a set of possible reactions, for which the likelihood depends on the culture, mood, and many more intrinsic and extrinsic factors. The final action which takes into account the hidden goals, intentions, anticipations, inner thoughts and the evaluation of different alternatives to fulfill the intention, would reveal a lot of information about norms and cultures, one's experience and his/her personality.

Once the action is executed, based on the previous assessment, a reaction may be fully anticipated, partially expected or totally unexpected, thus raising surprise. In conversation envisioning we follow this methodological rubric to clarify the prior step involved in decision making and action execution process. The envisioning process involves participants as the executors of the actions to disclose, explain, and visualize their mental states and assumptions as well as meta-participants to augment the conversation from a third-person viewpoint and to notify the ambiguous or conflicting assumptions. To this end, we leverage virtual reality and graphic recording to perform the envisioning process.

3   Virtual Reality for Conversation Envisioning

In an attempt to elicit obscure aspects of conversations, we introduce conversation envisioning as a powerful methodology, which strives to unveil the hidden structures of our conversation by exploiting the recent advances of virtual reality (VR) and artificial intelligence. We introduce virtual reality graphic recording (VRGR) as a platform to allow both observers and participants to disclose their thoughts in the conversation. We use such interpretations to augment the conversation and to build a training tool for the learners in order to acquire inter-cultural interactions. VRGR not only allows the investigators to annotate the conversation and fully share the situation from different perspectives (e.g., first or third-person view), but also endows the participants with a tool to envision their thoughts by themselves, thus learning from each other or from interpretations provided by the observers. Furthermore, VRGR serves as a training platform to provide timely assistance, thereby realizing smoother inter-cultural communication.

Figure 3 shows the building blocks of our proposed VRGR platform. In this framework, the video and audio of the participants are captured using camera and microphone, and their motions are alternatively captured by visual or markerless motion capture systems. Such input is fed to the system through the human-computer interface (HCI) unit and is used to reconstruct participant's motion and the speech by the character manager unit.

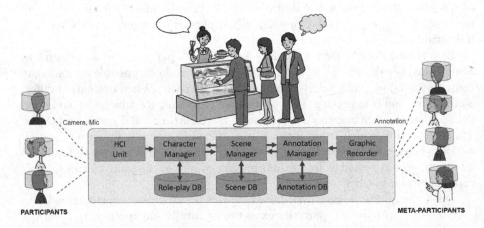

Fig. 3. VRGR framework

Character manager is in charge of generating participants' role-plays from a live input by transferring the motion data to the 3D avatar, transforming the input voice, and storing the play into role-play database. This can be further used as a source of training data for an autonomous agent that can produce realistic speech and action. This module is also in charge of replaying saved plays or synthesizing avatar behavior for an arbitrary input script.

The main character(s) as well as (optional) NPC characters populate a virtual world created by the Scene manager. The scene manager is in charge of providing context for the scenario, allowing for interaction of the avatars with the environment, providing different views of the scene such as first and third-person and free-form views, enabling traverse through time to facilitate the annotation task, allowing for playing/analyzing alternative role-plays, and enabling timely appearance of the desired annotations in conjunction with Annotation Manager.

The Annotation Manager is the main unit of gathering, combining and storing the annotations from different sources. The annotations may come from the participants themselves during a revision of the role-play to express their first-hand experience, their mental states, emotions, construal, thoughts, expectations, and explanations of the reason why they selected a specific course of actions throughout the scenario and what else they could do or they could think of. This process can potentially highlight the root-cause of miscommunications, conversation breakdowns, or possible negative feelings caused by different interpretations of the situation mainly due to the norm and cultural differences. Furthermore, third-party viewers from stakeholders, experts, people with similar experiences, or even interested ordinary people, can annotate the interaction to demystify why a behavioral pattern emerges in specific situations, or why the role-play has this particular trajectory rather than another one. This unit stores the annotations and retrieves them based on point-of-view, keywords, or even credibility of the annotations that using an internal up-voting mechanism.

To gather a thorough set of annotations, the Graphic Recording unit facilitates the annotation process of the participants/meta-participants by allowing them to place pre-built annotation templates such as text-boxes, arrows, highlight spheres, images, voice memos. Also, time controls (play, pause, slow-down, skip forward or backward) is provided to ease the task. This process can be done either within or outside of the virtual world (using a computer). In VR, an automatic speech recognition system is employed to execute commands and transcript annotator's input while wearing a VR head-mount display. The GR unit also enables editing and commenting on annotations to group relevant feedbacks in the same place.

4 Virtual Reality Graphic Recording for Training Inter-cultural Conversation

VRGR regenerates a played scenario and includes participants/observers in the annotation procedure while providing them with a friendly interface to add information to the interactions, thus making tacit information visible through graphic recording. With such functionality, the participants can explain their thoughts, play aloud to better disclose the underlying information, and revise the play on-the-fly. To highlight the vast amount of tacit information, we focused on a bargaining scenario as an interesting piece of social interaction, which represents many cultural points. We used VRGR as a tool to make an in-depth analysis of the scenario, provide useful interpretations to the participants, analyze the underlying reasoning, and uncover the mental processes that induced the conversational artifacts.

In a target multi-cultural bargaining scenario, first two or several participants play their roles that are captured by HCI unit. Meanwhile, character manager transforms their voices and motions to their corresponding avatars and interact with scene manager to update the virtual world based on this play. The scenario is then saved and replayed by the scene manager, while the participants along with extra third-party meta-participants discuss their experience and annotate the saved role-play accordingly.

During the discussion, several topics may be attended to or elaborated annotation may be considered necessary, which provides a rich source for situated analysis. The scenario may be replayed by other meta-participants later to augment the conversation and provide more insights on the situation. Through this reiteration, participant and meta-participants learn about the inter-cultural differences, practice how to resolve a specific situation or learn how to avoid unpleasant ones.

Using such tool learners can make a detailed analysis of the scenario and experience how different interpretations can lead the conversation into different branches. In this view, VRGR permits participants or observers to explicitly express their thought processes and virtual referents, hence experience or gain insights on the alternative situations that may result from different interpretations. Moreover, choosing topics such as bargaining, enable us to provide opportunities to develop the knowledge of other cultures in a game-like platform to motivate the learners. Using VR also allows us to design the environments that provide immersion in the target culture [16] and allow for cross-cultural experiential learning [17].

VRGR uses interpretations to augment the conversation and provides learners with a useful training tool, representing the interpretations of their actions and the expected outcomes induced from different perspectives or backgrounds.

5   Experimental Analysis

To evaluate our framework from the participants' viewpoints, we conducted a preliminary experiment, focusing on cross-cultural experiences gained from interactions using VRGR. The purpose of this experiment was to elicit learners' feedback on the effectiveness of this platform and to gain insight on improving the framework for future use and experiments. Through this interaction, the participants received interpretations on the cultural aspect of the conversation.

5.1   Participants

The participants of this study were 20 students with the majority being females (13) from different nationalities including Japanese, French, Thai, Korean, Palestinian, Chinese, American, etc. There were students of different majors including doctoral, master and undergraduate students.

5.2   Procedure

We asked the participants to play the role of the customer, in a conversation with the AI-shopkeeper who had a different cultural background. The goal of this conversation was to bargain with the shopkeeper and try to make a deal. The participants were asked to choose their next utterance from the available options. When doing so, the participants could hear their transformed voice and their avatar would perform the appropriate gestures.

The participants interacted with the AI agent two times, each time on a different scenario. On the first round, they did not receive any interpretations along with the conversation, whereas in the second round we provided culture-related interpretations to all the participants. We compare the participants' choices, monitored their interactions and analyzed the outcome of their conversation in bargaining with the agent for both conditions i.e. before and after receiving GR. After the experiment we asked the participants to explain if VRGR was helpful in understanding the situation, learning about cultural differences and leading to smoother conversation when the two participants have different cultural backgrounds and limited common ground.

5.3 Experimental Results

Participants' Interaction Using VRGR. Figure 4 shows how participants' interaction with the AI shopkeeper differed with and without GR. As the figure suggests, the majority of the participants tended to have more positive interaction with the agent when they learned about the actual intention of what the agent said based on the interpretations derived from the analysis of target cultural background and collected interpretations presented by VRGR.

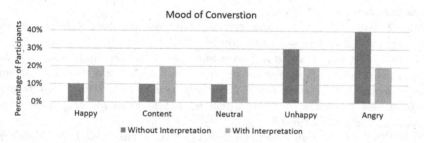

Fig. 4. Analysis of the mood in participants' conversations with the agent shopkeeper before and after receiving interpretations presented by GR

Figure 5 shows how VRGR helped the participants make a happy deal with the shopkeeper. Results demonstrate that VRGR not only helps the participant to reach the goal (make a deal) by conversation but also helps to raise participants' awareness about the other culture, thus bring the participants closer to the agent and induce a positive atmosphere. It should be noted that even when participants had a bad start in their conversations, later in the interaction they were given another chance to change the conversation's direction with the help of interpretations to reach a deal, although the negative mood arose at the beginning of the conversation may persist (see the rightmost bar in Fig. 5).

Participants' Feedback on VRGR. The followings are the participants' feedback on the use of VRGR given at the open-ended questions.

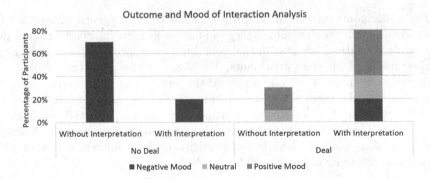

Fig. 5. Analysis of the outcome of participants' conversation with the agent shopkeeper (i.e. reaching a deal or not) in conjunction with the mood analysis before and after receiving interpretations presented by GR

Learning Cultural Points: The cultural points introduced in the interpretations were interesting and new to almost all of the participants (except one whose cultural background was close to the agent's):

- P3: "*I think GR is very helpful. It helps understanding why the shopkeeper was answering in a certain way.*"
- P12: "*I think GR can be helpful, especially to those who have different cultures or haven't experienced such situations.*"
- P5: "*Without GR, I would not have understood that the shopkeeper was just being polite when offering goods for free, for example, and I would not have asked again for the item price. In situations like this, I think GR is most helpful and I can see that this would be helpful when learning about other social norms and cultures.*"
- P19: "*The information about cultural background was helpful. If I didn't get the information about it, I'd have made another choice.*"

Building Common Ground: GR was mentioned as a beneficial tool for building and expanding common group between people, according to the participants:

- P2: "*GR was very helpful in bringing the two shopkeeper and customer to a common ground or understanding of what was going on. It made clear what the shopkeeper was intending with each phrase, gesture, or intonation.*"
- P14: "*I believe it is a very good idea to provide GR to help understand the flow of the conversation.*"

Serving the Purpose of Education: Participants particularly noted that the interpretations provided by GR sound promising for educational purposes:

- P7: "*I would consider this an effective method to practice social interaction in specific contexts with discrete education goals since the response choices are prescribed and the behavioral analysis and interpretation is discretely defined*" *during the conversation, which are conditions that do not appear in current reality."*

Our preliminary results received positive participant feedback on the usefulness of VRGR to raise understanding in cross-cultural interactions. Furthermore, the results suggested that VRGR augmented the conversation and assisted the participants to have a smoother cross-cultural communication. Therefore, the interpretations provided by GR were rated effective in developing cross-cultural competence and understanding the differences.

6 Conclusions

We introduced VRGR as a medium for training cross-cultural interactions. Benefiting from VR technology, participants are allowed to interact with a virtual agent in a simulated environment. The platform allows participants and meta-participants to revisit the conversation, and make an in-depth analysis of the situation. Following a specific rubric, participants and meta-participants use VRGR to specify the belief and intention that induce one's action and the expectations about a forthcoming reaction. We evaluated the proposed framework by conducting experiments with participants from different cultural backgrounds. VRGR has received positive participant feedback and is found as a promising tool to enhance cultural competencies and facilitate inter-cultural interactions. Future directions include the incorporation of facial expressions into avatars, the extraction of surface information from the conversation for meta-participants, and the design of content-specific training sessions for the participants.

References

1. Tversky, B.: Visualizing thought. Top. Cogn. Sci. **3**(3), 499–535 (2011)
2. Nishida, T., Nakazawa, A., Ohmoto, Y., Mohammad, Y.: Conversational Informatics: A Data-Intensive Approach with Emphasis on Nonverbal Communication. Springer, Japan (2014). https://doi.org/10.1007/978-4-431-55040-2
3. Miranda, A.H.: Best practices in increasing cross-cultural competency. Best Pract. Sch. Psychol. Found. **4**, 49–60 (2014)
4. Jones, J.: Best practices in providing culturally responsive interventions. Best Pract. Sch. Psychol. Found. **1**, 353–362 (2002)
5. Storti, C.: Cross-Cultural Dialogues: 74 Brief Encounters with Cultural Difference. Nicholas Brealey, Boston (2017)
6. Nishida, T.: Human-Harmonized Information Technology: Horizontal Expansion, vol. 2. Springer, Japan (2017). https://doi.org/10.1007/978-4-431-56535-2
7. Hofstede, G.J., Jonker, C.M., Verwaart, T.: An agent model for the influence of culture on bargaining. In: Proceedings of the 1st International Working Conference on Human Factors and Computational Models in Negotiation, pp. 39–46. ACM (2008)
8. Kim, J.M., Hill Jr., R.W., Durlach, P.J., Lane, H.C., Forbell, E., Core, M., Marsella, S., Pynadath, D., Hart, J.: Bilat: a game-based environment for practicing negotiation in a cultural context. Int. J. Artif. Intell. Educ. **19**(3), 289–308 (2009)

9. Lane, H.C., Ogan, A.E.: Virtual environments for cultural learning. In: Second Workshop on Culturally-Aware Tutoring Systems in AIED 2009 Workshops Proceedings. Citeseer (2009)
10. Fussell, S.R., Zhang, Q. (eds.): Workshop on Culture and Collaborative Technologies (CHI 2007), Proceedings (2007). http://www.cs.cmu.edu/~fussell/CHI2007/overview.shtml
11. Chi, M.: Self-explaining expository texts: the dual processes of generating inferences and repairing mental models. Adv. Instr. Psychol. 5, 161–238 (2000)
12. Hempel, C.G., et al.: Aspects of Scientific Explanation. Free Press, New York (1965)
13. Lave, J., Wenger, E.: Situated Learning: Legitimate Peripheral Participation. Cambridge University Press, Cambridge (1991)
14. Wellman, H.M.: Making Minds: How Theory of Mind Develops. Oxford University Press, New York (2014)
15. Wellman, H.M.: The Child's Theory of Mind. The MIT Press, Cambridge (1992)
16. Stanley, G., Mawer, K.: Language learners & computer games: from. TESL-EJ 11(4), n4 (2008)
17. Cushner, K., Brislin, R.W.: Intercultural Interactions: A Practical Guide, vol. 9. Sage Publications, Thousand Oaks (1996)

Gamification Design Framework for Mobile Health: Designing a Home-Based Self-management Programme for Patients with Chronic Heart Failure

Hoang D. Nguyen[1]([✉]), Ying Jiang[2], Øystein Eiring[3],
Danny Chiang Choon Poo[1], and Wenru Wang[2]

[1] Department of Information Systems and Analytics, School of Computing,
National University of Singapore, Singapore, Singapore
{hoangnguyen, dpoo}@comp.nus.edu.sg
[2] Alice Lee Centre for Nursing Studies, Yong Loo Lin School of Medicine,
National University of Singapore, Singapore, Singapore
{nurjiy, nurww}@nus.edu.sg
[3] Norwegian Institute for Public Health, Oslo, Norway
oystein.eiring@fhi.no

Abstract. Gamification is the design nexus between psychology and technology; thus, the ensemble of game design concepts and mobile health is promising for a far-reaching impact in public health. This paper presents a gamification design framework for mobile health as a unified, structured representation of activity systems aiming towards better health-related outcomes. It provides a valuable guideline for researchers and designers to model and gamify complex interventions into mobile health design with four steps: (i) defining activity systems, (ii) modelling, (iii) transforming, and (iv) designing. The framework is demonstrated for gamification of a home-based self-management programme for patients with chronic heart failure.

Keywords: Gamification · Design framework · Mobile health
Activity theory · Self-management · Chronic disease · Heart failure

1 Introduction

Gamification has increasingly become a hot topic over the past few years. The adaptation of game design concepts in non-gaming contexts is a new, promising paradigm for reshaping user engagement [1, 2]. Gamification design has been studied as useful in many areas such as e-commerce [3], education [4, 5], and health care [6]. It is the combination of psychology and technology point towards business objectives [7]; however, the role of technology has been significantly evolved in gamification. With unique technological capabilities, mobile health is the next suitable wave of technologies for modifying patient engagement and activities, thereby transforming gamification design in health care. Hence, the focal research question of this paper is: "how to gamify mobile health to address public health problems?"

© Springer International Publishing AG, part of Springer Nature 2018
G. Meiselwitz (Ed.): SCSM 2018, LNCS 10914, pp. 81–98, 2018.
https://doi.org/10.1007/978-3-319-91485-5_6

Chronic Heart Failure (CHF) has become a major problem of public health with its high and increasing prevalence worldwide [8]. The risk of CHF increases sharply along with patients' age and gender [9]. It is estimated that over 30% of individuals who are aged 55 and above will likely develop heart failure during the remaining course of their life [10]. CHF occurs as a clinical endpoint of many cardiovascular disorders, especially those that impair cardiac function or strain the cardiac workload [11]. It has become a heavy burden of healthcare systems; hence, self-management has been long recommended as an integral component of treatment for patients with CHF [12]. Nevertheless, many older patients who would potentially benefit from self-management interventions do not participate them [13]; moreover, a past study has shown that feelings of hopeless and powerlessness are common among them [14]. Therefore, it is imperative to explore the ensemble of gamification and mobile health for a simple, convenient means of self-management with a right amount of motivational affordances. It has capabilities to empower patients with continuous psychological and social support from care givers, family members, and friends.

This study investigates gamification design as a holistic structure of activity systems. We propose a gamification design framework for mobile health with four steps: (i) defining activity systems, (ii) modelling, (iii) transforming, and (iv) designing.

This paper contributes to the cumulative theoretical development of gamification design and mobile health in several folds. The proposed design framework provides a structured procedure for researchers and designers to model and to gamify complex interventions into mobile health design. We also present a gamification of home-based self-management programme for patients with CHF.

The structure of the paper is as follows. Firstly, we discuss the background of our research in the next section. Secondly, we describe a gamification design framework for mobile health in great details in the Sect. 3. And then, the paper demonstrates the use of the design framework for a home-based heart failure self-management programme. Lastly, we concluded our paper with findings and contributions of the research in the final section.

2 Background

2.1 Gamification Design

Gamification refers to the use of game design elements in non-gaming [1]. Such design elements are motivational affordance contexts for engagement purposes aim towards specific goals [15]. Recent studies have shown positive, psychological effects of gamification in many areas such as e-commerce [3], education [4, 5], and health care [6]. Table 1 summarizes the typical game design elements, which are adapted from Hamari et al. (2014) [16].

The design process of gamification is about how to incorporate various gamification elements into different contexts [17]. Several gamification design framework have been developed as guidelines for researchers and designers in general and business-specific domains [15]. Werbach and Hunter (2012) proposed the most popular design framework in Six Steps to Gamification (6D) [18]. The 6D framework consists

Table 1. Typical game design elements (Adapted from Hamari et al. 2014)

Game design element	Description
Reward	An obtainable object in recognition of participation, efforts, or performance such as monetary incentive, status, or self-development benefit
Point	Users earn points by joining the activity
Clear goal	The object of users' efforts to achieve in the activity
Leaderboard	A visual board for displaying and comparing user's performance with others
Achievement/Badge	Users can be awarded with virtual badges or recognitions, or merits for participation and performance
Level/Progress/Feedback	An indicator that provides reflections to users on where they are in the activity
Story/Theme	A mechanic that draws the users' interest into the activity. They provide guidance for users to move forward
Challenge	A pre-defined task that motivates users to action
Loop	An engagement mechanic that integrates positive reinforcement into repeated activities to keep users engaged

of six steps: (1) define business objectives, (2) delineate target behaviors, (3) describe your players, (4) devise activity cycles, (5) don't forget the fun!, and (6) deploy the appropriate tools. In a similar vein, the GAME framework developed by Marczewski (2012) suggests breaking the design process into two phases: planning and designing, to derive a solution in the gamification context [19]. In additions, a number of business-specific frameworks for designing gamification has been introduced to activate game elements and techniques in complex business processes [20, 21]. Nonetheless, there is a dearth of a structured framework for gamification design that is capable of modelling and translating activities in different contexts into practical design. Furthermore, new technologies are constantly reshaping human engagement; thus, gamification is not without considering the enabling role of technologies.

2.2 Mobile Health

The growth spur of mobile technologies has paved the path for new health interventions, "mobile health" or "mHealth" [22]. It is broadly defined as "the use of mobile computing and communication technologies in health care and public health" [23] which is capable of delivering health services to a huge number of people. Mobile health technologies have transformed a variety of health-related activities such as disease management and prevention [24–28], care surveillance [29–31] and instructional interventions [26, 32].

With the significant advantages of usability and mobility, mobile health apps are the next suitable wave of interventions for enhancing healthcare [33, 34]. They offer a wide range of capabilities from displaying and reporting to real-time sensing and social media sharing. Table 2 shows a list of common mobile health capabilities, which are typically used in combination.

Table 2. Common mobile health capabilities

Capability	Description
Interactive touch	A smartphone is typically equipped with a touch screen which allows users to control through simple or multi-touch gestures
Connectivity and social media	Mobile and internet connectivity allow smartphone users to maintain constant connections with social support actors or care givers
Displaying/Reporting/Visualization	Rich displaying and reporting contents such as websites, images, and videos provide multiple means for health intervention delivery
Wireless health sensing and monitoring	Wireless technologies such as Bluetooth, Near-Field Communication (NFC), and Wi-Fi allow users to connect medical devices for continuous sensing and monitoring
Push notification/Reminder	A mechanism to bring information to smartphone users for their attention

Mobile health interventions in their current form, however, are limited in psychological capabilities, which possibly render their effects temporary on health-related behaviors [35]. Hence, gamification of mobile health is a promising paradigm towards transforming patients' engagement for better outcomes.

3   Gamification Design Framework for Mobile Health

Gamification is defined by the Oxford dictionary as "the application of typical elements of game playing (e.g. point scoring, competition with others, rules of play) to other areas of activity" [36]. The core of gamification leverages on the principles of game design theory [15]; however, the design process of gamification is conceptually distinguished from game design, which is typically for entertainment purposes. The main purpose of gamification design is to enhance activities in different contexts towards specific, shared outcomes. Hence, the design process does not only entail the use of game design elements or techniques, but also revolves around human activities with the use of technology. In this paper, gamification design is viewed as the investigation of activity systems based on multiple factors such as tools, controls, contexts and communications. Exploring activities as dynamic phenomena is the key to obtaining extensive understanding of the design process, and supporting expressive human interactions in gamification.

The engagement between people and technology develops and shares meanings and objectives in activity systems. Mobile health, in which smartphones are interactive agents, has continuingly transformed human-technology interactions in an unprecedented way. It is necessary for the design process to capture the operationalization of the dialectal relationship between people and mobile technology. Therefore, a design framework for gamification of mobile health is proposed as a unified structure of activity systems, which encompasses various aspects of human-technology interactions, game design elements, and mobile health capabilities. It consists multiple steps for researchers and designers to extend the framework structure into practical design artefacts. Four steps in the gamification design framework are: (1) defining activity systems, (2) modelling, (3) transforming, and (4) designing.

Step 1: Defining Activity Systems

The first step of the gamification process is to define the activity systems in mobile health. It involves decomposition of the activity systems into several components for analysis and design from both psychological and technological perspectives. Grounded in the activity theory [37], we propose a holistic structure of activity systems for gamification of mobile health based on our previous thereotical framework [27]. It consists of seven components: (1) outcome, (2) subject, (3) object, (4) tool, (5) context, (6) control, and (7) communication, as shown in Fig. 1,

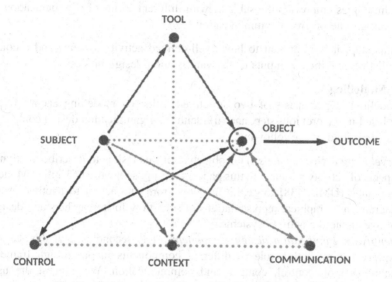

Fig. 1. Activity systems for gamification of mobile health

(1) *Outcome.* A well-developed understanding of the outcome in activities is vital for the gamification design process. In mobile health, improvements in knowledge and skills, as well as health-related behaviors are commonly defined as the collective outcome.

(2) *Subject.* The "subject" component depicts both individual and social nature of the activity. The stakeholders of health interventions are patients and psychological/social support actors.

(3) *Object.* The "object" or "objective" component reflects the purposes of activities that allow the manipulation of subjects' actions. In mobile health, the object is typically a collection of health-related behaviours and outcomes.

(4) *Tool.* The "tool" component shows the mediation aspects where in-teractions between subjects are not direct, but mediated through the use of technological capabilities and instructional materials. Mobile health apps have excellent capa-bilities as discussed in Sect. 2.2 to facilitate health interventions.

(5) *Control.* The "control" component refers to how activities are conducted and managed based on technological and psychological elements. Game design ele-ments such as stories, clear goals, or challenges are some examples of gamifi-cation controls.

(6) *Context.* The "context" component highlights the situation and environment, in which the subjects perform activities. For instance, locational, temporal, and social constraints have been removed in mobile health; thus, it enables subjects to participate in activities anywhere anytime. Social media is also a technological context that motivates subjects to action.

(7) *Communication.* The "communication" component refers to how subjects engage and what are their role structure in the activity systems. The use of mobile technologies empowers the subjects with different forms of communication such as one-to-one or group communication.

In this step, it is important to have well-defined activity systems and a compre-hensive list of activities as inputs of the gamification design process.

Step 2. Modelling
The modelling step consists of two structured tasks: (i) modelling each and every activity listed in the previous step, and (ii) identifying gamification design components. They are described as the following.

(i) *Developing activity models.* This objective of this task is to describe the dynamic aspects of activity systems. In this paper, we propose the use of Unified Modelling Language (UML) [38] to depict the flow from one action to another. Activity diagrams are graphical representations of such flow for researchers and designers to investigate the activity systems.

(ii) *Identifying gamification design components.* The second task of this step is to explore the mediating role of different components in our design framework including tool, control, context, and communication. We suggest the use of Mwanza's techniques [39] to generate possibile gamification opportunities. In Fig. 1, the arrows run from the subject through a mediating component towards the object, in which gamification opportunities can be recognized in the activity systems. There are four questions to identify gamification elements: what tool can help the subject to achieve the objective? what control can affect the subject to transform the object? what context can influence the subject to reach the object? or what communication can change the way the subject reach the object?

This step develops activity descriptions with a list of gamification design components.

Step 3. Transforming
Psychological/social support actors, including doctors, nurses, care givers, family members, or friends, play a critical role to intrigue patients to participate in health

interventions. Based on activity diagrams in the step 2, the activities can be transformed by supplementing or substituting such actor with technological subject in this step. In the former scenario, smartphones can be explored as an ancillary resource for enhancing patient engagement; while, the latter scenario portrays mobile devices as the primary actor, as known as mobile support actor. Hence, the activity diagrams and the list of gamification design components can be revised with the new role of mobile support actors. There are additional questions to describe gamification and design mechanics: how does tool help the subject to achieve the objective? how does control affect the subject to transform the object? how does context influence the subject to reach the object? or how does communication change the way the subject reach the object?

For example, how do wireless monitoring devices assist patients to improve weight management? Or how does reward points motivate patients to participate in a health exercises? Or how do displaying and visualization capabilities enable patients to recognize symptoms?

In additions, the inter-connections among tool, control, context, and communication can also be investigated to enhance the gamification models.

Step 4. Designing

The objective of the designing step is to translate gamification models in to a practical design product. In mobile health, mobile apps are commonly referred as the end product of the design and development process. Hence, this step involves the creative design of user interfaces of mobile apps for touch-sensitive display on smartphones. We suggest the use of UML-based use case diagrams and use case descriptions to describe design concepts. These concepts serve as the basis for user interface/ experience (UI/UX) designers to create mobile user interfaces.

This gamification design framework presents a unified, structured guideline for game designers, UX/UI designers, researchers, and theorists to gamify mobile health interventions. The use of the framework in a health-specific application is illustrated in the next section.

4 Designing Home-Based Heart Failure Self-management Programme

This paper discusses a home-based self-management programme for patients with chronic heart failure (CHF). The primary aims of the programme are to increase self-care behaviors, self-efficacy, and social support, as well as, improving health-related quality of life. It is imperative to equip CHF patients and family members/care givers with knowledge to improving heart failure self-management skills. In a long run, the programme is promising for a far-reaching impact, extending beyond the individual's physical and financial costs and affecting society at large. With integration of mobile health technologies, a larger proportion of the patients with CHF would be able to receive the rehabilitation in a more acceptable, easily accessible and fun way, especially for older patients. Hence, gamification design of the programme has been deliberated at the very beginning at the study.

The design process has spanned over six months by a group with more than ten members from many disciplines including clinicians, researchers, designers, and software developers. We employed the proposed gamification design framework as an effective means for brainstorming and communicating among team members to reach the final, practical design for the mobile-based self-management intervention. The process of gamification design is clearly documented in great details in subsequent sections.

4.1 Identifying Activity Systems

Self-care management is an integral component of heart failure treatment [12]; thus, it is critical to motivate patients, family members, and care givers to mandate tailored interventions on self-care maintenance and management with suitable practical guidelines and activities. Therefore, the home-based heart failure self-management programme is developed with several objectives in accordance to international practical recommendations for patients with CHF [40]. Figure 2 summarizes the objects of the activity systems under the programs towards achieving the primary aims.

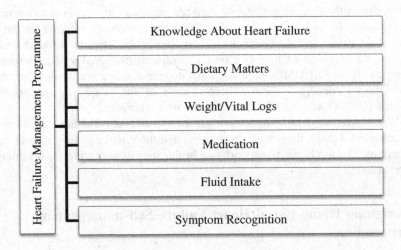

Fig. 2. The home-based heart failure self-management programme

(1) *Knowledge About Heart Failure.* The set of education activities for improving patients' knowledge and skills for several areas: (i) understanding heart failure, (ii), managing heart failure, and (iii) living with heart failure.

(2) *Dietary Matters.* Enhancing patients' self-care behaviors in selecting healthy food, and taking low salt diet. It is advisable for patients to have 2000 mg sodium nutrition plan per day.

(3) *Weight/Vital Logs.* Monitoring of lean weight, blood pressures, and heart rates on regular basis. It is critical for patients to respond to a sudden unexpected weight gain of more than 2 kg in 3 days.

(4) *Medication.* Understanding of medications and their uses/side-effects. The patients are instructed to take medications timely in accordance to doctors' recommendations.

(5) *Fluid Intake.* Focusing on optimal fluid management for patients with heart failure which is recommended at 1500 mL of fluid intake over 24 h. Participants can join interactive activities to keep track of their fluid consumption per day.

(6) *Symptom Recognition.* The patients are advocated to detect and recognize warning symptoms quickly to take appropriate action.

The activity systems of the home-based heart failure self-management programme, hence, are described in Table 3.

Table 3. Activity systems of the programme for CHF patients

Component	Elements
Outcome	• Self-care behaviors, self-efficacy, social support, and health-related quality of life
Subject	• Patients, Psychological/social support actors (e.g., family members, friends, caregivers, or clinicians)
Object	• Knowledge about heart failure, dietary matters, weight/vital logs, medication, fluid intake, symptom recognition
Tool	• Instructional materials • Mobile health capabilities: interactive touch, connectivity and social media, displaying/reporting/visualization, wireless health sensing and monitoring, push notification/reminder
Control	• Guidelines, protocols and recommendations • Gamification controls: reward, point, clear goal, leaderboard, achievement/badge, level/progress/feedback, story/theme, challenge, loop
Context	• Virtual communities: social media or health communities • Location-based • Time-based
Communication	• Face-to-face, one-to-one, group discussion, role structures

The activity systems consist of several high-level activities as below: (1) transferring knowledge and skills, (2) nutrition planning, (3) weight and vital signs monitoring, (4) medication management, (5) fluid management, and (6) symptom recognition.

4.2 Modelling

For each high-level activity, we use UML activity diagrams to model the activity with the structured flow of actions. This step enables us to identify the opportunities for gamification design. For brevity, we use the "nutrition planning" activity as the example of the modelling step; because it has been identified as one of complex activities in the programme. The nutrition planning activity requires patients' knowledge and patience to learn and manage their daily diets to avoid excessive salt intake. Figure 3 depicts the UML activity diagram of the nutrition planning.

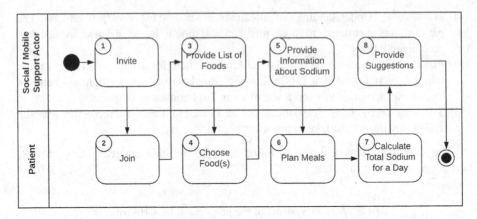

Fig. 3. The UML activity diagram of nutrition planning

There are eight (8) sub-activities in the nutrition planning activity. For each sub-activity, the questioning technique is employed to develop the activity descriptions as shown in Table 4.

Table 4. Gamification design elements in nutrition planning

Sub-activity	Elements
1. Invite	• Tool: push notifications/reminders • Control: challenge • Context: time-based
2. Join	• Control: point
3. Provide list of foods	• Tool: displaying & visualization, instructional materials
4. Choose food(s)	• Tool: interactive touch • Control: clear goal
5. Provide information on sodium	• Tool: displaying & visualization • Control: achievement
6. Plan meal	• Tool: interactive touch, and reporting • Control: achievement, reward • Context: time-based
7. Calculate total sodium	• Tool: displaying & visualization • Control: challenge, loop, progress
8. Provide suggestions	• Tool: displaying & visualization, connectivity and social media • Control: reward, level/progress, point, feedback • Context: social media

4.3   Transforming

In this step, the activity systems are transformed into mobile health by introducing the mobile support actor. Each component of the activity systems is investigated through a structured information gathering technique to determine how it can mediate the subject to achieve the objective. The interactions between people and technology are documented in use case descriptions, which are not included in this paper for brevity.

In Fig. 4, the design concepts for nutrition planning are illustrated in the left panel. There are several elements designed based mobile health capabilities and game design elements: (i) an information box for providing list of foods, selecting food, and providing information; (ii) time-based boxes for planning different meals of the day, (iii) an information box for a total sodium of the day, and (iv) an information box for providing suggestion and feedback.

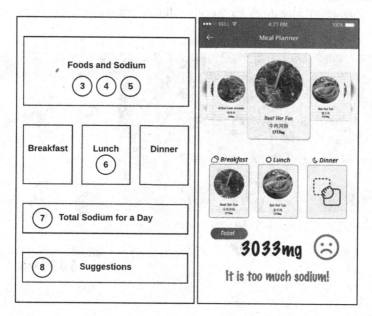

Fig. 4. Design concepts for nutrition planning

The deliverables of the transforming step are the revised activity diagrams, the updated use case descriptions, and the design concepts in words or graphical representations.

4.4   Designing

The last step of the gamification design framework is to translate the design concepts into practical user interfaces. This step heavily involves creative work from UI/UX designers to follow the information architecture and flow concepts for each activity. Mobile and design constraints are also considered to produce the final design of mobile health apps.

In the home-based heart failure self-management programme, several prototypes are developed iteratively and are improved over time with the updates of design concepts. The final design prototype is reviewed by a group of clinical experts and patients. It encompasses a number of gamification design elements and mobile health capabilities. The following highlights the key features of our mobile apps for patients with CHF. In Fig. 4, the final design of nutrition planning is manifested based on the design concepts in the right panel.

Rewards. The source of motivational affordances is embedded in the gist of the mobile app as shown in Fig. 5. It provides the gamification backbone for every activity participated by the users in the activity systems. An intelligent point system is introduced with different numbers of points awarded for participation and performance in different activities. For instance, users earn more points for adherence to medication management than following educational activities. In this study, the rewards feature is linked with monetary incentives for weekly self-care behaviors.

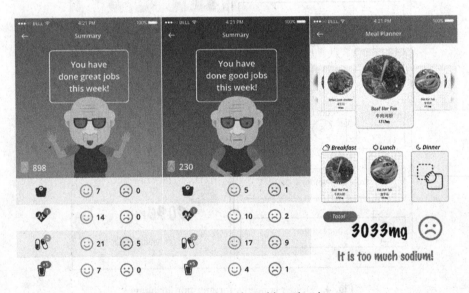

Fig. 5. Rewards and nutrition planning

Nutrition Planning. A mini-game is designed for patients to memorize healthy foods with sodium information as illustrated in Fig. 5. CHF patients are encouraged to play this game repeatedly to receive more reward points. It has over 40 common local foods, which potentially modify the dietary behaviors of the patients.

My Schedule. The mobile app offers clear goals and the day journey of self-care management to guide patients to perform activities as shown in Fig. 6. This feature is tightly integrated with push notification and reminder capabilities to constantly prompt for patients' actions. In additions, it also provides a one-stop screen for events related to patients' self-management activities such as play and learn, as well as, home visits by care givers.

Fig. 6. My schedule and weight monitoring

Weight Monitoring. CHF patients are advised to monitor their weights in a regular basis. If there is a weight change of more than 2 kg in 3 days, an alert will be sent to care givers for timely interventions. Furthermore, Bluetooth-enabled weighting scales are supported by the mobile app to streamline to burden of manual inputs for patients.

Vital Sign Monitoring. The mobile app allows patients to monitor their blood pressures, and pulse rates several times per day as shown in Fig. 7. The design of this

Fig. 7. Vital sign monitoring and fluid intake management

features follows conventional blood pressure monitoring devices; thus, it is friendly for older patients to avoid confusions for entering vital sign measurements.

Fluid Intake Management. The overloaded volume of fluid intake can worsen the symptoms of heart failure; and the patients are recommended to keep track of fluid intake daily as shown in Fig. 7. The optimal fluid management is 1500 mL over 24 h; the mobile app allows patients to enter estimated fluid intake with instant feedback for self-management.

Symptom Reporting. Symptoms have negative impacts on physical and psychological aspects of patients' daily life. The app allows patients to view the images of symptoms and to report warning symptoms quickly when they are worsening as shown Fig. 8. An alert will be sent to necessary psychological/social support actors to take appropriate action.

Fig. 8. Symptom reporting, education and social connectivity

Social Connectivity. The app allows the ability to connect with friends and family members as well as care givers via major social networking sites such as Facebook and Twitter. Once connected with appropriate permissions, the platform ensures the interactions and information exchange amongst the users in real-time. With the social support, the engagement between users and the mobile app would strengthen the frequent usage leading to a healthier lifestyle. In additions, in-app messaging between patients and psychological/social support actors is incorporated for timely recommendations and responses as illustrated in Fig. 8.

As illustrated in the methodological steps, the design process for gamification of mobile health entails the comprehensive understandings of activity systems towards achieving health-related outcomes. By exploring activities supported by technological capabilities and game design elements, the design framework is capable of translating complex health interventions into practical design artefacts.

5 Conclusion

This paper redefines gamification beyond the use of game design elements and techniques in non-gaming contexts. It is the nexus between psychology and technology; where activities can be transformed by motivational affordances, as well as, technological capabilities. We propose a gamification design framework for mobile health for translating activity systems into practical design. The framework provides a structured procedure of steps: (i) identifying activity systems, (ii) modelling, (iii) transforming, and (iv) designing; which are clearly demonstrated through a mobile health self-management programme for patients with chronic heart failure.

This study contributes to the cumulative theoretical development of gamification and mobile health in three folds. First, the dialectal engagement between people and technology is highlighted in gamification to make clear the requirements for a theoretical discourse. Second, the proposed design framework is capable of transforming health interventions supported by mobile technologies towards better outcomes. Last but not least, the study provides a holistic guideline how to incorporate psychological and technological elements in domain-specific gamification.

There are multiple implications for designers and developers of mobile health. The paper introduces a unified representation of gamification design for developing gameful mobile health applications. Furthermore, its findings on design concepts of mobile-based self-management interventions are well-informed and practical for creating new programmes for other chronic diseases.

Acknowledgment. This study is funded by a grant from the Singapore National Medical Research Council (grant number: HSRGWS16Jul007).

References

1. Deterding, S., O'Hara, K., Sicart, M., Dixon, D., Nacke, L.: Gamification: using game design elements in non-game contexts. In: Proceedings of 2011 Annual Conference on Human Factors in Computing Systems (2014).
2. Kankanhalli, A., Taher, M., Cavusoglu, H., Kim, S.H.: Gamification: a new paradigm for online user engagement. In: ICIS 2012 Proceedings, pp. 1–10 (2012)
3. Hamari, J.: Transforming homo economicus into homo ludens: a field experiment on gamification in a utilitarian peer-to-peer trading service. Electron. Commer. Res. Appl. **12**, 236–245 (2013)
4. Denny, P.: The effect of virtual achievements on student engagement. In: Proceedings of the SIGCHI Conference on Human Factors in Computing Systems - CHI 2013. p. 763. ACM Press, New York (2013)
5. Cheong, C., Cheong, F., Filippou, J.: Quick quiz: a gamified approach for enhancing learning. In: PACIS 2013 Proceedings, pp. 1–14 (2013)

6. Hamari, J., Koivisto, J.: Social motivations to use gamification: an empirical study of gamifying exercise. In: Proceedings of the 21st European Conference on Information Systems Society, pp. 1–12 (2013)
7. Zichermann, G., Cunningham, C.: Gamification by Design: Implementing Game Mechanics in Web and Mobile Apps. O'Reilly Media Inc., Sebastopol (2011)
8. Heidenreich, P.A., Trogdon, J.G., Khavjou, O.A., Butler, J., Dracup, K., Ezekowitz, M.D., Finkelstein, E.A., Hong, Y., Johnston, S.C., Khera, A., Lloyd-Jones, D.M., Nelson, S.A., Nichol, G., Orenstein, D., Wilson, P.W.F., Woo, Y.J.: Forecasting the future of cardiovascular disease in the United States: a policy statement from the American Heart Association. Circulation 123, 933–944 (2011)
9. Dickstein, K., Cohen-Solal, A., Filippatos, G., McMurray, J.J.V., Ponikowski, P., Poole-Wilson, P.A., Stromberg, A., van Veldhuisen, D.J., Atar, D., Hoes, A.W., Keren, A., Mebazaa, A., Nieminen, M., Priori, S.G., Swedberg, K., Vahanian, A., Camm, J., De Caterina, R., Dean, V., Dickstein, K., Filippatos, G., Funck-Brentano, C., Hellemans, I., Kristensen, S.D., McGregor, K., Sechtem, U., Silber, S., Tendera, M., Widimsky, P., Zamorano, J.L., Tendera, M., Auricchio, A., Bax, J., Bohm, M., Corra, U., della Bella, P., Elliott, P.M., Follath, F., Gheorghiade, M., Hasin, Y., Hernborg, A., Jaarsma, T., Komajda, M., Kornowski, R., Piepoli, M., Prendergast, B., Tavazzi, L., Vachiery, J.-L., Verheugt, F. W.A., Zamorano, J.L., Zannad, F.: ESC guidelines for the diagnosis and treatment of acute and chronic heart failure 2008: the task force for the diagnosis and treatment of acute and chronic heart failure 2008 of the European Society of Cardiology. Developed in collaboration with the heart. Eur. Heart J. 29, 2388–2442 (2008)
10. Bleumink, G.S., Knetsch, A.M., Sturkenboom, M.C.J.M., Straus, S.M.J.M., Hofman, A., Deckers, J.W., Witteman, J.C.M., Stricker, B.H.C.: Quantifying the heart failure epidemic: prevalence, incidence rate, lifetime risk and prognosis of heart failure The Rotterdam Study. Eur. Heart J. 25, 1614–1619 (2004)
11. Ng, T.P., Niti, M.: Trends and ethnic differences in hospital admissions and mortality for congestive heart failure in the elderly in Singapore, 1991 to 1998. Heart 89, 865–870 (2003)
12. Lainscak, M., Blue, L., Clark, A.L., Dahlström, U., Dickstein, K., Ekman, I., McDonagh, T., McMurray, J.J., Ryder, M., Stewart, S., Strmberg, A., Jaarsma, T.: Self-care management of heart failure: practical recommendations from the patient care committee of the Heart Failure Association of the European Society of Cardiology. Eur. J. Heart Fail. 13, 115–126 (2011)
13. Yu, D.S.F., Lee, D.T.F., Woo, J.: Improving health-related quality of life of patients with chronic heart failure: effects of relaxation therapy. J. Adv. Nurs. 66, 392–403 (2010)
14. Mahoney, J.S.: An ethnographic approach to understanding the illness experiences of patients with congestive heart failure and their family members. Heart Lung J. Acute Crit. Care 30, 429–436 (2001)
15. Mora, A., Riera, D., Gonzalez, C., Arnedo-Moreno, J.: A literature review of gamification design frameworks. In: 2015 7th International Conference on Games and Virtual Worlds for Serious Applications (VS-Games), pp. 1–8. IEEE (2015)
16. Hamari, J., Koivisto, J., Sarsa, H.: Does gamification work? – A literature review of empirical studies on gamification. In: 2014 47th Hawaii International Conference on System Sciences, pp. 3025–3034. IEEE (2014)
17. Brathwaite, B., Schreiber, I.: Challenges for Game Designers. Charles River Media Inc., Rockland (2008)
18. Werbach, K., Hunter, D.: For the Win: How Game Thinking Can Revolutionize Your Business. Wharton Digital Press, Philadelphia (2012)

19. Marczewski, A.: Gamification: A Simple Introduction. Andrzej Marczewski (2013)
20. Marache-Francisco, C., Brangier, E.: Process of gamification. From the consideration of gamification to its practical implementation. In: CENTRIC 2013, Sixth International Conference on Advances in Human-oriented and Personalized Mechanisms, Technologies, and Services, pp. 126–131 (2013)
21. Kumar, J.: Gamification at work: designing engaging business software. In: Marcus, A. (ed.) DUXU 2013. LNCS, vol. 8013, pp. 528–537. Springer, Heidelberg (2013). https://doi.org/10.1007/978-3-642-39241-2_58
22. Kumar, S., Nilsen, W., Pavel, M., Srivastava, M.: Mobile health: revolutionizing healthcare through transdisciplinary research. Computer **46**, 28–35 (2013)
23. Fiordelli, M., Diviani, N., Schulz, P.J.: Mapping mHealth research: a decade of evolution. J. Med. Internet Res. **15**, e95 (2013)
24. Hervás, R., Fontecha, J., Ausín, D., Castanedo, F., Bravo, J., López-de-Ipiña, D.: Mobile monitoring and reasoning methods to prevent cardiovascular diseases. Sensors **13**, 6524–6541 (2013)
25. Walton, R., DeRenzi, B.: Value-sensitive design and health care in Africa. IEEE Trans. Prof. Commun. **52**, 346–358 (2009)
26. Van Woensel, W., Roy, P.C., Abidi, S.S.: A mobile and intelligent patient diary for chronic disease self-management. In: MEDINFO 2015 eHealth-enabled Heal, pp. 118–122 (2015)
27. Nguyen, H.D., Poo, D.C.C., Zhang, H., Wang, W.: Analysis and design of an mHealth intervention for community-based health education: an empirical evidence of coronary heart disease prevention program among working adults. In: Maedche, A., vom Brocke, J., Hevner, A. (eds.) DESRIST 2017. LNCS, vol. 10243, pp. 57–72. Springer, Cham (2017). https://doi.org/10.1007/978-3-319-59144-5_4
28. Nguyen, H.D., Poo, D.C.C.: Analysis and design of mobile health interventions towards informed shared decision making: an activity theory-driven perspective. J. Decis. Syst. **25**, 397–409 (2016)
29. Prociow, P.A., Crowe, J.A.: Towards personalised ambient monitoring of mental health via mobile technologies. Technol. Health Care **18**, 275–284 (2010)
30. Magill, E., Blum, J.M.: Personalised ambient monitoring: supporting mental health at home. In: Advances in Home Care Technologies: Results of the Match Project, pp. 67–85 (2012)
31. Paoli, R., Fernández-Luque, F.J., Doménech, G., Martínez, F., Zapata, J., Ruiz, R.: A system for ubiquitous fall monitoring at home via a wireless sensor network and a wearable mote. Expert Syst. Appl. **39**, 5566–5575 (2012)
32. Junglas, I., Abraham, C., Ives, B.: Mobile technology at the frontlines of patient care: understanding fit and human drives in utilization decisions and performance. Decis. Support Syst. **46**, 634–647 (2009)
33. Carroll, A.E., Marrero, D.G., Downs, S.M.: The HealthPia GlucoPack Diabetes phone: a usability study. Diabetes Technol. Ther. **9**, 158–164 (2007)
34. Istepanian, R.S.H., Zitouni, K., Harry, D., Moutosammy, N., Sungoor, A., Tang, B., Earle, K.A.: Evaluation of a mobile phone telemonitoring system for glycaemic control in patients with diabetes. J. Telemed. Telecare **15**, 125–128 (2009)
35. AlMarshedi, A., Wills, G., Ranchhod, A.: Gamifying self-management of chronic illnesses: a mixed-methods study. JMIR Serious Games **4**, e14 (2016)
36. OED: Oxford English Dictionary. Oxford University Press, Oxford (2017)
37. Mursu, A., Luukkonen, I.: Activity Theory in information systems research and practice: theoretical underpinnings for an information systems development model. Inf. Res. **12**, 1–21 (2006)

38. Rumbaugh, J., Jacobson, I., Booch, G.: Unified Modeling Language Reference Manual, 2nd edn. Pearson Higher Education, London (2004)
39. Mwanza, D.: Towards an activity-oriented design method for HCI research and practice (2002)
40. Toukhsati, S.R., Driscoll, A., Hare, D.L.: Patient self-management in chronic heart failure – establishing concordance between guidelines and practice. Card. Fail. Rev. 1, 128 (2015)

The Impact of Gamification in Social Live Streaming Services

Katrin Scheibe[✉]

Department of Information Science, Heinrich Heine University,
Düsseldorf, Germany
katrin.scheibe@hhu.de

Abstract. Gamification – the implementation of game mechanics and dynamics in non-game contexts – has become a central part in designing websites and mobile applications. It is used to improve the motivation and engagement of users and should change their behavior. The general social live streaming service YouNow, which is mostly used by adolescents, offers its users many gamification and motivating elements. Therefore, it is used as a case study in this investigation. Some of the gamification elements are, for example, coins, levels, and different kinds of rankings. What gamification elements does You-Now offer to its users? To what extend are YouNow's users motivated by the game mechanics (gratifications sought)? What are the most motivating elements for producers, participants, and consumers? To what extend are the game elements perceived as a reward (gratifications obtained)? What are the most rewarding elements for specific user groups? To answer the research questions an online survey was conducted. It was available in five different languages and had 211 YouNow users as attendees.

The results show, that the producers are the most motivated as well as rewarded user group by obtaining actions of gamification elements. For them, getting fans is the best action. The participants are the most motivated by commenting streams as well as the most rewarded by making premium gifts. The best ones from the game mechanics for all users are coins as well as levels. All gamification elements, except for moments and "The Top Moments" ranking, which are both perceived as neutral, are highly leading users to gratifications sought as well as obtained.

Keywords: Social live streaming services · Gamification
Uses and gratifications · Motivation · User behavior · YouNow

1 Introduction

Games have always been a fundamental part of our society. Even our everyday life is influenced by games. Who does not know the "airplane landing" method, used by many desperate parents to lead their children to eat unwanted food by making it fun. Adding the promise to give the child a treat (reward) after finishing the meal makes the perfect combination to achieve the influence and change of the child's behavior [1]. This phenomenon called "gamification" is already applied in various situations, be it in school, or at work, in how we stay fit, or the way we travel. The implementation of

© Springer International Publishing AG, part of Springer Nature 2018
G. Meiselwitz (Ed.): SCSM 2018, LNCS 10914, pp. 99–113, 2018.
https://doi.org/10.1007/978-3-319-91485-5_7

game mechanics and dynamics in non-game contexts is used to increase one's engagement, motivation and activity. Therefore, it is no surprise that social media services and mobile applications already utilize it [2].

Social media, also known as "Web 2.0" [3], are web based open access services "which are predicated upon the active participation of broad masses of users" [4, p. 259]. Users generate a large number of data while using such services [5], thereby a user may produce information, as a producer, as well as consume the published information, as a consumer. Toffler [6] named this shared characteristics and behavior of users as "prosumers."

One arising kind of social media are social live streaming services (SLSSs). Live streaming is described as a synchronous function – users are producing live videos and viewers are able to interact in real-time with a broadcaster. This happens via chat messages or likes, rewards, or other gratifications, e.g. becoming a fan. The person being live is able to react immediately. Some SLSSs are known for being topic-specific, like Twitch for virtual games and electronic sports events, or Picarto for art, but most of them do not have any thematic context and are considered as general SLSSs, e.g. YouNow, Periscope, Ustream [7]. On YouNow, most of the users are highly motivated by the applied gamification elements [8, 9]. Therefore, YouNow was used as a case study in this paper.

To describe the impact of gamification and motivating elements in an SLSS like YouNow, this investigation refers to the model of users' information behavior on a gamified social live streaming service [10]. The model applies various theoretical aspects, such as the sender-centered Communication Formula by Lasswell [11] as well as the audience-centered Uses and Gratifications Theory by Blumler and Katz [12] with the differentiation between Gratifications Sought and Gratifications Obtained from Palmgreen et al. [13] as well as, additionally, the Self-Determination Theory of human motivation [14].

Users apply a certain service, because they are searching for gratifications [12] and to satisfy certain needs [14]. Palmgreen et al. [13] discuss the aspects of gratifications sought and gratifications obtained in relation to the uses and gratification theory – "since a gratification is sought it must necessarily be obtained" [13, p. 183]. Additionally, uses and gratifications are related to different information production, as well as reception behaviors. Users have a certain motivation and therefore are searching for gratifications. When users are getting a reward, certain gratifications are obtained. Gamification is considered as rewarding as well as motivating factor, regarding to Deterding's explanation: "gamification's guiding idea is to use elements of game design in non-game contexts, products, and services to motivate desired behaviors" [15, p. 14].

Users of social network services, respectively SLSSs, can be divided in three different user groups, namely producers, participants, and consumers [16]. Producers are users who are streaming live. They are producing content in a live stream. Participants are watching and are taking part by commenting, liking, or rewarding. Finally, consumers are users who are watching streams and reading comments, but are not producing any content and are not participating at all. Each user group has different motives for the usage of a social live streaming service. Consumers are using a service for entertainment as well as for information. Participants have the same motives as consumers and, additionally, the aim of social interaction. And finally, producers have additionally the goal to achieve self-realization and self-presentation.

With the theoretical backgrounds on uses and gratifications, the three user groups, and the difference between sought and obtained gratifications in mind, there are occurring three research questions about gamification on YouNow:

- *RQ1:* What gamification elements does YouNow offer its users?
- *RQ2:* To what extend are YouNow's user groups motivated by the game mechanics (gratifications sought)?
- *RQ3:* To what extend are the game elements on YouNow perceived as a reward (gratifications obtained) by the special user groups?

There are already several studies about live streaming services, about YouNow [17–20] as well as on gamification in social media, and gamification in general. Also, the worldwide popular video game live streaming platform Twitch.tv earns lots of attention from researchers [21]. Actually one investigation about giving and taking gratifications on YouNow [22] as well as another by Wilk et al. [23] about gamification on live streaming services, in particular, about gamification influencing the user behavior of mobile live video broadcasting users could be found. The researchers developed a mobile live broadcasting application in three different versions. The first version (A) was applied as base version. It consists of a simple overview page, as well as a view for watching the live video, and one for recording a live broadcast. The second version (B) includes the opportunity of leveling and the overview of one's process to the next level. The last version (C) applies all functions of version B as well as the performance of challenges and the chance of receiving badges. Each version was evaluated by different users. Some users have streamed over a longer duration with the implementation of levels, and users are significantly more motivated by the challenges and rewards, than using the base version.

2   YouNow and Its Gamification Elements as a Case Study

As YouNow is an SLSS which is mostly applied by teenagers and young adults between the age of 13 and 22 who are bored and want to have fun [8], it strikes that YouNow offers many gamification features as motivating factors. All gamification elements that are shown by the action of viewing a stream as a recipient are highlighted and listed in Fig. 1.

On Facebook there are friends, on Twitter as well as on Instagram there are followers, and on YouNow there are fans. Users connect with other users and stay up-to-date by following them via a so called fanning function (blue button above the live stream). There is also the opportunity to become a subscriber (white button above the live stream) of selected broadcasters, if you are willing to pay a monthly fee. Subscribers have special and additional functions: "Subscription entitles you to Super Chat privileges [...], you will have access to a Super Gift" and one will receive a "special" and "unique badge" that will "identify you as a subscriber" as well [24]. Speaking about badges, there are three different types on YouNow. The first one, the "Subscription Badge", was already mentioned. The second one is the "Broadcast

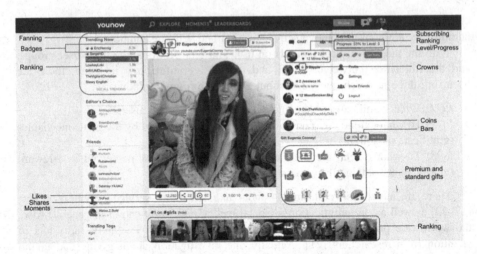

Fig. 1. YouNow live stream with marked gamification elements.

Badge". Only producers (streamers) are able to get this badge. It represents the users' broadcasting skills. There are nine different levels of this badge; each can be reached by different challenges. The levels are, namely and by order: Novice, Rookie, Junior, Captain, Rising Star, Boss, Ace, Superstar, Pro and finally, Partner. Users are moving up to a next status, if they reach certain goals (e.g. getting a determined number of fans or likes). Coming to the last badge which is called "Crowns", those badges are symbolizing top fans, who are supporting a streamer with bars. The more bars a user spends, the higher is his or her "Crown level". The "Crown level" is represented by one to five red or golden crowns. Those are shown on the user's profile, as well as beside one's username by commenting in the chat.

The virtual currencies on YouNow are coins as well as bars. Bars have to be bought with real money. They make it possible for users to bestow a broadcaster with so called premium gifts. The other virtual currency (coins) may be collected during certain activities on YouNow (e.g. broadcasting live). Coins are needed for site actions, such as bestowing a broadcaster with gifts. A user even needs coins for the action of liking a stream. Likes can be considered as some kind of feedback function by users or some kind of reward. Besides to likes, there are also shares and captured moments, which are presented below the live video (Fig. 1). YouNow live streams can be mentioned on other social media sites by sharing it. There is the opportunity to share it on Facebook, Twitter, or Tumblr. One may even invite fans to a live stream to support the broadcaster. By capturing a moment, the previous 15 s of a live stream will be saved to one's profile as well as to the "Moments Feed", and may be shared on social media platforms as well. Another function of YouNow is to be guest in a live broadcast of another streamer (there is normally a tab in the chat box, it is covered by the level, as well as level progress bar). The host has to accept the guest request first. This feature offers the opportunity to collaborate with other streamers. Returning to gifts as well as premium gifts (below the chat box), most of them are like stickers or icons in the chat, but some of them even have an influence on the stream (e.g. applause). They have varying prices, serve as a reward for streamers, and "they are a symbol of your dedication and appreciation" [25]. Furthermore, levels show the user's experience on YouNow and the level progress bar should motivate one to reach the next level. YouNow offers its users many leaderboards to compare their performance and accomplishment to the one of other users. The most utilized one is the "Trending Now" ranking (top left corner). The list displays the broadcasters being live and having the most viewers. The greater the audience of a broadcaster at the moment, the better he is ranked. Consequently, the one with the most viewers is ranked on top. While watching a stream, there is also the "Top Fans by Streamer" ranking (above the chat) as well as the "Trending by Hashtag" ranking (below the stream). The "Top Fans by Streamer" ranking shows which fan spent the most bars regarding to one streamer. The other ranking shows other streams with the same hashtags, regarding to the number of audience. Furthermore, other rankings are displayed on a special leaderboard site (Fig. 2).

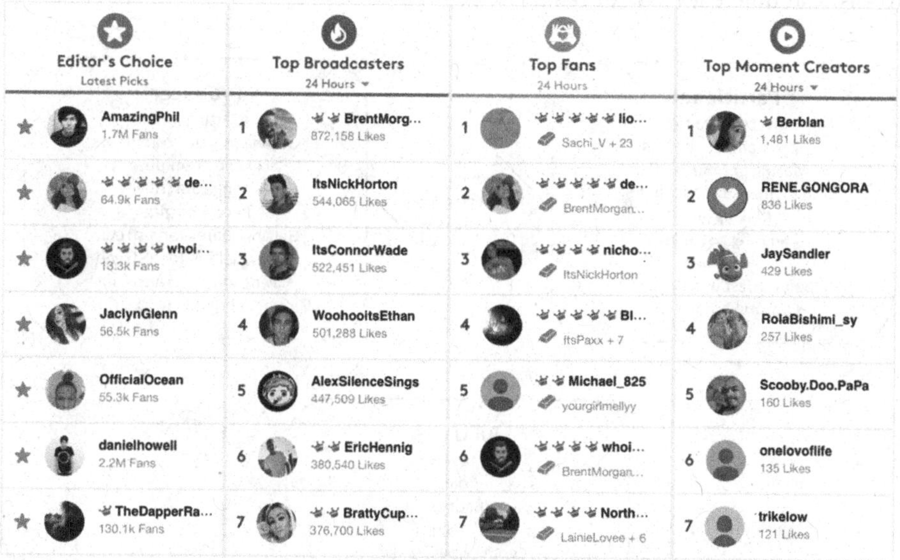

Fig. 2. Leaderboards on YouNow. (Color figure online)

On the one hand, there is the "Editor's Choice" ranking (green). It shows streamers who are discovered as talented by the editors of YouNow. "Editor's Choice is awarded on a rotating basis and will be removed after a few weeks" [26]. Following, there is the "Top Broadcaster" ranking (purple), it lists "broadcasters with the highest number of likes in a particular broadcast" [27]. Furthermore, the "Top Fans" ranking (pink) "shows fans who have supported broadcasters with the greatest value of gifts in the past 24 h" [27], and finally, the "Top Moment Creators ranking" (blue) displays captured moments that have been liked by users. Concluding, YouNow offers many gamification elements, whereof seven items are rankings and there are three different kinds of badges. Nearly all of the elements allow the interaction or comparison with other users. In this investigation, the currency of bars as well as subscribing have been considered as further motivational features, because they cost real money. Also, commenting streams as well as sharing streams have been added as further motivational element. Furthermore, the "Broadcast Badge" has not been considered in the online survey, because it is a new element that has been added in the time frame of this investigation. Other SLSSs like Periscope or the live streaming function on Instagram only offer their users the opportunity to show attention during flying hearts.

The mentioned gamification elements and their related actions can be associated with the three different user groups (Fig. 3). Producers (blue) can receive or get something (e.g. comments, gifts) during the action of streaming live. The action one user receives during a stream has to be taken or sent by another user, the participant (yellow). Therefore, there are opposite actions for producers as well as participants. If a viewer participates in a stream, the producer will consequently perceive the action. There are some gamification elements, which can be used by all three user groups (white) – producers, participants as well as consumers. These are levels, coins, badges, crowns, and different kinds of rankings.

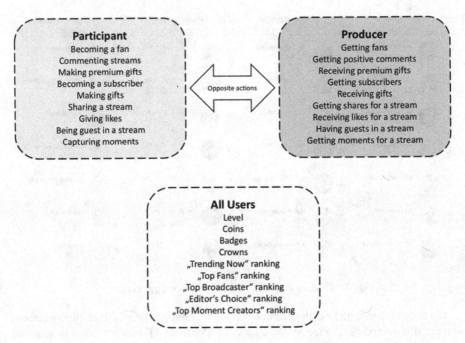

Fig. 3. User groups and the related game elements and actions on YouNow. (Color figure online)

3   Methods

As investigative method and to collect the required data, an online survey was conducted on umfrageonline.com. It was available in five different languages, namely English, German, Spanish, Arabic, as well as Turkish. The German survey was translated to English and Spanish, and the English version was translated to Arabic as well as Turkish. The main part of the survey consists of pre-formulated statements about the game mechanics and dynamics, and motivational elements on YouNow. As shown in Fig. 4, a picture of the respective element was presented. For every statement a seven-point Likert-scale, from "strongly disagree" (1) to "strongly agree" (7) [28] with equidistance between neighboring numbers was prepared. The Likert-scale had an uneven number to have the opportunity of a neutral (4) point. Additionally, the statements could be answered by "I don't know". The pre-formulated statements in the survey were based on different theoretical backgrounds:

- Motivation (gratification sought [1, 2, 13])
- Reward (gratification obtained [1, 12, 13])

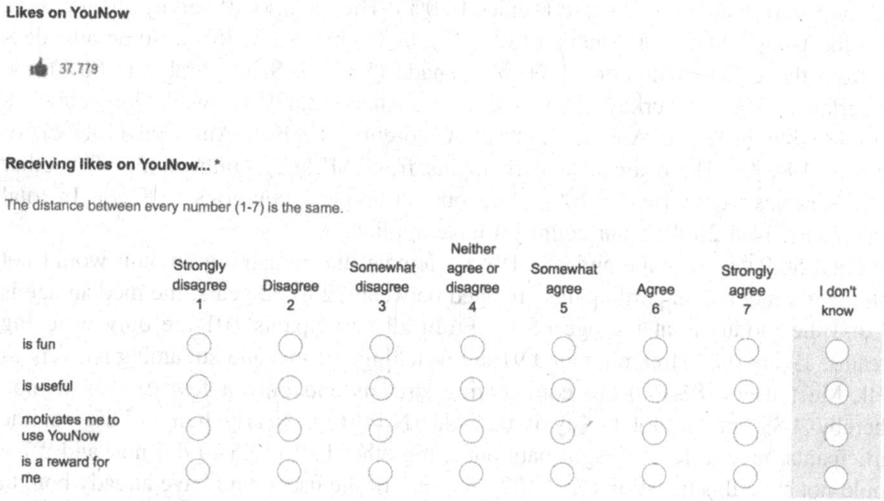

Likes on YouNow

👍 37,779

Receiving likes on YouNow... *

The distance between every number (1-7) is the same.

	Strongly disagree 1	Disagree 2	Somewhat disagree 3	Neither agree or disagree 4	Somewhat agree 5	Agree 6	Strongly agree 7	I don't know
is fun	○	○	○	○	○	○	○	○
is useful	○	○	○	○	○	○	○	○
motivates me to use YouNow	○	○	○	○	○	○	○	○
is a reward for me	○	○	○	○	○	○	○	○

Fig. 4. Online survey item (statement about likes).

The first question asked the attendee what he or she uses YouNow for, with the options to choose between "Only streaming", "Only watching streams", "Both: streaming and watching streams", or "I do not use YouNow". Either the attendee was only a producer (streaming), only a consumer (watching streams), or both, producer and consumer (streaming and watching streams). If the attendee was not a user of YouNow, the survey was finished. Furthermore, the attendee was asked with different questions if and with what actions he or she participates in streams. Some of the next survey items (questions and statements) were separated by the given answer of this question. Therefore, the survey items have a varying number of N. Finally, the survey participants were asked about age, gender as well as country. The survey was available from August 30, 2016 until March, 2017 and reached 211 YouNow users as participants.

Since the data is ordinal scaled and not normally distributed, the median was considered as first benchmark. Furthermore, the mean as well as the standard deviation (not mentioned in the table) where added as second and third sorting criterion. For the analysis of YouNow's ease of use, its usefulness, trustability, and giving users the experience of flow, we calculated the median and the interquartile range (IQR). In order to analyze the correlations between the positions of the actions of two rankings, Spearman's Rho rank coefficient has been calculated via SPSS. The common thresholds were used, namely two stars (**) for 99% as well as "ns" for "not significant."

4   Results

Considering Alexa's online traffic statistics [29] about YouNow.com, most users are from the United States of America (30.1%), followed by Germany (11.6%), Turkey (8.1%), Saudi Arabia (6.5%), and Mexico (4.0%). The majority of survey attendees are from the United States of America (29.25%), or Germany (20.75%). Some attendees are from the United Kingdom (6.60%), Canada (5.66%), Saudi Arabia (5.66%), the Netherlands (4.72%), Turkey (2.83%), or New Zealand (2.83%) as well. Only a few are from Mexico (1.89%), Algeria (1.89%), Colombia (1.89%), Australia (1.89%), or Austria (1.89%). There are also participants from MENA countries (3.76%), other Latin American countries (3.76%), and other European countries (2.82%). In total participants from 26 different countries have applied.

Total 50.94% are male and 48.11% are female, the remaining amount would not state their gender. The participants are aged between 12 to 62 years, the median age is 23 and the modus is at the age of 17. From all participants 101 are only watching streams, 19 are only streaming, and 91 are watching streams and streaming actively as well. Most users (86.1%) are commenting streams and only a few (9.1%) do not, whereby 4.8% prefer not to say if they do (N = 165). Nearly half (47.8%) of the participants have at least bought bars once, the other half (49.5%) did not, and 2.7% would not state this question (N = 182). 54.42% of the users who have already bought bars think it was money well spent, only 27.93% think the money was not well spent. The remaining 17.65% points have a neutral point of view to this. With 48.9 percent points, slightly more users have already subscribed to someone, while 47.8% have not subscribed yet, and 3.3% prefer not to say if they do (N = 182). Already 52.7% have been a guest in a live stream on YouNow, and 45.6% have never been guest in a stream, only few attendees (1.6%) would not answer this question (N = 182).

Regarding to what users think about the SLSS YouNow (Fig. 5), for most users (min. 75%) it is easy and funny to use, and they have also experienced the feeling of flow (min. 75% as well) while being on YouNow (median is 6, each and IQR from 5 to 7). With the median at 5.5, users also noticed YouNow as a useful system. But, some users seem to have doubts whether YouNow is trustable, or not (median: 4).

Table 1 shows two different rankings regarding to gratifications sought (top ranking) and gratifications obtained (bottom ranking) through the different gamification actions for producers. In both rankings, we can see a similarity at the positions and the median of the actions of "getting fans" (first rank, median of 7), "receiving premium gifts" (third rank, median of 7), "receiving gifts" (fifth position and median of 6),

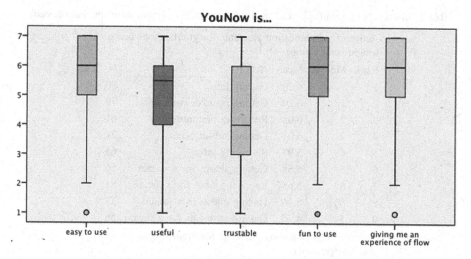

Fig. 5. What YouNows users think about the service (N = 160).

"having guests in a stream" (eighth position, median of 6) as well as "capturing moments for a stream" (ninth and last position, median of 5). One other action, "getting positive comments" is on the second position in the first ranking and on the fourth position in the second ranking, both with a median of 7. The action of "getting subscribers" is on the fourth place in the ranking for gratifications being sought and on the second position in the ranking for gratifications being obtained. Therefore, the positions for these actions have switched in both rankings. In the first ranking, the action of "getting shares" is on the sixth rank and "receiving likes" is on the seventh rank. Looking at the ranking about gratifications obtained, "receiving likes" is on the sixth rank and "getting shares" on the seventh rank. As mentioned, on the last two positions of both rankings is the action of "having guests in a stream" followed by the action of "getting moments." The first four positions of both rankings have a median of 7, the following positions, from five to eight, have a median of 6 and the last position has, in both cases, a median of only 5. The correlation between the positions of the actions of the two rankings, according to Spearman's Rho rank coefficient, is 0.917**.

Looking at the rankings of actions by gamification elements being sought as well as obtained for participants (Table 2), it strikes that the action on the first position of the first ranking, "commenting streams", has a median of 6 and the first position of the second ranking (gratifications obtained), "making premium gifts", has a median of 6.5. Comparing it to the previous table, the first rank starts with a smaller median for participants. "Commenting streams" is on the sixth place for gratifications obtained and "making premium gifts" is on the fifth position for gratifications sought. The second element in the ranking about gratifications sought is "being guest in a stream". Participants are motivated by being guest in a stream. Comparing it to the ranking about the reward, "being guest in a stream" is on the eighth ranking. Nevertheless, they both have a median of 6 and a similar mean value (5.48 for sought; 5.40 for obtained). Moving on with the third place, there is the action of "becoming a fan" in both

Table 1. Rankings of gamification elements for producers (gratification sought/obtained).

Actions by gamification elements for gratifications being sought (motivation)

Rank	Median	Mean	Action	N
1.	7	6.20	Getting fans	80
2.	7	6.20	Getting positive comments	79
3.	7	6.05	Receiving premium gifts	61
4.	7	6.01	Getting subscribers	74
5.	6	5.97	Receiving gifts	65
6.	6	5.68	Getting shares for a stream	56
7.	6	5.65	Receiving likes for a stream	81
8.	6	5.46	Having guests in a stream	57
9.	5	4.70	Getting moments for a stream	56

Actions by gamification elements for gratifications being obtained (reward)

Rank	Median	Mean	Action	N
1.	7	6.30	Getting fans	79
2.	7	6.12	Getting subscribers	75
3.	7	6.10	Receiving premium gifts	61
4.	7	6.04	Getting positive comments	78
5.	6	6.12	Receiving gifts	65
6.	6	5.75	Receiving likes for a stream	79
7.	6	5.70	Getting shares for a stream	57
8.	6	5.19	Having guests in a stream	57
9.	5	4.71	Getting moments for a stream	55

rankings. "Becoming a subscriber" is on the fourth position at gratifications sought (median of 5.5) and on the second ranking (median of 6) for gratifications obtained. "Sharing a stream" has a median of 4 in the first ranking (8th rank) and a median of 6 in the second ranking (7th rank). Again, "capturing moments" is at the last position in both rankings. For the positions of the two rankings in Table 2, the Spearman's Rho rank coefficient correlation is 0.233 and it is statistically not significant.

Considering the median values of the rankings about gamification elements concerning all user groups (Table 3), only the first rank of the motivating ranking (gratifications sought) has a median of 6, the second to eighth positions have a median of 5 and the last place has a median of 4. The other ranking, about gamification elements are experienced as a reward (gratifications obtained), has a median of 6 from the first to the fifth ranking. The sixth place has a median of 5.5 and the others a median of 5. Therefore, the standard game mechanics are generally experienced as more rewarding than motivating. Looking at the ranking positions, levels are on the first place for gratifications sought and on the second place for gratifications obtained (median of 6, each). The second position of the first ranking (sought) displays the gamification

Table 2. Rankings of gamification elements for participants (gratification sought/obtained).

Actions by gamification elements for gratifications being sought (motivation)

Rank	Median	Mean	Action	N
1.	6	5.60	Commenting streams	116
2.	6	5.48	Being guest in a stream	58
3.	6	5.22	Becoming a fan	131
4.	5.5	5.24	Becoming a subscriber	72
5.	5	5.03	Making premium gifts	68
6.	5	4.94	Making gifts	114
7.	5	4.63	Giving likes	133
8.	4	4.49	Sharing a stream	100
9.	4	4.03	Capturing moments	101

Actions by gamification elements for gratifications being obtained (reward)

Rank	Median	Mean	Action	N
1.	6.5	6.00	Making premium gifts	68
2.	6	6.04	Becoming a subscriber	72
3.	6	5.99	Becoming a fan	128
4.	6	5.74	Giving likes	134
5.	6	5.71	Making gifts	112
6.	6	5.68	Commenting streams	114
7.	6	5.49	Sharing a stream	101
8.	6	5.40	Being guest in a stream	58
9.	5	4.55	Capturing moments	101

element coins and has a median of 5. In the second ranking (obtained), coins are on the first place and have a median of 6.

The "Trending Now" ranking is in both rankings on the third place and has a median of 5 for gratifications sought and a median of 6 for gratifications obtained. Placed on the fourth position of motivating gamification elements, is the "Editor's Choice" ranking, it has a median of 5. On the fifth and sixth position are crowns (5th) as well as badges (6th), both with a median of 5 as well. In the ranking about rewarding gamification elements, the "Editor's Choice" ranking can be found at the seventh rank with a median of 5, crowns are on the fourth position, and badges are on the fifth position, both with a median of 6. The "Top Fans" ranking and the "Top Broadcasters" ranking are on the seventh and eighth place of the first rankings and on the sixth and eighth place of the second ranking. With a median of 5, the "Top Fans" ranking is on the eighth position for gratifications obtained. Finally, the "Top Moment Creators" ranking is on the last rank for gratifications sought as well as for obtained. It has a median of 4 for motivating and a median of 5 for rewarding. The Spearman's Rho rank coefficient correlation for the positions of the elements in the two rankings (Table 3) is 0.850**.

Table 3. Rankings of gamification elements for consumers, producers, and participants as well as (gratification sought/obtained).

Gamification elements for gratifications being sought (motivation)

Rank	Median	Mean	Element	N
1.	6	5.21	Levels	113
2.	5	5.15	Coins	124
3.	5	5.03	"Trending Now" ranking	105
4.	5	4.86	"Editor's Choice" ranking	101
5.	5	4.86	Crowns	98
6.	5	4.84	Badges	93
7.	5	4.79	"Top Fans" ranking	104
8.	5	4.64	"Top Broadcaster" ranking	104
9.	4	4.52	"Top Moment Creators" ranking	103

Gamification elements for gratifications being obtained (reward)

Rank	Median	Mean	Element	N
1.	6	5.48	Coins	123
2.	6	5.36	Levels	108
3.	6	5.25	"Trending Now" ranking	101
4.	6	5.13	Crowns	100
5.	6	5.10	Badges	93
6.	5.5	5.05	"Top Broadcaster" ranking	96
7.	5	4.97	"Editor's Choice" ranking	95
8.	5	4.97	"Top Fans" ranking	100
9.	5	4.65	"Top Moment Creators" ranking	97

5   Discussion

This investigation presented a first insight about game mechanics and their rewarding, respectively motivating aspect on general live streaming services, and, if and through what gamification elements users are searching for as well as obtaining gratifications. On YouNow, users are confronted with many types of gamification elements. Every registered user has a level as well as a level process bar to compare their experience with other users and to be motivated to reach a next status. While watching a stream the audience is able to reward the streamer with likes and gifts. There is the opportunity to share a stream on other social media services as well as to capture moments (15 s) of a stream. To collaborate, one is able to request to be guest in a stream of a producer. Users stay up-to-date through the fanning as well as subscribing function and collecting coins through several site activities. The other currency on YouNow, besides coins, that has to be bought with real money, is called bars. Bars are needed for special premium gifts. Moreover, YouNow offers its users seven different leaderboards to compare the performance towards other users and three different kinds of badges.

The online survey asked YouNow's users if they perceive the particular gamification elements as rewarding and motivating. According to the results, the actions a user perceives while producing a stream are experienced as the most rewarding as well as motivating. The rank order from the ranking about gratifications sought (motivating) is similar to the ranking about gratifications obtained (rewarding) for producers (correlation of 0.917**), even the median values of each action is the same in both rankings. For producers, getting fans is the best way to search as well as to obtain gratifications, whereas getting moments is the least.

The actions users perceive while participating in a live stream are conceived as slightly less rewarding as well as motivating as for the producers. They are gently more rewarded trough the different gamification actions than motivated and they are not searching for gratifications as much as they are obtaining them. For participants, commenting streams is the most motivating action and making premium gifts is perceived as the most rewarding. The least action for participating is in both circumstances capturing moments.

Coming to the general gamification elements for all users, the most motivating one are levels and the most rewarding are coins. At the last ranking positions of both rankings was the "Top Moment Creators" ranking. In general, the standard gamification elements for all users of YouNow are perceived as the least rewarding as well as motivating, but all are at least perceived as neutral and the majority as thoroughly motivating, respectively rewarding. Users are more motivated by the actions they are able to perform on the service. Moments as well as the "Top Moment Creators" ranking are rated rather low, because YouNow's users want to replay and watch the full video instead of the captured moments (15 s). Additionally, the results show that YouNow is easy as well as fun to use, YouNow's users are experiencing flow while using the platform, and they think the information service is somehow useful as well.

As limitations of this investigation, one can mention the rather small number of participants (N = 211); also not every survey attendee has answered all questions concerning the high number of survey items. As alternative, qualitative interviews with producers as well as the audience will be more accurate than pure quantitative data. The interviews could be performed live on YouNow. Moreover, 50% of the survey participants are 23 years old and older. If more users from generation Z [30] had participated, the data will be more accurate. The common users of YouNow are mainly teenagers and adolescents. For further research, a comparison of other live streaming services' game mechanics would be helpful to have data on different live streaming services (as, e.g., Periscope is mainly used by generation Y and generation X and older people are mainly using Ustream). Moreover, an investigation about comparing the extent of gamification elements a service applies should be conducted. Also, an investigation about the distinction of users by gender, age, and culture will be an interesting research topic.

All in all, YouNow's game mechanics are accepted very well. The very young users of YouNow do really enjoy the gamification elements of the service.

References

1. Zichermann, G., Cunningham, C.: Gamification by Design: Implementing Game Mechanics in Web and Mobile Apps. O'Reilly, Sebastopol (2011)
2. Deterding, S., Dixon, D., Khaled, R., Nacke, L.: From game design elements to game-fulness: defining "gamification". In: Proceedings of the 15th International Academic MindTrek Conference: Envisioning Future Media Environments, pp. 9–15. ACM, New York (2011)
3. O'Reilly, T.: What is Web 2.0. (2005). http://www.oreilly.com/pub/a/web2/archive/what-is-web-20.html. Accessed 08 Feb 2018
4. Linde, F., Stock, W.G.: Information Markets. A Strategic Guideline for the I-Commerce. De Gruyter Saur, Berlin, New York (2011)
5. O'Reilly, T., Battalle, J.: Web Squared: Web 2.0 Five Years on (2009). https://conferences.oreilly.com/web2summit/web2009/public/schedule/detail/10194. Accessed 08 Feb 2018
6. Toffler, A.: The Third Wave. Morrow, New York (1980)
7. Scheibe, K., Fietkiewicz, K.J., Stock, W.G.: Information behavior on social live streaming services. JISTaP **4**(2), 6–20 (2016)
8. Friedländer, M.B.: And action! Live in front of the camera: an evaluation of the social live streaming service YouNow. Int. J. Inf. Commun. Technol. Hum. Dev. **9**(1), 15–33 (2017)
9. Scheibe, K., Zimmer, F., Fietkiewicz, K.J.: Das Informationsverhalten von Streamern und Zuschauern bei Social Live-Streaming Diensten am Fallbeispiel YouNow [Information behavior of streamers and viewers on social live-streaming services: YouNow as a case study]. Inf. Wiss. Prax. **68**(5–6), 352–364 (2017)
10. Zimmer, F., Scheibe, K., Stock, W.G.: A model for information behavior research on social live streaming services (SLSSs). In: Meiselwitz, G. (ed.) SCSM 2018. LNCS, vol. 10914, pp. xx–yy. Springer, Cham (2018)
11. Lasswell, H.D.: The structure and function of communication in society. In: Bryson, L. (ed.) The Communication of Ideas, pp. 37–51. Harper & Brothers, New York (1948)
12. Blumler, J.G., Katz, E.: The Uses of Mass Communications: Current Perspectives on Gratifications Research. Sage, Newbury Park (1973)
13. Palmgreen, P., Wenner, L.A., Rayburn II, J.D.: Relations between gratifications sought and obtained: a study of television news. Commun. Res. **7**(2), 161–192 (1980)
14. Ryan, R.M., Deci, E.L.: Self-Determination Theory. Basic Psychological Needs in Motivation, Development, and Wellness. Guildford Press, New York, London (2017)
15. Deterding, S.: Gamification: designing for motivation. Interactions **19**(4), 14–17 (2012)
16. Shao, G.: Understanding the appeal of user-generated media: a uses and gratification perspective. Internet Res. **19**(1), 7–25 (2009)
17. Zimmer, F., Fietkiewicz, K.J., Stock, W.G.: Law Infringements in social live streaming services. In: Tryfonas, T. (ed.) HAS 2017. LNCS, vol. 10292, pp. 567–585. Springer, Cham (2017). https://doi.org/10.1007/978-3-319-58460-7_40
18. Friedländer, M.B.: Streamer motives and user-generated content on social live-streaming services. JISTaP **5**(1), 65–84 (2017)
19. Honka, A., Frommelius, N., Mehlem, A., Tolles, J.N., Fietkiewicz, K.J.: How safe is YouNow? An empirical study on possible law infringements in Germany and the United States. J. MacroTrends Soc. Sci. **1**(1), 1–17 (2015)

20. Fietkiewicz, K.J., Scheibe, K.: Good Morning… Good Afternoon, Good Evening and Good Night: adoption, usage and impact of the social live streaming platform YouNow. In: Proceedings of the 3rd International Conference on Library and Information Science, 23–25 August 2017, Sapporo, Japan, pp. 92–115. International Business Academics Consortium, Taipei (2017)
21. Gros, D., Wanner, B., Hackenholt, A., Zawadzki, P., Knautz, K.: World of streaming. Motivation and gratification on twitch. In: Meiselwitz, G. (ed.) SCSM 2017. LNCS, vol. 10282, pp. 44–57. Springer, Cham (2017). https://doi.org/10.1007/978-3-319-58559-8_5
22. Scheibe, K., Göretz, J., Meschede, C., Stock, W.G.: Giving and taking gratifications in a gamified social live streaming service. In: Proceedings of the 5th European Conference on Social Media. Academic Conferences and Publishing International, Reading (2018)
23. Wilk, S., Wulffert, D., Effelsberg, W.: On influencing mobile live broadcasting users. In: 2015 IEEE International Symposium on Multimedia, pp. 403–406. IEEE, Washington, DC (2015)
24. YouNow: What is a Subscription? https://younow.zendesk.com/hc/en-us/articles/206320655-What-is-a-subscription. Accessed 14 July 2017
25. YouNow: What are Stickers and Bar-Based Gifts? https://younow.zendesk.com/hc/en-us/articles/206320625-What-are-stickers-and-bar-based-gifts. Accessed 14 July 2017
26. YouNow: What is Editor's Choice? https://younow.zendesk.com/hc/en-us/articles/206320675-What-is-Editor-s-Choice. Accessed 14 July 2017
27. YouNow: How Do I Get Featured on YouNow? https://younow.zendesk.com/hc/en-us/articles/212230083-How-Do-I-Get-Featured-on-YouNow. Accessed 14 July 2017
28. Likert, R.: A technique for the measurement of attitudes. Arch. Psychol. **140**, 5–55 (1932)
29. Alexa: Younow.com Traffic Statistics. https://www.alexa.com/siteinfo/younow.com. Accessed 30 June 2017
30. Fietkiewicz, K.J., Baran, K.S., Lins, E., Stock, W.G.: Other times, other manners: how do different generations use social media? In: 2016 Hawaii University International Conferences, pp. 1–17. Arts, Humanities, Social Sciences and Education, Honolulu (2016)

Proposal of Learning Support SNS Utilizing Gamification

Syun Usami[1]([⊠]), Kohei Otake[2], and Takashi Namatame[3]

[1] Graduate School of Science and Engineering, Chuo University,
1-13-27 Kasuga, Bunkyo-ku, Tokyo 112-8551, Japan
al3.nnma@g.chuo-u.ac.jp
[2] School of Information and Telecommunication Engineering, Tokai University,
2-3-23 Takanawa, Minato-ku, Tokyo 108-0074, Japan
otake@indsys.chuo-u.ac.jp
[3] Faculty of Science and Engineering, Chuo University,
1-13-27 Kasuga, Bunkyo-ku, Tokyo 112-8551, Japan
nama@indsys.chuo-u.ac.jp

Abstract. Recently, a learning support system using the internet is often used in the field of education. However, many of them focus only on improving academic ability and communication among students. Also, as education problem in Japan, there is a lack of voluntary and learning motivation. In this research, we aim to propose a system to promote student's self-learning motivation to learn and establishment of learning habits. The proposed system enables well communication not only among students but also between students and teachers. We also utilize gamification for the proposed system. We believe that propose system can effectively promote student's self-learning motivation and establishment of learning habits. Finally, we describe an experiment plan using the propose system.

Keywords: E-learning · Gamification · Social networking service
Motivation management

1 Introduction

Recently, as a problem of education in Japan, it is often mentioned that voluntary and sustained learning motivation is lacking [1]. Lack of voluntary and learning motivation leads to a decrease in motivation for self-learning. Currently, there is "assignment" as a method for cram schools and schools to participate in self-learning at home. However, with "assignment" alone, it is not possible to solve the problem of a decrease in motivation for student's self-learning. In this situation, junior high school and high school students who are using social networking services (SNSs) such as Twitter and Instagram for self-learning are on the rise. From the survey by the school corporation Takamiya Gakuen Yoyogi Seminar, 57% of preparatory students have shown that they use SNSs for self-learning (Fig. 1) [2]. According to a questionnaire survey conducted by Asahi Student Newspaper in 2017, the main reason why junior high school students and high school students use SNS for self-learning are "information gathering" and "solving unknown things". Besides that, "Increase motivation" accounts for about 50%

© Springer International Publishing AG, part of Springer Nature 2018
G. Meiselwitz (Ed.): SCSM 2018, LNCS 10914, pp. 114–125, 2018.
https://doi.org/10.1007/978-3-319-91485-5_8

(Fig. 2) [3]. From these surveys, it is common for junior high and high school students to use SNSs for self-learning. Moreover, it can be seen that SNSs functions not only as a solving unknown things and information gathering purpose but also as a system for increasing motivation. Besides SNSs, there are many researches on systems aiming at learning support [4–7]. However, many of these systems are often focused on improving academic ability [4]. We think that improvement of academic ability is the result of improvement of motivation for self-learning and establishment of self-learning habit. Therefore, in this research we propose a system focusing on improvement of motivation for self-learning and establishment of self-learning habits for student.

This paper is organized as follows. In Sect. 2, we describe the purpose of our research. In Sect. 3, we introduce the previous research and gamification. In Sect. 4, we describe preliminary surveys for teachers who are instructing junior high school students and clarify more detailed subjects. In Sect. 5 we explain the outline of SNSs

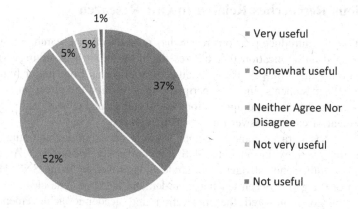

Fig. 1. Question about effectiveness of SNSs for self-learning [3]

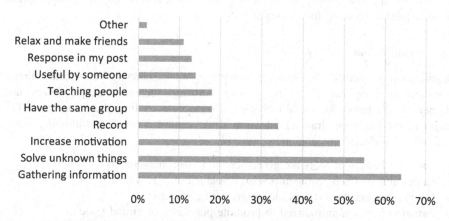

Fig. 2. Purpose of utilizing SNSs for study (multiple answers) [3]

utilizing the gamification propose in our research. In Sect. 6, we explain the outline of experimental design using proposed system. Finally, we describe conclusion and future works with Sect. 7.

2 Purpose of Our Research

In this research, we propose a system to increase student's motivation to self-learning and establishment of self-learning habits. The system is created based on gamification methods and SNSs elements. We explain gamification methods in Sect. 3.1. First of all, we conduct questionnaire survey for teachers on the proposed system and problems. Next, we propose a system. Finally, we propose an experimental plan for junior high school students.

3 Previous Researches Related to Our Research

In this section, we introduce previous researches related to this research. Hatsugai and Iyoda developed an application that utilizes gamification such as giving experience points by solving question [4]. They conducted experiments from junior high school students to college students. In the experiment, they verified the effect on computing capacity by using the proposed application. The result gained educational effect, such as improvement of correct answer rate in calculation test.

Sasaki and Sasakura examined whether learning opportunities and student satisfaction are improved by introducing SNS into university lessons [5]. As a result, learning opportunities and satisfaction of students improved by using SNS. However, the problem of the gap between teacher's burden and merit became clear.

Hanes and Fox compared the motivation and academic achievement in the university class with the group using gamification and the group not using it [6]. Results were showed effect in the group utilizing gamification. However, there were cases where the game elements did not lead to essential motivation and the satisfaction degree was lowered in some cases. The conclusion suggested that it is necessary to use appropriate game elements according to the purpose.

3.1 Gamification

Gamification is method that induces voluntary and sustainable behavioral change. Gamification is defined as "to use gaming elements, such as concept, design, and mechanics of a game, for social activities or services other than the game itself" [7]. This method has been drawing attention since around 2010. The following seven methods are included as the Gamification [7].

- Honorific badges or titles are given according to achievements
- Names and scores of competitors are displayed on a real-time basis
- The graphic interface shows the progress of each task
- Virtual currency is introduced to promote purchases of virtual goods
- Rewards such as coupons or gifts are provided

- Assignments that encourage users to collaborate together are presented
- Simple games are prepared between activities in order to keep users from being bored

Gamification is now applied in a wide range of fields including marketing and education. Foursquare is a representative service using Gamification [8]. Foursquare is a web service based on location information that allows registered users to connect with friends and update their locations. There are few Gamification such as badge, which is given when checking in to a specific place (Fig. 3). Foursquare makes users compete with badges and scoring points. This has triggered the user to visit various places.

Fig. 3. Foursquare's play screen showing the specific place

4 Preliminary Survey

In order to clarify current problems about student's self-learning, we conducted questionnaires to 16 teachers who are teaching at junior high school. The question items are shown in Table 1.

Table 1. Questions for preliminary survey

Question	Contents of question
1	Do you think self-learning is necessary?
2	How do you feel about students' awareness of self-learning?
3	Do you think students are studying hard while they are in a cram school?
4	What is the reason why students can study hard while in a cram school?
5	Are students working hard to self-learning?
6	What do you think is the reason why students do not self-learning hard?
7	Can you able to grasp and manage the student's learning situation outside the class?

From the results of questionnaires, 87.6% of teachers answered that "Strongly Agree" or "Agree" in response to the question "Do you think self-learning is necessary?" (Fig. 4). However, no one answered that "Very high" or "High" in response to the question "How do you feel about students' awareness of self-learning?" (Fig. 5). From the result, we think that there is a present situation that students are not positively working on self-learning which is thought to be necessary.

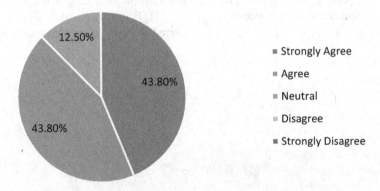

Fig. 4. Result of question 1: Do you think self-learning is necessary?

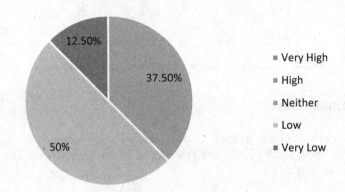

Fig. 5. Result of question 2: How do you feel about students' awareness of self-learning?

No one answered "Disagree" or "Strongly Disagree" in response to the question "Do you think students are studying hard while they are in a cram school?" (Fig. 6). Also, in response to the question "What is the reason why students can study hard while in a cram school?", About 50% of teachers answered "teacher encouragement" or "teacher's commentary for study" for the reason. From this result, we think that prompt feed-back from the teachers on learning can be considered as a reason for studying hard in the cram school. On the other hand, only 31.3% of teachers answered "Agree" response to the question "Are students working hard to self-learning?" (Fig. 7). The

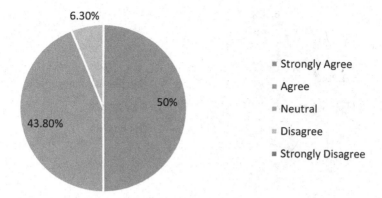

Fig. 6. Result of question 3: Do you think students are studying hard while they are in a cram school?

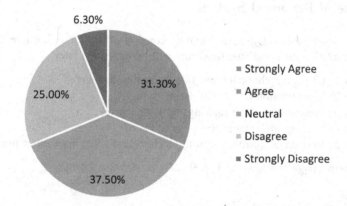

Fig. 7. Result of question 5: Are students working hard to self-learning?

answer to the question "The reason why the students are not working hard to self-learning" is that "Students do not think that have to study", "Students cannot see other students studying", "Students cannot enjoy self-learning".

31.3% answered "Not be both", 62.5% answered "Only grasping" in the question "Are you able to grasp and manage the student's learning situation outside the class?" (Fig. 8). It is difficult for the teachers to make appropriate guidance in the class in a situation where students' self-learning is not grasp and manage.

Based on these results of questionnaires, there are four problems at present such as "motivation for self-learning is weak", "No prompt feedback for self-learning", "situation other than self cannot be grasped" and "The teacher cannot grasp/manage the student's self-learning". In this research, we propose a system to solve these problems.

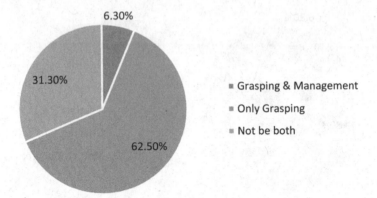

Fig. 8. Result of question 7: Can you grasp and manage the learning situation of the students outside the class?

5 Outline of Proposed System

The system proposed in this research mainly consists of the following five functions. A detailed explanation of the five functions will be described later.

Function 1: Changing character and title according to level
Function 2: Visualization of system usage
Function 3: Present and share rankings by user's level
Function 4: Sharing information on other students
Function 5: Encourage comments from teachers and likes from other users.

The system image is shown in Fig. 9.

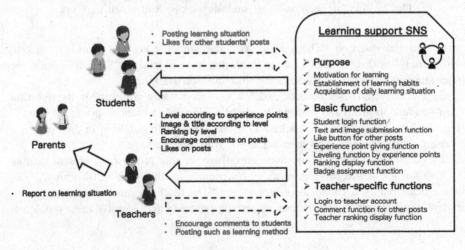

Fig. 9. Image of the propose system

5.1   Changing Character and Title Using Experience Points

We implemented the function that the level is improved by accumulating the experience value given by each action, and the character and title displayed on the user's My Page are changed according to the level (Fig. 10). The purpose of this function is to let the user feel visual growth as well as numerical values.

This function is expected to improve the problem of "motivation for self-learning is weak", "No prompt feedback for self-learning".

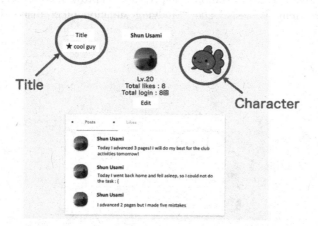

Fig. 10. Changing characters and title using experience points

5.2   Visualization of System Usage

The function to display the total number of logins and likes, the current level, my posts, posts that pushed favorites on the user's My Page were implemented in the proposed system (Fig. 11). The purpose of this function is visualization of system usage. As the

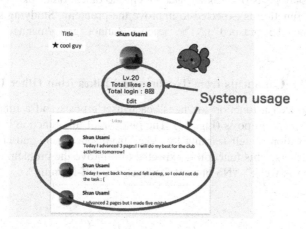

Fig. 11. Visualization of system usage

result, we think that users can obtain self-learning situation and learning situation of other students.

This function is expected to the improve the problem of "motivation for self-learning is weak" and "Studying situation other than self cannot be grasped".

5.3   Present and Share Ranking by User's Level

By competing with other users, we implemented a function that encourages users to self-learning (Fig. 12). This function is expected to improve the problem of "motivation for self-learning is weak" and "Studying situation other than self cannot be grasped".

Fig. 12. Present and share ranking by user's level

5.4   Sharing Information on Other Students

In order to reduce the sense that students study independently, we introduced the function to display posts of all other users in chronological order like Twitter timeline (Fig. 13). This function is expected to improve the problem "Studying situation other than self cannot be grasped", "The teacher cannot grasp/manage the student's self-learning".

5.5   Encourage Comments from Teachers and Likes from Other Users

We have introduced a function to "likes" each user's posts and a function to allow students to comment on posts (Fig. 14). The purpose of this function is student's user maintains motivation of self-learning by "like" from other users and encourage comments from teachers. This function is expected to improve the problem of "motivation for self-learning is weak", "No prompt feedback for self-learning".

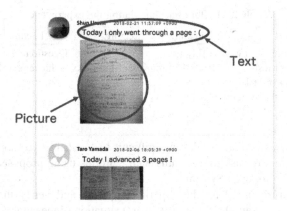

Fig. 13. Sharing information on other students

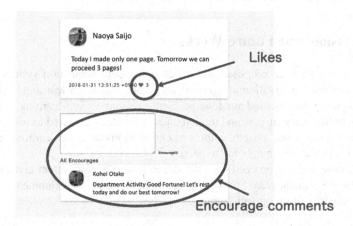

Fig. 14. Encourage comments from teachers and likes from other users

6   Outline of Experimental Plan

We will conduct a verification experiment at a cram school using the system pro-posed in Sect. 5. We describe outline of the experimental plan. This experiment covers junior high school students. Subjects are divide into "Groups using proposed Sys-tem" and "Groups not using proposed System". This experiment carries out for 2 months. During the experiment period, subjects are gathered once a week and lecture are held face to face. The content of the lecture is as shown in Table 2.

In the first lecture, we will conduct a questionnaire survey on current learning situations and awareness of study, academic ability test, presentation of tasks during the experiment, and explanation of how to use the system. For the second and subsequent lecture, we conduct weekly tests and investigate the progress of the task. In the last lecture, we conduct academic performance test and questionnaire again.

Table 2. Lecture schedule during the experiment

First Lecture	2nd Lecture $\sim 7^{th}$ Lecture	Last Lecture
Questionnaire survey	Weekly test	Questionnaire survey
Achievement test	Check the progress of the tasks	Achievement test
Presentation of tasks		
How to use the system		

The task presents challenges that are difficult to finish in two months. Then, we compare the progress of task in 2 months with "Groups using proposed System" and "Groups not using proposed System".

Then, we analyze based on the experimental data and verify the effect of the proposed system. For example, we clarify the use situation of the student's system from the access log, and clarify the relation between system using situation, progress of the task and score of the academic ability test can be seen.

7   Conclusion and Future Works

In this research, we have proposed and developed a communication system aimed at improving student's motivation for learning and establishing self-learning habits. In the proposed system, we focused on four problems related to self-learning, which was clarified by preliminary survey, and implemented a function expected to solve them. we think using the proposed system, improvement of motivation for learning and establishment of self-learning habits are expected.

In the future we will proceed with the development of the system and experiment based on the experiment plan. From the experimental results, we examine the influence of the developed system on self-learning.

References

1. Ministry of Education: Culture, Sports, Science and Technology: Direction of improvement. http://www.mext.go.jp/b_menu/shingi/chukyo/chukyo0/toushin/attach/1346331.htm. Accessed 20 Feb 2018 (in Japanese)
2. School corporation Takamatsu Gakuen Yoyogi Seminar: https://www.yozemi.ac.jp/sapixgroup/__icsFiles/afieldfile/2017/03/28/20170328_release_2.pdf. Accessed 20 Feb 2018 (in Japanese)
3. Asahi Student Newspaper: https://prtimes.jp/main/html/rd/p/000000052.000021716.html. Accessed 20 Feb 2018 (in Japanese)
4. Hatsugai, T., Iyoda, M.: A method that gamification aimed at increasing mathematical ability. In: Proceedings of the 76th National Convention of IPSJ, vol. 1, pp. 633–635 (2014). (in Japanese)
5. Sasaki, Y., Sasakura, C.: Practical training of computer literacy using SNS to support learning: practices and evaluation. Jpn. J. Educ. Technol. **33**(3), 229–237 (2010). (in Japanese)

6. Michael, D.H., Jesse, F.: Assessing the effects of gamification in the classroom: a longitudinal study on intrinsic motivation, social comparison, satisfaction, effort, and academic performance. Comput. Educ. **80**, 152–161 (2015)
7. Otake, K., Shinozawa, Y., Sakurai, A., Uetake, T., Oka, M., Sumita, R.: A proposal of SNS to improve member's motivation in voluntary community using gamification. Int. J. Adv. Comput. Sci. Appl. **6**(1), 82–88 (2015)
8. Foursquare. https://ja.foursquare.com/. Accessed 20 Feb 2018

Because It's Good for My Feeling of Self-worth: Testing the Expanded Theory of Planned Behavior to Predict Greek Users' Intention to Review Mobile Apps

Charalampos Voutsas, Ardion Beldad[(✉)], and Mark Tempelman

University of Twente, Enschede, The Netherlands
babisvoutsas@gmail.com,
{a.d.beldad, m.h.tempelman}@utwente.nl

Abstract. Mobile apps, just like traditional products (e.g. books, electronic goods) and services (e.g. hotels) sold and marketed online, are increasingly being subjected to after-use evaluations. While the factors influencing people's intention to write reviews for product and services have been increasingly understood, the mechanisms behind people's willingness to review mobile apps, which often can be used without any cost, are not yet fully explored. Using the Theory of Planned Behavior and a set of functions for writing reviews identified in previous studies, a model was tested with survey data from 214 Greek mobile app users to identify the factors that influenced their intention to write reviews for mobile apps. Results of a hierarchical regression analysis shows that app review writing intention is influenced by a positive attitude towards the act, perceived behavioral control, descriptive social norms, and ego-defensive function.

Keywords: Mobile app reviews · Theory of Planned Behavior
Online review writing functions

1 Introduction

In a post-Web 2.0 era, consumers have increasingly gained the ability to be involved in the commodity production process, which is realized partly by providing them with the chance to say something about their experience with a product. Aptly termed as online reviews, customers' assessment of their interaction with a certain product has been known to either increase or decrease other people's inclination to use that product.

Online reviews are also significantly affecting the market for mobile apps. Often when people have no prior experience with or information about a certain app, their decision to download it could be hinged on some factors visible to them at the moment of download decision. User reviews are often used as one of the bases for a download decision.

In the literature on online reviews, it has been noted that several factors influence people's proclivity to provide reviews for specific products or services. For instance, people have both rational (e.g. knowledge sharing) and emotional (e.g. making friends) motivations for writing reviews [14]. Additionally, when one subscribes to the Theory

© Springer International Publishing AG, part of Springer Nature 2018
G. Meiselwitz (Ed.): SCSM 2018, LNCS 10914, pp. 126–136, 2018.
https://doi.org/10.1007/978-3-319-91485-5_9

of Planned Behavior (TPB) [1], it can also be argued that people's willingness to provide reviews for certain apps could be influenced by factors such as their evaluation of the review writing act (attitude), their perceived ability to perform the act (behavioral control), and what they expect people within their networks expect them do (subjective norm).

As studies into the predictors of people's intention to write reviews for mobile apps are virtually non-existent, this research aimed at determining the factors that prompt people to publish narratives of their experiences with certain mobile apps. The hypotheses proposed for the study were tested using data collected from 214 mobile users from Greece through an online survey.

2 Theoretical Framework

Online reviews of products and services are regarded important sources of information for customers who are still in the process of deciding whether or not to purchase or acquire a specific product or service. As a form of user-generated contents, online reviews benefit not only customers but also companies that either sell or produce a product or provide a service. For online companies or vendors, online reviews can be a low-cost form of advertisement (especially if consumer reviews are positive) and an effective quality control approach, as reviews can provide companies with insights into customers' reactions and levels of satisfaction [10]. Such benefits could explain why online companies are incentivizing, in one way or another, customers who write reviews for products or services acquired.

Although the factors that motivate people to write online reviews, in general, have already been identified in previous studies [e.g. 11, 23], the relative newness of mobile apps as commodities for mass consumption, unlike more established products (e.g. books, electronic goods) and services (e.g. hotels, restaurants), implies that the factors influencing users' motivations to write reviews for certain apps are not yet adequately understood. Additionally, the fact that some apps can be downloaded for free, while others can be purchased for a relatively low price, would also signify that the mechanisms governing people's decision to review a mobile app could be different, especially if the app to be reviewed does not have a high price tag.

In this study, a model for the determinants of users' intention to write mobile app reviews was tested using Ajzen's [1] Theory of Planned Behavior (TPB). However, questions pertaining to the sufficiency of the model [8] to explain variance in people's behavioral intention has prompted calls for the inclusion of context-relevant variables. For instance, in a study into the factors influencing the use of ICT in classrooms [19] and instant messaging [15], TPB was employed alongside factors such as perceived usefulness and perceived ease of use.

2.1 The Effects of TPB Factors: Attitude, Subjective Norm, and Perceived Behavioral Control

The central proposition of TPB is that an individual's actual behavior is a function of his or her intention to perform the behavior, which, in turn, is predicated on three factors, namely attitude towards the behavior, subjective norm, and perceived behavioral control [1]. Attitude refers to the 'degree to which a person has a favorable or unfavorable evaluation or appraisal of the behavior in question' [1, p. 188], while subjective norm is defined as the person's perception of 'social pressure to perform or not to perform the behavior' [1, p. 188]. Perceived behavioral control refers to the 'ease or difficulty of performing the behavior' [1, p. 188].

In a several studies into the effects of these three TPB factors on the intention to use a technology, it has been shown that attitude, perceived behavioral control, and subjective norm contribute to people's decision to use specific types of technology such as instant messaging [15] and mobile devices for learning [5]. Additionally, the three TPB factors have also been found to be significant predictors of computer-mediated behaviors such as online shopping [12].

Nonetheless, Ajzen's [1] 'subjective norm' concept might be limited by its focus on a person's expectation of how his or her relevant contacts expect him or her to behave. This point prompts the decision to re-conceptualize the role of social influence in people's decision to write reviews by taking into account the impact of two types of social norms, namely injunctive social norms and descriptive social norms. While injunctive social norms refer to expectations of what other people approve (hence, conceptually similar to subjective norm), descriptive social norms refer to beliefs in the acceptability of an act because it is something typically performed by others [6]. Both injunctive [4, 21] and descriptive [2] social norms have been found to influence people's decision to share various types of personal information online.

Results of previous studies into the effects of TPB variables on behavioral intention, therefore, precipitated the first set of research hypotheses.

Hypothesis 1: Mobile users' positive attitude towards writing reviews for mobile apps positively influences their intention to write app reviews.

Hypothesis 2: (a) Injunctive social norms and (b) descriptive social norms positively influence mobile app users' intention to write app reviews.

Hypothesis 3: Perceived behavioral control positively influences mobile app users' intention to write app reviews.

2.2   The Functions of Writing Reviews

Daugherty et al. [9], using Katz's functional theory, claim that people's decision to create user-generated contents, such as online reviews, is predicated on four functions, namely (a) utilitarian, (b) knowledge, (c) ego-defensive, and (d) value-expressive. From a utilitarian standpoint, UCG creation is motivated by the availability of incentives; whereas, from a knowledge standpoint, UCG creation is prompted by people's need to understand their environment and themselves. The ego-defensive function of UCG creation is hinged on people's need to reduce self-doubt, increase their sense of belongingness, and minimize feelings of guilt for not contributing; while the value-expressive function of UCG creation is triggered by a feeling of gratification for being able to create something and by a degree of validation of who they are upon engagement in the creation act [9].

Ambiguity in the operationalization of the 'knowledge function', however, spurred the decision within this study to drop the concept from the model and to replace it with 'social function', which is approximately similar to the notion that writing online reviews provides reviewers with social benefits (e.g. writing reviews allows a person to meet others) [11]. Additionally, the 'value-expressive' concept is further extended (and referred to as 'emotional expression') to refer to people's desire to voice out the feelings that emerged from either their positive or negative experience of using a product. In the current study, the authors argue that this function enables product users not only the possibility to fully express themselves but also to inform others of their subjective experiences of using a certain product. In a way, hence, emotional expression also assumes the function of knowledge sharing from one consumer to another.

In a previous study, economic incentives (or the utilitarian function), social benefits, concern for other consumers, and positive self-enhancement (or ego-defensive) have been found to increase online word-of-mouth behavior [11]. Hence, it can also be hypothesized that the four functions of writing reviews could influence mobile app users' intention to write reviews for certain mobile apps. The second set of research hypotheses is presented below.

Hypothesis 4: The utilitarian function of writing reviews positively influences mobile app users' intention to write reviews for mobile apps.

Hypothesis 5: The social function of writing reviews positively influences mobile app users' intention to write reviews for mobile apps.

Hypothesis 6: The ego-defensive function of writing reviews positively influences mobile app users' intention to write reviews for mobile apps.

Hypothesis 7: The emotional expression function of writing reviews positively influences mobile app users' intention to write reviews for mobile apps.

3 Methods

3.1 Research Design and Procedure

An online survey was implemented to collect the necessary data to test the hypotheses proposed for the study. A link to the electronic questionnaire was sent to Greek mobile app users, who were approached through social networking sites, e-mails, and online discussion platforms.

A snowball sampling approach was used to reach as many survey respondents as possible. Despite the limitation of this sampling strategy (e.g. non-representativeness of the sample), it enables the researchers to collect data within a short timeframe and with less financial costs. After a ten-day collection period, completed questionnaires from 214 respondents were collected.

3.2 Survey Respondents

Of the 214 Greek respondents whose data were used for analysis, 123 (57%) were females. Majority of the respondents (n = 129, 60%) fall under the age cluster '25 to

34', with another 20% of the total number of respondents belonging to the age cluster '18 to 24'. Exactly 80% (n = 172) of the respondents have obtained higher education (e.g. a four-year bachelor's or a master's degree). Moreover, 67% (n = 143) of 214 respondents are primary users of social networking apps.

3.3 Measurements

The different research constructs were measured using previously validated scales. The 'attitude' construct was measured with five items (e.g. 'Writing a review for a mobile app is....good/bad, pleasant/unpleasant') on a semantic differential scale originally formulated by Daugherty et al. [9] and Moon and Kim [16]. Injunctive (e.g. 'My close social contacts approve of me writing mobile app reviews.') and descriptive (e.g. 'A lot of people around me write mobile app reviews.') social norms were measured with four and three items by White et al. [22], respectively. Five items, mostly from Netemeyer, Burton, and Johnston [17], were used to measure 'perceived behavioral control' (e.g. 'If I wanted to, I could easily write a review for a mobile app.').

Five items (e.g. 'Submitting an online review for a mobile app benefits me personally.') by Daugherty et al. [9] were used to measure 'utilitarian function', while four items (e.g. 'writing an online review for a mobile app makes me feel part of a community') by Clary et al. [7] measured 'social function'. Additionally, three items (e.g. 'Writing an online review for a mobile app makes me feel important.') by Clary et al. [7] were selected for 'ego-defensive function', while four newly formulated items (e.g. 'Writing a review provides me with the opportunity to express my opinion about the app.') were used to measure 'emotional expression function'. Finally, three newly formulated items were used to measure 'intention to write reviews for mobile apps'.

3.4 Measurement Validity and Reliability

To determine the validity of the constructs, an exploratory factor analysis, using principal component analysis (PCA), was performed with the 35 items measuring the nine constructs. For this analysis, the Kaiser-Meyer Olkin Measure of Sampling Adequacy value is .850 (higher than the recommended value of .60) [13], while the Bartlett's Test of Sphericity X^2 (561) = 3,885.75 is significant ($p < .001$), which means that the correlation among the 35 items is high enough for PCA. However, analysis revealed that only eight factors (instead of nine) had eigenvalues higher than 1. Inspection of the rotated component matrix indicated that items measuring 'social function' loaded with items measuring both 'injunctive' and 'descriptive' social norms. Hence, the 'social function' construct, considering its questionable validity, was removed from further analysis.

A second exploratory factor analysis was subsequently executed with the remaining 31 items intending to measure 8 constructs. Kaiser-Meyer Olkin Measure of Sampling Adequacy value is .827, while the Bartlett's Test of Sphericity X^2 (435) = 3,331.31 is also significant ($p < .001$). The analysis resulted in seven factors (instead of eight) having eigenvalues higher than 1. Items measuring 'injunctive social norms' had problematic loadings, as they loaded with 'intention' items. Hence, 'injunctive social norms' was also excluded from analysis.

A third exploratory factor analysis without the items for 'injunctive social norms' was performed (Kaiser-Meyer Olkin Measure of Sampling Adequacy = .815; Bartlett's Test of Sphericity X^2 (351) = 2,897.36, $p < .001$), which resulted in seven factors having eigenvalues higher than 1. With the removal of 'injunctive social norms' and 'social function' from the research model, hypotheses 2a and 5, respectively, will not be tested.

Cronbach alpha's values of the seven constructs were calculated to determine their reliability. The reliability for all constructs ranges from acceptable to good, as alpha values for the seven constructs are higher than .70. Table 1 presents the Cronbach's alpha, mean, and standard deviation (SD) values for the seven constructs.

Table 1. Cronbach's alpha, mean, and standard deviation (SD) values for the seven constructs

Construct	No. of items	Cronbach's α	Mean	SD
Intention to write reviews (INT)	3	.76	2.76	0.75
Attitude (ATT)	4	.76	3.35	0.71
Descriptive social norms (DES)	3	.74	2.54	0.69
Perceived behavioral control (PBC)	5	.73	3.58	0.64
Utilitarian function (UTI)	5	.87	2.45	0.84
Ego-defensive function (EGO)	3	.89	2.03	0.88
Emotional expression function (EMO)	4	.87	3.93	0.70

4   Results

4.1   Test for Multicollinearity

To determine if there are multicollinearity issues among the dependent variables, a correlation analysis was executed using all the research constructs. Correlation analysis reveals that there are no multicollinearity issues as correlation values among the constructs are remarkably lower than .70, the minimum for high correlation [3].

Furthermore, the tolerance level and the variance inflation factor (VIF) values were also calculated to fully ensure that, indeed, multicollinearity issues are not present. Tolerance level values for the six predictors ranged between .73 and .91 (higher than the prescribed limit of .10 for high correlation to exist), while the VIF values for the construct ranged between 1.08 and 1.37 (lower than the prescribed value of 10 to indicate multicollinearity) [3]. The absence of multicollinearity clearly denotes that the six predictors can be included in the regression analysis. Table 2 shows the inter-correlations among the seven research constructs.

4.2   Hypotheses Testing

A hierarchical regression analysis, which enabled the sequential determination of the five predictors [3] on the intention to write reviews for mobile apps was performed to test the final set of research hypotheses (1, 2a, 3, 4, 6, and 7). In the first block, the TPB factors – attitude, perceived behavioral control, and descriptive social norms (instead of

Table 2. Inter-correlations among the 7 seven research constructs

	INT	ATT	DES	PBC	UTI	EGO	EMO
INT	1						
ATT	.42**	1					
DES	.36**	.23**	1				
PBC	.41**	.33**	.23**	1			
UTI	.23**	.22**	.23**	.07	1		
EGO	.38**	.29**	.22**	.11	.24**	1	
EMO	.24**	.30**	.16*	.46**	.07	.17*	1

$**p < .01; *p < .05$.

subjective norm or injunctive social norms as the construct has poor validity) – were entered resulting in an adjusted R^2 of .29 ($F_{3, 210} = 31.31, p < .001$).

In the second block, the three functions with high validity were entered (utilitarian, ego-defensive, and emotional expression), which resulted in an increase in the adjusted R^2 (.34; $F_{3, 207} = 19.94, p < .001$). The adjusted R^2 value for the final model indicates that 34% of the variance for Greek mobile app users' intention to write mobile app reviews could be explained by the six predictors included in the analysis.

The final model further indicates that four of the six hypothesized predictors of mobile review app writing intention have significant effects on the dependent variable of interest. The four variables include the three TPB variables, namely attitude ($b = .22, p < .01$), perceived behavioral control ($b = .28, p < .001$), and descriptive social norms ($b = .18, p < .01$), and one functional predictor – ego-defensive ($b = .24, p < .001$). These results mean that hypotheses 1, 2a, 3, and 6 are supported, whereas hypotheses 4 and 7 could not be supported.

Table 3 presents the unstandardized and the standardized coefficients of the different predictors of Greek mobile app users' intention to write reviews for mobile apps.

Table 3. Unstandardized and standardized coefficients of the different predictors of Greek mobile app users' intention to write reviews for mobile apps

	B	Std. error	β	Adj. R^2 (ΔR^2)
(Constant)	.04	.29		
Attitude	.30	.07	.28***	.30
Descriptive social norms	.25	.07	.23***	(.31)
Perceived behavioral control	.31	.07	.26***	
(Constant)	−.08	.31		
Attitude	.23	.07	.22**	.35
Descriptive social norms	.20	.06	.18**	(.06)
Perceived behavioral control	.32	.08	.28***	
Utilitarian function	.06	.05	.06	
Ego-defensive function	.20	.05	.24***	
Emotional expression function	−.03	.07	−.02	

$***p < .001; **p < .01; *p < .05$.

5 Discussion and Future Research Directions

5.1 Discussion of Results

A review of the literature on the factors influencing people's word-of-mouth intention in the offline setting and an empirical study into the determinants of people's decision to write electronic reviews [11] clearly demonstrate how diverse people's motivations are for verbalizing their views and feelings for products and services they have used. Online reviews, as a specific form of word-of-mouth, benefit not only a general population of consumers (e.g. reviews as primary sources of necessary information) but also companies that sell or produce commodities being reviewed (e.g. reviews as low-cost advertisements).

The benefits app creators can derive from app user ratings and review could sufficiently explain why those creators are incentivizing app users to either rate or review apps they have downloaded (e.g. gaming apps promising game points to users who will decide to rate or review those apps). From a business standpoint, then, it helps to understand which factors would prompt people to write reviews for mobile apps. The fact that a multitude of mobile apps can be downloaded for free (hence, emotions arising from instances when apps will not meet prior expectations might be less intense), prompts the question on whether or not the mechanism behind people's propensity to write reviews for paid products and services would also translate into the context of reviewing mobile apps.

Results of an online survey with 214 Greek mobile app users show that their inclination to write reviews for mobile apps are predicated on their beliefs that they are capable (either because they have the know-how and the time) of writing reviews. Although writing reviews does not require a specialized expertise, the act can only be performed when one (a) has basic knowledge of how to post a review, (b) has an adequate understanding of the product's pros and cons, (c) has elementary knowledge of writing in a specific language, and (d) the time to write the review. The likelihood that an individual will write a review for a mobile app is higher when he or she could meet some of the prerequisites just mentioned.

A plethora of research into attitude-behavioral intention relationship has also shown that people will engage in an action that is positively viewed. In fact, it has been noted that behavioral intentions are higher when they are anchored on attitude (an autonomous belief in something) than on subjective norm (socially influenced belief in a thing) [18]. This particular result strongly suggests that mobile app users will have to be fully convinced of the positive features of writing a mobile app review before they will decide to engage in the act of writing.

As writing reviews is a social act that could be performed without serious demands for secrecy and confidentiality, people might be highly predisposed to write reviews just because others within their immediate environment are doing it. People have a strong tendency to mimic behaviors by others [20], and this could explain why descriptive social norms have been found to increase people's intention to perform a specific action. The performance of an action by a certain number of individuals may give an indication of the worth, value, and acceptability of an act, and this might provide an individual with a reason to engage in a similar action. Hence, people who are aware that others within their social networks write reviews for mobile apps might also be encouraged to review apps they have already used.

Results also show that Greek mobile app users will be inclined to write reviews for mobile apps if they are convinced of the ego-defensive function that the act of review writing extends. This result somehow corresponds to what a previous study has found – that self-enhancement needs (e.g. feeling good about being able to tell others about one's success) have a strong impact on consumers' willingness to write reviews [11].

The absence of statistical support for the effects of utilitarian and emotional expression functions on mobile app review writing has a couple of implications. First, mobile app users' decision to write a review might have resulted from their calculation of the value of the compensation they are bound to receive in exchange for the effort and time they have to invest in writing a review. It is highly likely that the apps they have used do not offer attractive incentives for review writing.

Second, mobile app users' decision to write reviews might be strongly hinged on the intensity of emotions they have upon using a specific app. This point means that it might require a very high level of satisfaction and joy for mobile users to write reviews, just as users have to be extremely disappointed or must have a nightmarish experience with an app to invest time to review it. A disappointing experience with mobile apps that can be downloaded for free may not suffice to instigate users to publicly vent out their frustrations with those apps as users can just decide to uninstall those apps.

5.2   Implications and Future Research Directions

Results of this survey have a number of implications for mobile app designers. First, the finding that Greek mobile app users would be willing to write a mobile app review if people within their social networks are doing the same signifies that app developers should persistently explore ways to capitalize on the potential of social influence to motivate users to review apps they have used.

Second, the critical role that perceived behavioral control plays in nudging people to write reviews for mobile apps suggests that app designers should ensure that the act of review writing is something that will not cause people much time and effort to perform. However, it should be noted that if the act, indeed, would require some time investment from app users, attractive incentives should be offered to them. This point is proposed in relation to the premise that the statistically insignificant effect of utilitarian function on app review writing intention might be due to the absence or the unattractiveness of rewards or incentives for review writing.

Third, the effect of attitude towards review writing on people's intention to write reviews implies that mobile app developers have to look into strategies that would prompt their users to regard the act of reviewing apps in a positive way. Influencing app users' attitude towards app review writing might mean that mobile app developing should identity strategies to increase the salience of the functions of app review writing.

Results of the current study must be interpreted with caution. The cross-sectional nature of the study would limit any claim pertaining to the real causal relationships between the predictors and the dependent variable of interest. Future studies, hence, should consider resorting to an experimental approach to test the possible effects of variables such as incentives and levels of satisfaction (or dissatisfaction) on people's intention to write reviews for mobile apps.

The current study's reliance on a relatively small sample of Greek mobile users invited using a non-random sampling approach (e.g. snowball sampling) also means that the results may not entirely reflect the mechanisms behind app review writing among a wider population of Greek mobile app users. More importantly, the use of data from a specific cultural or national cluster would also signify that the results will not totally apply to mobile users from other cultural or national clusters. This point will certainly open up avenues for research into the factors influencing mobile app review writing intentions in a cross-cultural context.

Furthermore, as the current study opted not to take a more nuanced view on app review writing intention across various types of mobile apps (e.g. paid apps vs free apps), the results might be remarkably different in a study that focuses on the determinants of mobile app review writing intention within the context of mobile apps that people have to pay for. One can only surmise that the impact of emotional expression function on users' review writing intention, for instance, would be stronger when they have to review an app they have paid for compared to an app that was downloaded for free.

References

1. Ajzen, I.: The theory of planned behavior. Org. Behav. Hum. Decis. Process. **50**(2), 179–211 (1991)
2. Beldad, A., Hegner, S.: More photos from me to thee: Factors influencing the intention to continue sharing personal photos on an online social networking (OSN) site among young adults in the Netherlands. Int. J. Hum. Comput. Interact. http://www.tandfonline.com/doi/full/10.1080/10447318.2016.1254890?scroll=top&needAccess=true
3. Burns, R.B., Burns, R.A.: Business Research Methods and Statistics Using SPSS. Sage, London (2008)
4. Chang, C., Chen, G.: College students' disclosure of location-related information on Facebook. Comput. Hum. Behav. **35**, 33–38 (2014)
5. Cheon, J., Lee, S., Crooks, S.M., Song, J.: An investigation of mobile learning readiness in higher education based on the theory of planned behavior. Comput. Educ. **59**, 1054–1064 (2012)
6. Cialdini, R., Goldstein, N.: Social influence: compliance and conformity. Annu. Rev. Psychol. **55**, 591–621 (2004)
7. Clary, E.G., Snyder, M., Ridge, R.D., Miene, P.K., Haugen, J.A.: Matching messages to motives in persuasion: a functional approach to promoting volunteerism. J. Appl. Soc. Psychol. **24**(13), 1129–1146 (1994)
8. Conner, M., Armitage, C.J.: Extending the theory of planned behavior: a review and avenues for further research. J. Appl. Soc. Psychol. **28**(15), 1429–1464 (1998)
9. Daugherty, T., Eastin, M.S., Bright, L.: Exploring consumer motivations for creating user-generated content. J. Interact. Advert. **8**(2), 16–25 (2008)
10. Dellarocas, C.: The digitization of word-of-mouth: promise and challenges of online feedback mechanisms. Manag. Sci. **49**(10), 1407–1424 (2003)
11. Hennig-Thurau, T., Gwinner, K.P., Walsh, G., Gremler, D.D.: Electronic word-of-mouth via consumer-opinion platforms: what motivates consumers to articulate themselves on the internet? J. Interact. Mark. **18**(1), 38–52 (2004)

12. Hsu, M.H., Yen, C.H., Chiu, C.M., Chang, C.M.: A longitudinal investigation of continued online shopping behavior: an extension of the theory of planned behavior. Int. J. Hum Comput Stud. **64**, 889–904 (2006)
13. Kaiser, H.F.: An index of factorial simplicity. Psychometrika **39**(1), 31–36 (1974)
14. Krishnamurthy, S., Dou, W.: Note from special issue editors: advertising with user-generated content: a framework and research agenda. J. Interact. Advert. **8**(2), 1–4 (2008)
15. Lu, Y., Zhou, T., Wang, B.: Exploring Chinese users' acceptance of instant messaging using the theory of planned behavior, the technology acceptance model, and the flow theory. Comput. Hum. Behav. **25**, 29–39 (2009)
16. Moon, J.W., Kim, Y.G.: Extending the TAM for a world-wide-web context. Inf. Manag. **38**, 217–230 (2001)
17. Netemeyer, R.G., Burton, S., Johnston, M.: A comparison of two models for the prediction of volitional and goal-directed behaviors: a confirmatory analysis approach. Soc. Psychol. Q. **54**(2), 87–100 (1991)
18. Sheeran, P.: Intention-behavior relations: a conceptual and empirical review. Eur. J. Soc. Psychol. **12**(1), 1–36 (2002)
19. Teo, T.: Examining the intention to use technology among pre-service teachers: an integration of the technology acceptance model and theory of planned behavior. Interact. Learn. Environ. **20**(1), 3–18 (2012)
20. Van Baaren, R.B., Holland, R.W., Kawakami, K., Van Knippenberg, A.: Mimicry and prosocial behavior. Psychol. Sci. **15**(1), 71–74 (2004)
21. Van Gool, E., Van Ouytsel, J., Ponnet, K., Walrave, M.: To share or not to share? Adolescents' self-disclosure about peer relationships on Facebook: an application of the prototype willingness model. Comput. Hum. Behav. **44**, 230–239 (2015)
22. White, K.M., Smith, J.R., Terry, D.J., Greenslade, J.H., McKimmie, B.M.: Social influence in the theory of planned behaviour: the role of descriptive, injunctive, and in-group norms. Br. J. Soc. Psychol. **48**(1), 135–158 (2009)
23. Yoo, K.H., Ulrike, G.: What motivates consumers to write online travel reviews? Inf. Technol. Tourism **10**(4), 283–295 (2008)

Stay Connected and Keep Motivated: Modeling Activity Level of Exercise in an Online Fitness Community

Li Zeng[1]([✉]), Zack W. Almquist[2], and Emma S. Spiro[3]

[1] Information School, University of Washington, Seattle, USA
lizeng@uw.edu
[2] Departments of Sociology, School of Statistics, and Minnesota Population Center,
University of Minnesota, Minneapolis, USA
almquist@umn.edu
[3] Information School, Department of Sociology,
Center for Statistics and the Social Sciences,
University of Washington, Seattle, USA
espiro@uw.edu

Abstract. Recent years have witnessed a growing popularity of activity tracking applications. Previously work has focused on three major types of social interaction features in such applications: cooperation, competition and community. Such features motivate users to be more active in exercise and stay within the track of positive behavior change. Online fitness communities such as Strava encourage users to connect to peers and provide a rich set of social interaction features. Utilizing a large-scale behavioral trace data set, this work aims to analyze the dynamics of online fitness behaviors and network subscription as well as the relationship between them. Our results indicate that activeness of fitness behaviors not only has seasonal variations, but also vary by user group and how well users are connected in an online fitness community. These results provide important implications for studies on network-based health and design of application features for health promotion.

Keywords: Fitness behaviors · Social interaction
Event history analysis · Online fitness communities · Social media
Behavioral traces

1 Introduction

Peer influence and social interaction have been found to have positive health-related effects, such as helping people lose weight and increasing physical activities [1–5]. Recently, increased attention has been paid to promoting health habits through online social interaction in online fitness communities (e.g., running or cycling online groups) [6,7]. Studies show that social interaction features such as cooperation and competition provide participants with a group of peers and

G. Meiselwitz (Ed.): SCSM 2018, LNCS 10914, pp. 137–147, 2018.
https://doi.org/10.1007/978-3-319-91485-5_10

help motivate them to reach their fitness or health goals [2, 8]. In this work, we study the relationship between fitness behaviors and online social behavior via a subscription service where users can post and follow peers' activity feeds.

Traditionally, studies on physical activity and social interaction have relied on information that is self-reported (e.g., diary studies) or measured via (expensive) wearable sensors, which have been limited in terms of scale, granularity of activity and duration of observation period [9, 10]. Recent developments in smart phone GPS tracking and accessibility has provided for an increasing large adoption of mobile devices which track everyday physical activities. These resultant behavioral trace data allow for precise measurement of individual's activities and online social action in scale and cost that traditional survey-based methods for collecting data on physical activities cannot match, and provide a good alternative to classic sensor studies [11, 12]. Our work leverages these new data sources (e.g., behavioral trace data) to explore the relationship between fitness behaviors and online social interaction over time. We employ this novel data set along with event history methods to understand the relationship between online social interaction and activity levels within one of these app-based activity communities. Specifically, we focus on the large app-based community known as Strava where users have covered over 12 billion miles worldwide [13]. In this work we focus on two comparable major metropolitan areas in the US.

This work analyzes the dynamics of online fitness behaviors and network subscription as well as the relationship between them. We ask the following research questions: (1) how do users' activity levels of exercise change over time? (2) how do exercise activeness and network subscription vary among users? (3) how subscription magnitude is associated with activity occurrence? We find clear seasonal patterns of users' fitness behaviors and discuss the implications for fitness application designs and health prevention. We also show that paid-plan users exercise more actively and attract more followers than free-plan users. Last, our analysis shows a positive relationship between social subscription and physical activities, supporting the claim – "stay connected, keep motivated".

The remainder of this paper is organized as follows. We start by reviewing existing studies on online fitness communities and roles of social interaction in physical activities in online fitness communities. Next, we describe the behavioral trace data collected from Strava for this work. We then describe the methodology for this work, describing our analysis on activity levels and online social interaction via network subscription, as well as the methodology used to model physical activities given user characteristics and network subscription. We finally discuss and summarize our findings.

2   Related Work

With advances in pervasive technologies, activity tracking applications and online fitness communities such as Strava, RunKeeper, MapMyRun, etc. are attracting more and more users around the world. Online fitness communities usually have features of both activity logging and social networking [9].

Online fitness communities log activity-related data and help users analyze their performance. Online fitness communities also serve as an activity-based network that connect users and provide users with a series of social interactions features meant to encourage behavioral change and healthy life-style promotion. For example, on Strava, users can follow both recreational and professional athletes, view their activities and interact with them by making comments, giving kudos, etc.

Behavioral trace data archived by online fitness communities record large amount of data that is generated by users throughout their physical activity (e.g., running/cycling) and through their social interaction online. One can compare this data to the more often collected self-report surveys which acquire individual's perception of their workout routine and social engagement, rather than the behavioral trace data which records their exact physical activity, timing, distance, etc. and precise social interaction (e.g., running in groups or liking someones run activity). This data source also shows advantages in terms of scale, granularity and observation duration against data collected via expensive sensors [10]. Thus, it provides researchers with new opportunities for understanding the relationship between fitness behavior and social interaction.

Online fitness communities are increasingly attracting researchers from a variety of disciplines. Some studies revolve around incentives and interventions. For example, studying potential of health devices and applications for health-related behavioral change [2,14,15]. Another major body of work focuses on the technical potential of wearable sensors and human-computer interaction aspects of these technologies. For example, examining specific features in designs of fitness applications [16–18]. However, there are few studies on how online social network structure influence activity engagement, leaving a gap in our current understanding of the social dynamics in these settings.

Recent work have studied the relationship between social interaction and physical activity using fitness applications. Social interaction in online fitness communities may include cooperation and competition and sharing physical activities. Studies by [2,19–21] suggest that social interaction is essential to motivate users to perform physical activities. As mentioned above, online fitness communities enable users to connect and interact with a group of peers online. In the case of Strava, users can follow other athletes, view their profile and activities, and receive activity feed once their peer post a completed activity. They may also compare workout and network-based stats with each other, "like" others' posted activities and make comments under posts. The work by [9] examines how Strava users' social motives predict perceived usefulness of the platform based on survey responses collected from 394 Strava users. Three aspects of social motives are considered: staying informed on friends' activities, viewing progress made by friends, and receiving support from others via kudos and comments. The results show that social motives influence habitual Strava use directly, and when compared to novice users social motives are more important for experienced users.

3 Data

This work utilizes behavioral trace data collected from the activity-based network, Strava. Strava sits at the intersection of social media and activity tracking applications and is known colloquially as the "Facebook" of activity-based apps; users have the ability to not only track and log their activities, but also connect to and interact with a group of peers online. The platform continues to grow in popularity among cyclists and runners in recent years around the world.

In this work, we study 2,605,147 cycling and running activities from 11,245 anonymized users from an activity-based tracking platform known as Strava. Our data includes but is not limit to the following three main components: (1) user profile information, including gender, date of birth, location, user account status (e.g. free or paid plan), sign-up date, etc.; (2) logs of posted activity for users in the sample over time, including activity timestamp, location, type, performance stats, etc.; and (3) social subscription: who followed whom and the corresponding timestamp.

We focus on two major metro areas within the continental US, which have a large active set of Strava users. We have chosen to focus on San Francisco City/County, CA which is where Strava started and continues to be headquartered, and Boston/Suffolk County, MA. Both cities represent similar size metropolitan areas within the US. Boston metro is ranked 10th with about 4.8 million residents, and San Francisco metro is ranked 11th with about 4.7 million residents. Thus, these two areas represent comparable cities on the coasts within the US context, but with wildly different weather patterns. For example, Boston has an average high of 36° Fahrenheit in January, and San Francisco has an average high of 58° Fahrenheit in same month. Our analysis examines

Table 1. Data summary

	San Francisco County	Suffolk County
# of Activities	2312669	292478
# of Users	9831	1414
Gender Proportion		
- Male	0.777	0.760
- Female	0.223	0.240
User Type Proportion		
- Cyclist	0.695	0.623
- Runner	0.305	0.377
Age Proportion		
- 0–17	0.005	0.008
- 18–35	0.524	0.590
- 36–49	0.380	0.303
- 50+	0.091	0.099

and compares fitness behaviors and social subscription behaviors of users from these two counties. The key difference in these two areas is their weather patterns, and so we expect differences in community activities to stem from these seasonal differences.

Table 1 presents the summary of our data in terms of user group and activities. Male users are proportionally greater than female users in San Francisco and Suffolk County, and gender proportion is relatively similar in both locations. There exist more cyclists than runners in both groups – this is to be expected as Strava was originally developed by a group of cyclists. San Francisco users have 69.5% of self-reported cyclists and 30.5% of runners, while the proportion of cyclists and runners of Suffolk County users are 62.3% and 37.7%, respectively. Lastly, we observe that the major age group of users is between 18 and 35 for both locations, and the second majority age group is between 36–49.

4   Methods

First, we examine seasonal patterns of workouts done by users in our sample. Specifically, we look at total numbers of posted activities across all users ranging from September, 2009 to April, 2017. Theoretically, seasonality of physical activities may vary by activity type and location. Hence, this analysis focuses on the two major activity types - cycling and running, which accounts for 53% and 40% of total activities, respectively. Moreover, we do so for users from San Francisco County and Suffolk County separately in order to have a simple control setting for weather in the analysis.

As we are interested in the relationship between fitness behaviors and online social interaction over time, we examine users' activity level and network subscription by different user groups. We begin with a simple metric *activity level* measuring to what extent a user engages Strava to track/log exercise actively:

$$\text{Activity level} = \frac{\#\text{ of activities by a user}}{\#\text{ of days a user in Strava service}}$$

We focus on how users' activity level and number of their followers varies by gender, age group, training plan enrolled (i.e. free or paid plan) and athlete type (i.e. cyclist or runner). For gender differences in physical activity, prior work argues that physical inactivity is more prevalent among female. Indeed, male users are proportionally greater than female users in Strava. However, we are also interested in exploring gender differences in activeness in the online fitness community. Next, we observe a disproportionate distribution of users age in our sample data (i.e., the majority of Strava users aged between 18 and 49). Therefore, we want to examine how actively each user age group engages these activities. Further, we are interested in finding whether users from paid plan workout more than free-plan users. We are also interested understanding how workout activity differ by athlete type.

Last, we model user workout frequency over time with a particular interest in examining whether a user who is followed by more peers tend to exercise more.

In this work, we perform an event history analysis in order to characterize the occurrence of the repeated events - physical activities given a time-dependent variable - follower count along with time-independent variables - gender, age, training plan enrolled and athlete type (i.e. cyclist or runner). To control for seasonality of location we include a dummy variable to indicate whether a user comes from San Francisco County (indicator = 0) or Suffolk County (indicator = 1). We use the popular cox proportional hazards model where the unique effect of a unit increase in a covariate is multiplicative with respect to the hazard rate. In this model, the dependent variable is $h(t)$ - the hazard of an event at time t. Roughly speaking, $h(t)$ can be interpreted as the instantaneous probability that an event will occur at time t. The hazard function follows the form:

$$\lambda(t|X_i) = \lambda_0(t)\exp(\beta_1 X_{i1} + \cdots + \beta_p X_{ip} + \beta_{p+1} Z_i(t))$$

where $X_{ip}, ..., X_{ip}$ are the 1st to the pth time-independent covariate and $Z_i(t)$ is the time-dependent covariate at time t for the observation i. Users who churned the Strava service and hence deleted the account are censored in this analysis.

5   Results

5.1   Seasonality of Physical Activity

Figure 1 shows a time series of activity frequency in Strava ranging from 2009 to 2017. For both runs and rides, we first observe a clear seasonal pattern for both locations; number of activities done per year usually peak at late spring or summer and then drop to the local lowest point at winter. When comparing between the two locations, we observe Suffolk users' activity behaviors are more greatly influenced by the time of year. Unsurprisingly, this indicates that weather is indeed a great factor for exercise activeness especially for outdoor exercise. Further, it is interesting to notice that cycling is more subject to seasonal change with larger difference in frequency during different seasons when compared to running. Moreover, it is important to note an overall increasing trend for both running and cycling, which indicates an increasing popularity of exercise within the Strava app over time. This analysis suggests that it is important to consider temporal dimension in the analysis and modeling of user fitness behaviors.

5.2   Exercise Activeness and Network Subscription

Figures 2 and 3 show how users' activity level and follower count vary in terms of gender, training plan, athlete type and age. We compare the two locations with the entire set to see if there is any spatial difference. Overall, we find that distributions of activity level and follower count follow similar patterns for both locations.

One significant difference is between users with free plan and users with paid plan. We observe that paid-plan users tend to exercise more actively and have more followers. It is worth pointing out that on average female paid-plan users

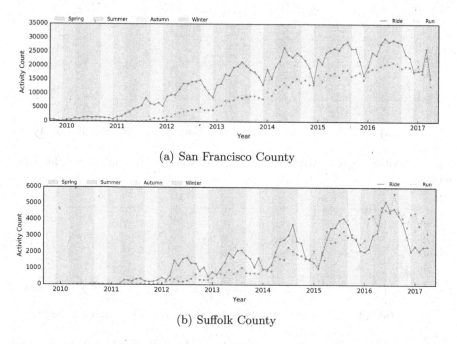

(a) San Francisco County

(b) Suffolk County

Fig. 1. Seasonal pattern of physical activity frequency.

tend to have more followers than their male paid-plan counterparts, while we do not see significant gender difference in follower count in free-plan user group.

We find that the activity level of cyclists is on average higher than runners, and male cyclists and runners tend to be more active than their female counterpart. Future study might examine users and their activities from other areas and countries for a generalization of these findings.

Last, we observe that activity level differs by age group and gender. Men who are over 50 tend to be more active than men in other age groups, and the pattern seems consistent across two locations. However, for female users in Suffolk County differences in activity level by age group is larger; women who are between 36 and 49 tend to be more active than women who are younger or older. For gender differences, we find that men tend to be exercise more than women across all age groups. While mid-aged or older-aged users tend to exercise more actively, the results show that younger users tend to receive more follower counts. However, note that the age group of 0–17 contains only a small number of observations and may not be representative of the larger sub-population.

5.3 Modeling Activity Occurrence

Overall, in the Figs. 2 and 3 we observe that activity level does not always align with social subscription level. For example, users who are over 50 exercise quite actively but have far fewer followers compared to other age groups. This suggests

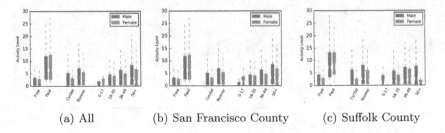

(a) All   (b) San Francisco County   (c) Suffolk County

Fig. 2. Distributions of activity level of workout by gender, training plan, athlete type and age

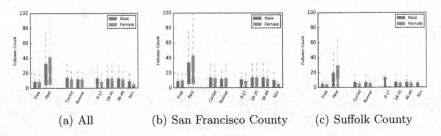

(a) All   (b) San Francisco County   (c) Suffolk County

Fig. 3. Distributions of user follower count by gender, training plan, athlete type and age

that there might exist a more complex relationship between activity level and online social interaction. Therefore, we move on to discuss the results of modeling users' physical activities over time, given variables of interest that are explored in our previous analysis.

Table 2 presents the results of this event history analysis. To recall, the variables in the model are follower count which is time dependent as well as age, gender, athlete type and location which are time independent. Time independent variables except for age are categorical variables in this analysis. The results of cox model show that gender of male, runner and paid plan are significantly related to increase in possibility of activity occurrence, supporting the findings from our previous exploratory data analysis.

Moreover, our cox model reveals that every one unit increase in follower count results in an increased 2.1% probability of an exercise occurrence ($P - value$ <0.001). For instance, the model suggests that a user who has 50 followers has an approximately 100% increase in the probability of performing a physical activity. Therefore, even though 2.1% appears to be a modest boost for activity occurrence, this could be a relatively large boost given that the followers is a count variable. This suggests that a greater follower count that a user has is thus correlated to a higher probability that the user exercises. Strava users who have more followers are experiencing more exposure of their posted activities to their followers and likely receiving more social feedback (i.e. comments, kudos) from them.

Table 2. Modeling activities using cox hazard models

Variable	Exponential	p-value
Age	1.014	<0.001
Gender (Male)	1.407	<0.001
Athlete Type (Runner)	1.807	<0.001
Plan (Paid)	4.345	<0.001
Follower count	1.021	<0.001
Location (San Francisco)	1.217	0.038

6   Discussion

In online fitness community such as Strava, a rich set of social interaction features starts with following other athletes and hence building users' activity-based social network. Therefore, our work aims to analyze the dynamics of online fitness behaviors and network subscription as well as the relationship between them. Specifically, we ask how users' activity levels of exercise change over time, how exercise activeness and network subscription vary among users, and how subscription magnitude is associated with activity occurrence. We utilize behavioral trace data from the online community Strava to answer these research questions. Data focus on two major U.S. metro areas that have a large number of active Strava users - San Francisco, CA and Suffolk, MA; data contain profile information of sampled users, user activity logs as well as network subscription logs.

We find that users' fitness behaviors display clear seasonal patterns. In general, late spring and summer are more attractive seasons for rides and runs, whereas winter appears to be less attractive. We also observe that compared to running, cycling is more sensitive to seasonality. Although strong seasonal patterns of physical activities (especially outdoors activities) are unsurprising in human behavior, results demonstrate that individual physical inactivity is likely to be aligned with seasonality in a systematic way. For designers of fitness applications, an implication of the analysis may be to take into account both individual exercise preference and optimal seasons for certain activity types. For example, Strava supports a great variety of activity type, but current practices in using and advertising the application are limited to outdoors activities, (mostly cycling and running).

Paid-plan users exercise more actively and attract more followers than free-plan users. We also observe significant gender differences in follower counts among paid-plan users; while activity levels of paid-plan users do not vary much by gender, active female users tend to have more followers than male users do. However, reasons behind the findings require a further investigation. It could be that active female users tend to connect to more users and hence receive more followers in return. Future work might also examine gender differences in the way that networks are structured in terms of symmetric and asymmetric ties for free-plan and paid-plan users.

The results demonstrate a positive relationship between social subscription and activity occurrence. Modeling individual activity occurrence using event history analysis enables us to quantify the "power" of gaining one follower for users to exercise more. In this work, we focus on characteristics of egos (eg. gender, training plan, age group, etc.). One analysis that may be worth to perform next is to take into account nodal covariates for both egos and alters. For example, users who have many active followers versus users who have many inactive followers; or users who are mostly followed by the same gender versus users who are mostly followed by users whose gender differs from them. Also, built upon the findings of this work, future work may further compare one-way connections with mutual connections to see which type of connections has a stronger association with user activity levels of exercise.

7   Conclusion

Our work analyzes the dynamics of online fitness behaviors and network subscription as well as the relationship between them. We utilize a large-scale behavioral trace data set from an online fitness community Strava. Our results indicate that fitness activity levels not only has seasonal variations, but also vary by user group. The results of event history analysis suggest that individual activity levels are significantly associated with how well users are connected in an online fitness community. The implications of these results for studies on network-based health and design of application features for health promotion are also discussed.

Acknowledgement. This material is based upon work supported by an Information School Strategic Research Award, University of Washington and the Office of the Vice President for Research, University of Minnesota.

References

1. Ahtinen, A., Isomursu, M., Mukhtar, M., Mäntyjärvi, J., Häkkilä, J., Blom, J.: Designing social features for mobile and ubiquitous wellness applications. In: Proceedings of the 8th International Conference on Mobile and Ubiquitous Multimedia, p. 12. ACM (2009)
2. Chen, Y., Pu, P.: Healthytogether: exploring social incentives for mobile fitness applications. In: Proceedings of the Second International Symposium of Chinese CHI, pp. 25–34. ACM (2014)
3. Wing, R.R., Jeffery, R.W.: Benefits of recruiting participants with friends and increasing social support for weight loss and maintenance. J. Consult. Clin. Psychol. **67**(1), 132 (1999)
4. Kulik, J.A., Mahler, H.I.: Social support and recovery from surgery. Health Psychol. **8**(2), 221 (1989)
5. Dishman, R.K., Sallis, J.F., Orenstein, D.R.: The determinants of physical activity and exercise. Public Health Rep. **100**(2), 158 (1985)
6. Centola, D., van de Rijt, A.: Choosing your network: social preferences in an online health community. Soc. Sci. Med. **125**, 19–31 (2015)

7. King, A.C., Glanz, K., Patrick, K.: Technologies to measure and modify physical activity and eating environments. Am. J. Prev. Med. **48**(5), 630–638 (2015)
8. Hamari, J., Koivisto, J.: "working out for likes": an empirical study on social influence in exercise gamification. Comput. Hum. Behav. **50**, 333–347 (2015)
9. Stragier, J., Abeele, M.V., Mechant, P., De Marez, L.: Understanding persistence in the use of online fitness communities: comparing novice and experienced users. Comput. Hum. Behav. **64**, 34–42 (2016)
10. Althoff, T.: Population-scale pervasive health. IEEE Pervasive Comput. **16**(4), 75–79 (2017)
11. Prince, S.A., Adamo, K.B., Hamel, M.E., Hardt, J., Gorber, S.C., Tremblay, M.: A comparison of direct versus self-report measures for assessing physical activity in adults: a systematic review. Int. J. Behav. Nutr. Phys. Act. **5**(1), 56 (2008)
12. Bernard, H.R., Killworth, P., Kronenfeld, D., Sailer, L.: The problem of informant accuracy: the validity of retrospective data. Ann. Rev. Anthropol. **13**, 495–517 (1984)
13. Strava, Strava just hit a huge activity milestone (2017). http://womensrunning. competitor.com/2017/05/news/strava-one-billion-activities_75359
14. Foster, D., Linehan, C., Kirman, B., Lawson, S., James, G.: Motivating physical activity at work: using persuasive social media for competitive step counting. In: Proceedings of the 14th International Academic MindTrek Conference: Envisioning Future Media Environments, pp. 111–116. ACM (2010)
15. Cavallo, D.N., Tate, D.F., Ries, A.V., Brown, J.D., DeVellis, R.F., Ammerman, A.S.: A social media-based physical activity intervention: a randomized controlled trial. Am. J. Prev. Med. **43**(5), 527–532 (2012)
16. Consolvo, S., McDonald, D.W., Toscos, T., Chen, M.Y., Froehlich, J., Harrison, B., Klasnja, P., LaMarca, A., LeGrand, L., Libby, R., et al.: Activity sensing in the wild: a field trial of ubifit garden. In: Proceedings of the SIGCHI Conference on Human Factors in Computing Systems, pp. 1797–1806. ACM (2008)
17. Lister, C., West, J.H., Cannon, B., Sax, T., Brodegard, D.: Just a fad? gamification in health and fitness apps. JMIR Serious Games 2(2)
18. West, J.H., Hall, P.C., Hanson, C.L., Barnes, M.D., Giraud-Carrier, C., Barrett, J.: There's an app for that: content analysis of paid health and fitness apps. J. Med. Int. Res. 14(3)
19. Campbell, T., Ngo, B., Fogarty, J.: Game design principles in everyday fitness applications. In: Proceedings of the 2008 ACM Conference on Computer Supported Cooperative Work, pp. 249–252. ACM (2008)
20. Choi, G., Chung, H.: Applying the technology acceptance model to social networking sites (sns): impact of subjective norm and social capital on the acceptance of sns. Int. J. Hum.-Comput. Interact. **29**(10), 619–628 (2013)
21. Consolvo, S., Everitt, K., Smith, I., Landay, J.A.: Design requirements for technologies that encourage physical activity. In: Proceedings of the SIGCHI Conference on Human Factors in Computing Systems, pp. 457–466. ACM (2006)

Social Network Analysis

Application of Social Network Analytics to Assessing Different Care Coordination Metrics

Ahmed F. Abdelzaher[1]([✉]), Preetam Ghosh[2], Ahmad Al Musawi[3],
and Ju Wang[1]

[1] Virginia State University, Petersburg, VA 23806, USA
{amohammed,jwang}@vsu.edu
[2] Virginia Commonwealth University, Richmond, VA 23084, USA
pghosh@vcu.edu
[3] Thi Qar University, Nasiriyah, Iraq
almusawiaf@utq.edu.iq
http://www.linkedin.com/in/ahmed-abdelzaher-322899119
http://www.linkedin.com/in/preetam-ghosh-5441502
http://www.linkedin.com/in/ahmad-al-musawi-577410141

Abstract. Social network analytic approaches have been previously proposed to identifying key metrics of physician care coordination. Optimizing care coordination is a primary national concern for which yields significant cuts in medical care costs. However, the proposed metric-termed 'care density' for estimating care coordination is not completely accurate. Our objective is to compare the accuracy of the previously proposed 'care density', with our proposed 'weighted care density', 'time varying care density', and 'time varying weighted care density' in terms of predicting the cost of care. Our proposed metrics are based on the former care density, however, takes other variables into consideration, mainly patient hospitalization time frames and number of physician visitations. Our findings suggest that physicians coordinating over short time spans spike the cost of care above normal.

Keywords: Social network analytics · 2 mode bipartite networks
Support vector regression · Incidence matrix

1 Introduction

Social networks belonging to a category of complex systems termed 'scale-free' [15], are known to follow a power law distribution for which the probability $p(K)$ for nodes to have neighbors is of the form $p(K) \sim K^{\gamma}$. The power law suggests that the evolution of such networks occurs in a sparse [7] manner, but more importantly, they exhibit "topological patterns". Similarly with other naturally occurring topologies, such as networks describing metabolic reactions in the animal cell, the World Wide Web and gene regulatory networks [4]. Essentially, such

© Springer International Publishing AG, part of Springer Nature 2018
G. Meiselwitz (Ed.): SCSM 2018, LNCS 10914, pp. 151–160, 2018.
https://doi.org/10.1007/978-3-319-91485-5_11

network interactions exhibit particular behaviors having inner construction that distinguish there structural properties, and are not to be equaled with randomly generated networks using the famous Erdös-Rényi model [8] and the small world properties model of Watts and Strogatz [17].

The concepts of social network analytics have been applied extensively in the medical field attempting to estimate the level of collaboration among physicians sharing patients [14] and the factors affecting their influential views of primary medical practices [10]. Patient sharing increases the chances for interactive communications as well as higher levels of information exchange [5] and increase their chances of receiving efficient synchronized care. In fact, care coordination has been identified as one of the nation's most concerned area of research [1,3,13]. This motivated our work here, for which we try to understand the correlation between different metrics that can be useful in evaluating the level of care coordination in the medical field.

In doing so, we examine the correlation between the previously proposed care coordination measure, 'care density' [14], and the costs of medical care for patients admitted for suffering from pneumonia –a disease for which the estimated deaths combined with influenza was 53,582 for 2009, and for which care coordination is likely important [2]. We try to enhance the accuracy of the proposed metric by considering additional variables to the measure, mainly, the time frames for which the patients have been admitted and the number of distinct patient-physician visits. We propose, namely: (1) the weighted care density, (2) the time varying care density, and (3) the time varying weighted care density, in addition to the former care density and examine their cost of care correlations as well. Though previously suggested that increased synchronized care yields significant cuts in costs of care [14], our findings suggest that this might not be the case when considering hospitalization time frames of the patients.

In order to determine the most accurate care coordination measure, we extracted additional data per patient as discussed in the Data section, which are relevant for constructing support vector regression (SVR) models as discussed in the Methods section. Our SVRs model the effects of the different patient features (extracted data) when combined with the care coordination metrics to predicting the cost of care. We construct 4 different SVRs, each considering different care coordination metrics as distinct features per patient and the corresponding accuracy of the models are recorded in the Results section.

2   Data

Our records for patients suffering from pneumonia were provided by The Medical Center of Virginia (MCV). The data contains two spread sheets: (1) one contains 33920 records of different patient-physician visitations, (2) the other indexes the different codes for the type of patient discharge. The data encompasses 2324 pneumonia patients and 1506 providers operating between the dates: the 30th of September 2007 and the 11th of April 2008. For the time being, we employ patients feature extractions from the first document and we consider all types of patient discharges the same. Computational programs have been constructed for parsing (and calculating if necessary) the different features pertaining to each patient, their notations and descriptions are portrayed in Table 1. Other information including race and gender of the patient, specialty of the provider, the patient's source of admission and the payer IDs are available within the spreadsheet but were not considered.

Table 1. Patient features

Data notation	Content description
#Doctors	Number of different physicians visited by patient
#Interventions	Number of different physician-patient visitations
#RX	Number of different medical prescription type visits
#LAB	Number of different lab orders type visits
#ADM	Number of other types of medical admitted visits
LOS	The length of hospitalization in days
Cost	The total cost of hospitalization
t	The time the patient admitted relative to t_0*

* t_0 is the time when patient was first admitted

3 Methods

3.1 2-Mode Bipartite Social Netwok

Calculating the care density requires forming a social network of patients and physicians known as the 2-mode bipartite social network- a network formed by assigning physicians as groups and patients they treat as subscribers to the groups [11,12,16]. Therefore, our network can be expressed by a set of doctors D and a set of patients P, together $(P \cup D) = N$, were N denotes the set of nodes of our network. We denote a binary relationship "Visits" as;

$$R_v = \{< p, d > \,|\, \exists\, p\,visits\,d\,or\,d\,provides\,for\,p\}, \tag{1}$$

wherein $p \in P$ and $d \in D$. Hence for $< i, j >\, \in R_v$, the value of the incidence matrix $G_{ij} = 1$ for a visit and 0 otherwise. In the weighted version of the 2-mode bipartite graph, the value of G_{ij} would be the number of times patient i visited doctor j. Note that G's number of rows equal to the number of patients and columns equal to the number of physicians.

3.2 Care Density

The Care Density (CD) [14] calculates an approximated value for the collaboration among doctors that a particular patient has visited during his entire stay in the hospital. For a particular patient p who has visited n_d doctors, CD can be computed as follows;

$$CD_p = \frac{\sum_{i=1}^{m} w_{p,i}}{n_d(n_d - 1)/2} \tag{2}$$

CD has proven to be correlated to the reduction in the mean charges of hospitalization as reported in a study involving patients with Diabetes and Congestive Heart Diseases [14]. Consider Fig. 1, patient $p2$ visits $d1$, $d2$ and $d3$ during his/her entire stay which yields a 3 possible provider combination, therefore the denominator is 3. Between $d1$ and $d2$, 2 patients are shared, similarly with $d2$ and $d3$, while $d1$ and $d3$ share 1 patient. Therefore the physicians share a total of 5 patients over 3 possible pairs of doctors, CD = 1.67. We hypothesize that CD is not an accurate measure for physician's collaboration as it does not consider the different time windows at which patients were admitted and discharged.

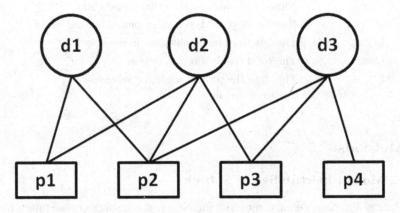

Fig. 1. A representation of a 2-mode bipartite graph between doctors (d1-d3) and patients (p1-p4).

3.3   Time Varying Care Density

Now, consider introducing two time windows $t_0 - t_1$ and $t_1 - t_2$ in Fig. 2, such that p1 and p2 have been admitted and discharged before t_1, and consider $p3$ and $p4$ being admitted after t_1. Doctors $d2$ and $d3$ share $p3$ after $p2$ was discharged and the effects of their collaboration post $p2$s treatment should not be incorporated in the CD of $p2$. Similarly with $p3$, CD should not account for the effects of the doctor's collaboration before $p3$ got admitted. Moreover, it makes no sense to try correlating the charges of a particular patient based on his provider's activity after a patient has been discharged and billed. Therefore the existing implementation of CD over estimates the collaboration of the doctors, which lead us to introducing the Time Varying Care Density (TCD). TCD excludes doctor's collaborations that occur outside the patient's time window. Essentially, $p2$ will lose the effect of doctors $d2$ and $d3$ sharing patient $p3$, which brings down the CD from 1.67 to 1.33, or TCD = 1.33.

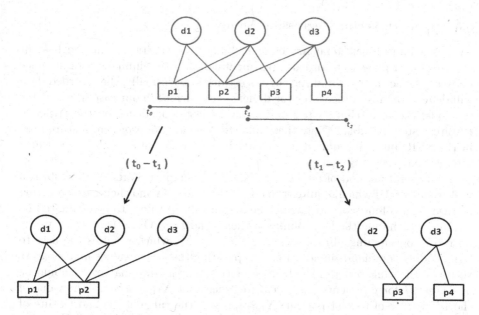

Fig. 2. A representation of the effects of introducing time windows to the social network of Fig. 1.

3.4 Weighted Care Densities

So far (in the above examples) we have considered the un-weighted incidence matrix, that is, we have yet to considered the number of visits patients have committed to doctors. In other words, we have considered CD to be the same for the following scenarios of 2 doctors sharing a patient: (Scenario 1) $p1$ visits $d1$ 9 times and $d2$ 1 time, and (Scenario 2) $p1$ visits $d1$ 5 times and $p1$ visits $d2$ 5 times. Both scenarios have a sum of visits equal to 10 but ideally $d1$ and $d2$ have a tendency to collaborate more in the second, i.e. CD should be higher in the first scenario.

To account for the weights (visits), we consider the collaboration between 2 physicians to be a maximum when both doctors have equal visits for all their common patients, and decreases as the difference between the numbers of visitations increase. The collaboration between a pair of doctors can be estimated using

$$w_{p,i} = 1 - 2 \left| \frac{1}{2} - \frac{v_1}{v_1 + v_2} \right| \tag{3}$$

where $w_{p,i}$ is the value of a single operation of the summation of the numerator of CD_p. v_1 and v_2 constitute the sum of visits for the common patients between $d1$ and $d2$ respectively. From here on we refer to the weighted care density as WCD and the time varying weighted density as TWCD.

3.5   Support Vector Regression Modeling

SVR is a multidimensional modeling technique which tries to fit the best fit curve or hyper plane to recorded experimental data with minimal error difference between the actual data and the curve's estimation. Normally, the recorded data will depend on a set of features (or dimensions) such as in our case, were we try to model the total charges per patient and their dependence on the patient's features listed in Table 1. Note that time windows are not a patient feature used in the SVR, but are required for calculating the different care densities which are relevant patient dimensions.

As a useful method of modeling, many tools support SVRs. WEKA [9] is a user friendly GUI which requires the user to list the data and their corresponding features in a column format having the actual values (values to be estimated by the hyper plane) on the last column of the spreadsheet. Other tools come in the form of programming libraries such as LibSVM [6], which requires the user to write coding commands in addition to separating the actual values in a separate vector in the same order of the feature matrix. The feature matrix X should be of size $p \times f$ for p patients and f features such that X_{11} represents the value of the 1st feature of the 1st patient; X_{12} represent the value of the 2nd feature of the 1st patient and so up to X_{pf}. The actual data should be stored in the same order of the patients in a vector y of size $1 \times p$. Ultimately the SVR gives a best fit with approximated values;

$$y_a = wX + b \tag{4}$$

were w is a $1 \times f$ weights vector, which represents the relative influence of each feature on the SVR, and b is the bias which represents a constant shift to the curve. The above process is referred to as training, while testing w and b on foreign data is referred to as prediction.

In order to test the SVR, we use a 10 k-fold cross validation technique. The data is divided into 10 different chunks of 90 percents and 10 percents. The training is performed on the 90% portion and the corresponding w and b are used for prediction of the remaining 10%. The error of the 10% is accumulated each time using;

$$e = e + \left[\frac{\|y - y_a\|}{y} \times 100 \right] \tag{5}$$

then averaged across the 10 folds to give the cross validation error. We compare the 4 SVR accuracies in terms of predicting the cost of hospitalizations using the patient features listed in Table 1 (excluding cost and t) plus an additional feature for each model, that is the different care densities (CD, WCD, TCD, TWCD).

4   Results

Results of the un-weighted and weighted mean care densities, recorded in Table 2, show that higher care density receivers have lower average costs of hospitalizations. Because the care densities are skewed, we divide the statistical brackets into tertiles of almost equal numbers of data points: lower, middle and upper. In both (CD and WCD) cases, there is no strong correlation between the average age of the patients and the average CD values.

Table 2. Features of pneumonia patients, stratified by care densities

Mean (SD)	Un-weighted			Weighted		
	Lower	Middle	Upper	Lower	Middle	Upper
N	768	765	751	739	755	790
Care density	3.32 (1.37)	7.53 (1.31)	20.22 (12.86)	2.68 (1.09)	6.03 (1.06)	13.94 (6.05)
Age	48.82 (23.26)	47.52 (24.50)	50.85 (21.93)	49.34 (22.89)	47.58 (24.52)	49.96 (22.5)
#Doctors	11.77 (9.08)	12.09 (7.94)	6.4 (4.57)	12.81 (10.97)	12.71 (8.30)	6.83 (4.83)
#interventions	16.74 (13.87)	17.01 (12.39)	8.63 (7.02)	18.47 (17.18)	18.00 (12.97)	9.22 (7.41)
LOS	15.39 (15.03)	13.01 (12.4)	6.22 (6.79)	20.19 (44.14)	14.09 (13.88)	6.84 (7.68)
Charges*	102K (119K)	96K (113K)	39K (64K)	141K (252K)	107K (137K)	44K (72.8K)

*K: ×1000

On the other hand, considering the time frames reverses every correlation mentioned above as depicted in Table 3. The mean costs of hospitalizations as well as mean number of interventions and mean LOS increase with respect to an increase in the TCD values. The higher the LOS, the higher the TCD simply based on the way TCD is calculated. Moreover, there is a direct correlation between the mean age and the mean TCDs, i.e. more care is required for elderly patients. It is also interesting to notice that the standard deviations of the CDs for each tile decrease as we add more variables to the metric, meaning $SD_{CD} > SD_{WCD} > SD_{TCD} > SD_{TWCD}$.

In order to determine the most suitable metric, we constructed an SVR using the 6 features: #Doctors, #Interventions, #RX, #LAB, #ADM, and LOS and compared it with the other 4 SVRs discussed above that have an additional CD as a 7th feature. As displayed in Table 4, All 4 SVRs show less cross validation error prediction than that of the SVR which excludes the CD. Though the difference in cross validation error is not so significant, we can still see a slight decrease in the cross validation errors as we consider more variables (time and weight) to affect the CD outcome.

Table 3. Features of pneumonia patients, stratified by time varying care densities

Mean (SD)	Un-weighted			Weighted		
	Lower	Middle	Upper	Lower	Middle	Upper
N	754	773	759	756	781	749
Care density	1.16 (0.3)	1.81 (0.21)	3.27 (1.07)	0.99 (0.26)	1.55 (0.18)	2.78 (0.9)
Age	46.7 (24.63)	48.3 (23.79)	52.1 (20.99)	47.31 (24.22)	48.47 (24.04)	51.42 (21.28)
#Doctors	8.64 (7.21)	9.45 (6.98)	12.25 (8.93)	8.65 (7.25)	9.67 (7.04)	12.05 (8.94)
#interventions	11.78 (10.91)	13.13 (10.63)	17.59 (13.85)	11.96 (10.97)	13.43 (10.75)	17.17 (13.9)
LOS	8.91 (10.07)	10.57 (11.92)	15.26 (14.4)	8.86 (10.06)	10.84 (12.22)	15.1 (14.24)
Charges*	49.6K (75.3K)	67.5K (90.9K)	120KK (130K)	50.1K (75.7K)	68.8K (94K)	119K (128K)

*K: ×1000

Table 4. Prediction errors for the different SVRs

SVR type	Cross validation error
Excluding CD	40.1%
Using CD	37.7%
Using WCD	37.7%
Using TCD	36.7%
Using TWCD	36.1%

5 Conclusion and Discussion

The original hypothesis by Pollack et al [14] is true when the care density values do not consider the time windows, however the time windows flaws the hypothesis. The original hypothesis considered an expanded time window that does not account for the stress exerted by the physicians, in fact, the effort exerted by a pair of doctors is always the same. However, collaborating in tight time spans reveals the urgency of the patient, and increases the stress amongst the physicians, which should normally bump up the cost of hospitalization.

The weighted time varying care density is the most accurate metric for assessing the physician collaboration. All 4 SVRs that consider one of the CDs as a 7th feature are more accurate than the 6 feature SVR, however, the SVR which considers the TWCD is most accurate. Moreover, TWCD accounts for the visitations as well as the time windows, which give a more accurate estimation of the relative efforts exerted by physician pairs. The final costs of hospitalization should not be accountable for the activities of the physician past the patients discharge dates. Furthermore, TWCD gives a fair correlation between the age of the patient and the urgency of the disease.

It is important to note that the weight vector w of the SVRs show that the relative effect of the LOS feature on the SVR supersedes the rest of the features by a huge margin (including CD). This explains the slight, but not significant decrease in errors as we add more variables to the care density metric.

A more accurate SVR modeling approach would account for many other features which can influence the patient expenses. Firstly, the type of payer or insurance policy can be grouped to fall under 3 or 4 types of insurance policy, for which we can add a feature per type of policy, or model the data separately according to which policy type they fall onto. However for this kind of analysis we would require 3 to 4 times the amount of patients at hand. Similarly with the type of discharges, a study can be made to add a feature for discharges that are similar, such as; (1) discharged to another short term hospital and (2) transferred to another type of inpatient care institution. Thirdly, a feature can be added to the SVR which assesses the severity of the Pneumonia, for instance, a moderate case can take a value of 2.0 and a severe case a 5.0.

Acknowledgements. This work started as a class assignment, initially supported under the leadership of John A. Palesis, Ph.D, of Virginia Commonwealth University, and Johnathan P. Deshazo, Ph.D, of The Medical Center of Virginia. Without their support and guidance, the execution of this work would have been unlikely.

References

1. National Priorities and Goals: Aligning Our Efforts to Transform America's Healthcare. National Quality Forum, November 2008
2. 2011 national healthcare quality report (2011)
3. Adams, K., Corrigan, J.M.: Priority Areas for National Action: Transforming Health Care Quality. The National Academies Press, Washington (2003)
4. Albert, R., Jeong, H., Barabasi, A.: Error and attack tolerance of complex networks. Nature **406**(6794), 378–382 (2000)
5. Barnett, M., Landon, B., O'Malley, A., Keating, N., Christakis, N.: Mapping physician networks with self-reported and administrative data. Health Serv. Res. **46**(5), 1592–1609 (2011)
6. Chang, C.-C., Lin, C.-J.: Libsvm: a library for support vector machines. ACM Trans. Intell. Syst. Technol. **2**(3), 27:1–27:27 (2011)
7. Del Genio, C.I., Gross, T., Bassler, K.E.: All scale-free networks are sparse. Phys. Rev. Lett. **107**, 178701 (2011)
8. Erdös, P., Rényi, A.: On the evolution of random graphs. Publ. Math. Inst. Hung. Acad. Sci. **7**, 17 (1960)
9. Hall, M., Frank, E., Holmes, G., Pfahringer, B., Reutemann, P., Witten, I.H.: The weka data mining software: an update. SIGKDD Explor. Newsl. **11**(1), 10–18 (2009)

10. Keating, N., Ayanian, J., Cleary, P., Marsden, P.: Factors affecting influential discussions among physicians: a social network analysis of a primary care practice. J. Gen. Intern. Med. **22**(6), 794–798 (2007)
11. Luke, D., Harris, J.: Network analysis in public health: history, methods, and applications. Annu. Rev. Public Health **28**, 69–93 (2007)
12. Newman, M.E.J.: Networks: An Introduction. Oxford University Press, Oxford (2010)
13. U. D. of Health and H. Services. 2011 report to congress: National strategy for quality improvement in health care (2011)
14. Pollack, C., Weissman, G., Lemke, K., Hussey, P., Weiner, J.: Patient sharing among physicians and costs of care: a network analytic approach to care coordination using claims data. J. Gen. Intern. Med. **28**, 459–465 (2012)
15. Vázquez, A., Dobrin, R., Sergi, D., Eckmann, J.P., Oltvai, Z.N., Barabási, A.L.: The topological relationship between the large-scale attributes and local interaction patterns of complex networks. Proc. Natl. Acad. Sci. U.S.A. **101**(52), 17940–17945 (2004)
16. Wasserman, S., Faust, K.: Social Network Analysis: Methods and Applications. Cambridge University Press, Cambridge (1994)
17. Watts, D., Strogatz, S.: Collective dynamics of 'small-world' networks. Nature **393**, 440–442 (1998)

Personality Based Recipe Recommendation Using Recipe Network Graphs

Ifeoma Adaji[✉], Czarina Sharmaine, Simone Debrowney,
Kiemute Oyibo, and Julita Vassileva

University of Saskatcehwan, Saskatoon, Canada
{ifeoma.adaji, czarina.sharmaine, simone.debrowney,
kiemute.oyibo}@usask.ca, jiv@cs.usask.ca

Abstract. There is usually a vast amount of information that people have to sift through when searching for recipes online. In addition to looking at the ingredient list, people tend to read the reviews of recipes to decide if it is appealing to them based on the feedback of others who have prepared the recipe, with some recipes having hundreds of reviews. Several researchers have proposed recipe-based recommendation systems using details such as the nutritional information of the recipe, however, such recommendations are not personalized to the characteristics of the user. To contribute to research in this area, we propose a personalized recommendation system that makes suggestions to users based on their personality. People of the same personality tend to have many similarities, and personality is a predictor of behavior, we thus propose that the use of personality types could make recommendations more personalized. In this paper, we present the result of a preliminary investigation into the use of the personality of reviewers of recipes and a recipe-based network graph in recommending recipes to users.

Keywords: Recipe recommendation · Personality · Recipe network graph

1 Introduction

The vast amount of information available to users online highlights the need for recommendation systems. Recommendation systems provide users with a list of suggestions based on their preferences (content-based filtering), preferences of other users they share similarities with (collaborative filtering), or a combination of both methods (hybrid filtering) [1]. Recommendation systems are popular commercially and in the research community and are being used in several domains including e-commerce [2] and social media [3]. Using the right recommendation strategy in a given scenario determines how precise the results of the recommendation will be. For example, using the collaborative filtering algorithm to recommend products to a person who has no similarity with others will likely produce suggestions that are not useful to the user. There have been several attempts to enhance the accuracy of recommendation systems. Guo et al. [4] propose the use of an improved Apriori algorithm to enhance a mobile e-commerce recommendation system. Similarly, Wang et al. [5] propose a deep learning approach to improve music recommendation systems. In the health domain,

Freyne and Berkovsky [6] explore the use of recipes and ingredients ratings in improving content based recommendations.

To contribute to ongoing research in this area, we propose a more personalized based recommendation system. In particular, we explore the use of the personality of reviewers and network graphs in making recommendations to users in a food and recipe domain, allrecipes.com. A person's personality are their notable characteristics or qualities such as attitudes, habits and skills in various social and personal context [7]. These characteristics or qualities differ among different personality types. There are various models that classify people based on their personality, with people in each group having a high tendency to behave in a similar way in a given context. In this paper, we adopted the Big Five Personality types [7] which describes a person's personality using five dimensions: openness to *experience, conscientiousness, extraversion, agreeableness* and *neuroticism*. We used this model because it has been used extensively in different domains such as e-commerce [8] and health [9]. In order to determine a person's personality, we adopted the Linguistic Inquiry and Word Count (LIWC) tool [10]. LIWC computes a value for each of the personality types (*openness, conscientiousness, extraversion, agreeableness* and *neuroticism*) for each user, with the personality type having the highest value being the dominant personality of the user.

The use of network graphs have also been used successfully for recommendation and forecasting [11]. Network graphs are made of nodes which in this case represent the *recipes* and *edges* which represent a connection between recipes. Recipes are connected if they have been reviewed by the same people. After identifying the personality type of the reviewers, we used a recipe network graph to plot the network of recipes with people of the same personality type being in the same community (cluster). From the communities formed, we could identify recipes (nodes) that were common among people of the same personality. We could also identify the commonly reviewed recipe pairs which could be recommended to people of similar personalities.

The main contribution of this paper is to propose a model that uses the personality of reviewers and network graphs to recommend recipes. To the best of our knowledge, this approach has not been used before. This study is still work in progress. In the future, we plan to develop an evaluation strategy that will compare the result of our recommendation to that of other systems in recipe and food-based applications.

2 Related Work

2.1 Recipe Based Recommendation Systems

Recipe based recommendation systems is an ongoing research area. Freyne and Berkovsky [6] investigate the use of recipe and ingredients ratings to improve content based recommendations. The aim of their study was to reduce the effort required by users to change their diets from unhealthy to healthier options by understanding their food preferences. In their study, the authors compared the performance of collaborative filtering, content-based filtering and hybrid filtering techniques in order to determine which one performed best. In implementing the content-based approach, the authors used the ingredient list from the recipes. The result of their analyses shows that the content-based approach which uses the ingredients of the recipes performed best. Their study differs from ours because we identify and use the personality of the reviewers of recipes and not the ingredient list.

Sobecki et al. [12] implemented a hybrid recommendation system for a web-based cooking assistant. One of the aims of the system was to recommend personalized user interfaces to users. Their recommendation system used the demographics data of users in clustering similar users together. Users in each cluster are then recommended similar user interfaces.

In their study of recipe recommendation using ingredient networks in a cooking site, Teng et al. [13] explored the relationship between ingredients using network graphs and how important various ingredients are in preparing the recipe. Their results show possible substitutes of ingredients as suggested by users in the cooking site. In addition, their results suggest ingredients that tend to co-occur frequently. Furthermore, the authors show that recipe ratings can be predicted using features of the ingredient list and nutrition information.

Despite ongoing research in the area of recipe-based recommendation, the use of personality to create a more personalized recommendation has not been explored. In this paper, we present the result of a preliminary investigation into the use of the personality of reviewers of recipes and a recipe-based network graph in recommending recipes to user.

2.2   Network Graphs

While the use of network graphs is an ongoing research area in social networking sites such as Facebook [14] and Twitter [15], they are also commonly used in recommendation systems. Odiete et al. explored the use of network graphs in identifying skill gaps in the question and answer social network, Stack Overflow. The authors investigated the relationship between experts in Stack Overflow and how this can be used to make recommendations to novices using a network graph.

Using network graphs, Teng et al. [13] explored the relationship between various ingredients in a recipe and how important these ingredients are in preparing the recipe. Their results show possible substitutes of ingredients as suggested by users in the cooking site. In addition, their results suggest ingredients that tend to co-occur frequently. Furthermore, the authors show that recipe ratings can be predicted using features of the ingredient list and nutrition information.

Silva et al. [16] used network graphs to develop a friend recommendation algorithm for social networks. To determine similarity among users, the authors used the friend-of-friends (FOF) concept separated by two degrees of separation.

2.3   Linguistic Inquiry and Word Count (LIWC)

In this paper, we explore the results of a preliminary investigation into the use of the personality of reviewers of recipes in developing a recipe-based recommendation system. In order to determine the personality of the reviewers, we used the LIWC tool [10]. LIWC computes a value for each of the personality types (*openness, conscientiousness, extraversion, agreeableness* and *neuroticism*) for each user, because according to the Big 5, everyone exhibits some traits of all personality types, however one trait is more dominant than the others [7]. The LIWC tool reads text and determines what percentage of words in the text reflect personality, emotions, thinking styles and social concerns of the writer. LIWC works by calculating the percentage of given words that match its built-in dictionary of words. The LIWC dictionary consists of about 6,400 words, word stems and emoticons. LIWC has been used extensively in analyzing users in social communities with success. Bazelli et al. [17] used the LIWC tool in exploring the personality traits of users in a popular question and answer social media, Stack Overflow. Their research suggests that top contributors in the community are extroverts. Romero et al. [18] also used the LIWC tool in their study of social networks. The authors explored how the personality traits and behavior of decision makers in a large hedge fund change based on price shocks. Based on the popularity and success of the LIWC tool as reported by other researchers, we chose to use it in this research in identifying the personality types of recipe reviewers.

3   Research Design and Methodology

The aim of this paper is to explore the use of personality types of reviewers of recipes in order to recommend recipes to other users who are of similar personality types. For example, if many users of personality type *openness to experience* have cooked and subsequently reviewed several recipes, we aim to answer the question; which of these recipes should be recommended to another user who is also of personality type *openness to experience*? In this paper, we present the result of a preliminary investigation carried out using a recipe-based network graph. In this section, we describe the methodology we adopted in carrying out our investigation.

3.1   Data Gathering

This study was carried out using data from the popular recipe site allrecipes.com[1]. Allrecipes.com is an online social networking site focused on sharing recipes and cooking. Users post recipes they have prepared in the past while other users cook those recipes and post reviews based on how useful the recipe turned out to be. Recipes are categorized by several factors such as meal type, ingredients, season (for example Christmas, Easter, Saint Patrick's Day) and cooking style (for example slow cooker, barbeque and grilling). We used the site's data about recipes such as the title of the recipe, the author, the rating received by the reviewers and the reviews written by other users. Because this study is part of a larger one which focuses on pulses, we only collected data about three popular pulses: lentils, soybeans and chickpea. This data was stored in a database for easy retrieval and manipulation.

---

[1] www.allrecipes.com.

4 Data Analyses and Result

4.1 Test for Personality

We crawled allrecipes.com and extracted the recipe title, ingredient list, reviews, category of recipe (vegan or meat based) and the ratings of recipes made using lentils, soybeans and chickpea[2]. We were particularly interested in the reviews because the reviews were used to identify the personality type of the user who reviewed a recipe. Because the LIWC tool performs better when the text it is analyzing is at least 200 characters long, we excluded short reviews with 200 or less characters. Table 1 summarizes the data we extracted that met the criteria above.

Table 1. Summary of data extracted for this study

	Number of recipes	Number of reviews
Lentils	184	15,504
Soybeans	49	1083
Chickpea	336	17,146

For each recipe, we identified the personality type of all reviewers using the Linguistic Inquiry and Word Count (LIWC) tool [10]. LIWC computes a value for each of the personality types (*openness, conscientiousness, extraversion, agreeableness* and *neuroticism*) for each user, because according to the Big 5 personality model, everyone exhibits some traits of all personality types, however one trait is more dominant than the others [7]. The dominant trait usually has the highest value. Thus, for each recipe, we identified the dominant personality trait for each reviewer. Figure 1 shows the average dominant personality for each of the types of recipe: lentils, soybeans and chickpea.

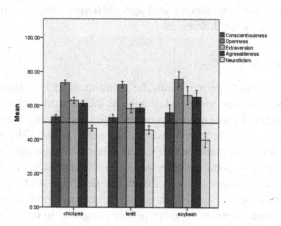

Fig. 1. Mean of dominant personality for each recipe type (lentil, soybeans and chickpea)

---

[2] We used these pulses only because this project is part of a larger one that studies pulses.

To ensure that there was significant difference between the five personality types for each recipe, and to determine any role the category of recipe (vegan or meat based) played in such differences, we carried out a two way mixed ANOVA with meal-category (vegan or meat based) as the between-subject factors and the five personality types (*openness, conscientiousness, extraversion, agreeableness* and *neuroticism*) as the within-subject factors.

Before carrying out the two way mixed ANOVA, we met the assumptions for mixed ANOVA: there were no outliers as assessed by the examination of studentized residuals for values greater than ±3, the five personality types were normally distributed as assessed by Normal Q-Q plot.

The result of two way mixed ANOVA revealed the following:

- Mauchly's test of sphericity indicated that the assumption of sphericity was violated for the two-way interaction, $\chi^2(2) = 185.829$, $p < .0001$. In addition, there was no statistically significant interaction between the personality types and the type of recipe, $F(4, 2236) = 2.354$, $p = .052$, partial $\eta^2 = .004$. We adjusted the degrees of freedom in order to correct this bias by calculating the Greenhouse-Geisse correction. There was still no statistical interaction between the persuasive strategies and the types of recipe, $F(3.387, 1893.386) = 2.354$, $p = .062$, partial $\eta^2 = .004$, $\varepsilon = .847$. **Hence we concluded that the category of recipes, meat based or vegan, had no influence on the personalities of the reviewers.** Based on this, we excluded the category of recipe from the network graph.
- There was however significant main effect of the different personality types, $F(4, 2236) = 242.246$, $p < .0005$, partial $\eta2 = 0.302$. To know where these differences exist, we carried out a pairwise comparison of the five personality types. All pairs of personality types showed significant differences, there was however no significant difference between *extraversion* and *agreeableness*. Based on this, we concluded that **there is no significant difference between the reviewers who are of personality types *extraversion* and *agreeableness***, while the other personality types showed significant differences.

4.2   Network Graph

In order to determine what recipes could be recommended based on the personality types of reviewers, we used a network graph. In the network graph, we represented nodes as recipes and edges (connection between nodes) as reviews. Two nodes (recipes) are connected if they have been reviewed by the same person. The size of the node represents the degree of the node, which is a measure of how popular that node is; the bigger the node, the more popular the recipe is and the more it has been reviewed. The thickness (width) of the edges between two nodes (recipes) shows how many people have reviewed both recipes; a very thick edge between two nodes (recipes) shows that both recipes have been reviewed by more people compared to other pairs of recipes.

Figure 2 illustrates the network graph of all the recipes in our dataset, with communities shown in various colors based on the personality types of the reviewers and the average rating of each recipe. In order to describe how a recommendation can be

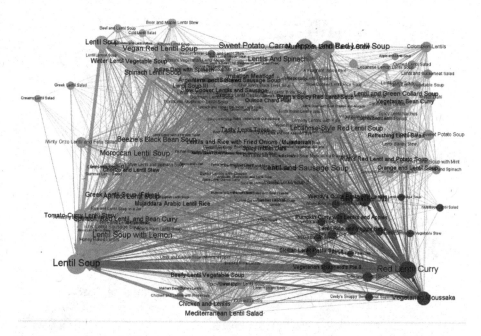

Fig. 2. Network graph of recipes showing various communities based on personality type and average ratings of each review

made using this graph, we isolated a community of recipes for reviewers with dominant personality type *openness to experience* as shown in Fig. 3.

4.3   Recommending Recipes

Figure 3 shows the community of recipes reviewed by users of personality type *openness to experience*. If a recommendation is to be made to a new user, the recipe(s) with the largest node will be the first recommendation. The largest node represents the most popular recipe in the network which has been reviewed by the highest number of people. In this case as shown in Fig. 3, lentil soup is the most popular recipe, thus will be the first recipe to be recommended to a user who is of personality type "*openness to experience*". This method will likely eliminate the cold start problem; a problem common to recommendation systems which is related to sparsity of information [19].

Because the edge between *lentil soup* recipe (marked A in Fig. 3) and *red lentil curry* recipe (marked B) has the highest weight (is thicker than all the other edges), if a user of personality type *openness to experience* has prepared and reviewed *lentil soup*, a recommendation to that user would be *red lentil curry* recipe. This is because most people that reviewed *lentil soup* also reviewed *red lentil curry* recipe and people typically review recipes they have tried out. Similarly, a user with personality type *openness to experience* who has made *Morrocan lentil soup* (marked C in Fig. 3) will be recommended the *lentil soup* recipe because of the weight of the edge between both recipes.

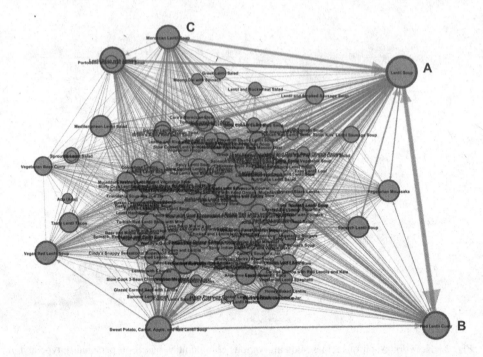

Fig. 3. Community of recipes reviewed by users who are of *"openness to experience"* personality type

5   Conclusion and Future Work

Recommendation systems have been important in online platforms because of the large amount of information that people have to sift through when trying to decide on an item such as a movie to watch, a product to purchase or a recipe to prepare. In online recipe-based systems such as allrecipes.com, in choosing a recipe, users may have to search through description of recipes and reviews written by other users to decide on what recipe to use. With some recipes having hundreds of reviews, it becomes almost impossible to sift through all this information. Several researchers have proposed recommendation systems using details such as the nutritional information of the recipe, however, such recommendations are not personalized to the personal characteristics of the user. Because personality has been shown to be a good method for personalization, we propose a more personalized recommendation system that uses the personality type of reviewers in making recipe-based recommendation. Because people of the same personality tend to have many similarities, we thus propose that the use of personality types could make recommendations more personalized.

In this paper, we present the result of a preliminary investigation into the use of the personality of reviewers of recipes and a network graph in recommending recipes to other users. Using data from allrecipes.com, a recipe sharing social networking site, we identified the personality types of reviewers using the well researched Linguistic Inquiry and Word Count (LIWC). We then carried out a two way mixed ANOVA to determine if there was any interaction between the personality types and the category of recipes. Our results show that the category of recipes, meat based or vegan, had no influence on the personalities of the reviewers. We thus excluded the category of recipes from the network graph. Finally, we developed a network graph using the recipes as nodes and reviews as edges. We identified clusters of communities in the network based on similarity of personality. Recommendations were then made based on the weight of the edges between nodes which shows pairs of recipes that have commonly been reviewed by the same user.

In the future, we plan to evaluate our findings by comparing the results we obtained here to that of other recommendation systems. In addition, we will determine if the ingredient list has any influence on the outcome of the recommendation.

Our result is limited in a few ways. First, we assume that reviewers prepare a recipe before reviewing it and that the review is based on their experience of using the recipe. This is however not always the case. In addition, the data set we worked on is only a fraction of the entire users on allrecipes.com. This is because we only selected recipes based on three not so popular ingredients: lentils, soybeans and chickpea. In the future, we plan to expand the types of recipes used in the investigation.

References

1. Shani, G., Gunawardana, A.: Evaluating recommendation systems. In: Ricci, F., Rokach, L., Shapira, B., Kantor, P.B. (eds.) Recommender Systems Handbook, pp. 257–297. Springer, Boston, MA (2011). https://doi.org/10.1007/978-0-387-85820-3_8
2. Li, S.S., Karahanna, E.: Online recommendation systems in a B2C E-commerce context: a review and future directions. J. Assoc. Inf. Syst. **16**(2), 72 (2015)
3. Guy, I., Zwerdling, N., Ronen, I., Carmel, D., Uziel, E.: Social media recommendation based on people and tags. In: Proceedings of the 33rd International ACM SIGIR Conference on Research and Development in Information Retrieval (2010). dl.acm.org
4. Guo, Y., Wang, M., Li, X.: Application of an improved Apriori algorithm in a mobile e-commerce recommendation system. Ind. Manage. Data Syst. **117**(2), 287–303 (2017)
5. Wang, X., Wang, Y.: Improving content-based and hybrid music recommendation using deep learning. In: Proceedings of the 22nd ACM International Conference on Multimedia (2014). dl.acm.org
6. Freyne, J., Berkovsky, S.: Recommending food: reasoning on recipes and ingredients. In: De Bra, P., Kobsa, A., Chin, D. (eds.) UMAP 2010. LNCS, vol. 6075, pp. 381–386. Springer, Heidelberg (2010). https://doi.org/10.1007/978-3-642-13470-8_36
7. Goldberg, L.R.: An alternative description of personality: the big-five factor structure. J. Pers. Soc. Psychol. **59**(6), 1216 (1990)
8. Huang, J., Yang, Y.: The relationship between personality traits and online shopping motivations. Soc. Behav. Pers. Int. J. (2010)
9. Orji, R., Nacke, L.E., Di Marco, C.: Towards personality-driven persuasive health games and gamified systems. In: Proceedings of the 2017 CHI Conference on Human Factors in Computing Systems - CHI 2017, pp. 1015–1027 (2017)
10. Pennebaker, J.: Linguistic inquiry and word count: LIWC 2001 (2001). downloads.liwc.net. s3.amazonaws

11. Odiete, O., Jain, T., Adaji, I., Vassileva, J., Deters, R.: Recommending programming languages by identifying skill gaps using analysis of experts. a study of stack overflow. In: Adjunct Publication of the 25th Conference on User Modeling, Adaptation and Personalization, UMAP, pp. 159–164 (2017)

12. Sobecki, J., Babiak, E., Słanina, M.: Application of hybrid recommendation in web-based cooking assistant. In: Gabrys, B., Howlett, R.J., Jain, L.C. (eds.) KES 2006. LNCS (LNAI), vol. 4253, pp. 797–804. Springer, Heidelberg (2006). https://doi.org/10.1007/11893011_101

13. Teng, C.Y., Lin, Y.R., Adamic, L.A.: Recipe recommendation using ingredient networks. In: Proceedings of the 4th Annual ACM Web Science Conference (2012). dl.acm.org

14. Traud, A.L., Mucha, P.J., Porter, M.A.: Social structure of Facebook networks. Phys. A Stat. Mech. Appl. **391**(16), 4165–4180 (2012)

15. Ediger, D., Jiang, K., Riedy, J., Bader, D.A., Corley, C.: Massive social network analysis: mining Twitter for social good. In: 2010 39th International Conference on Parallel Processing, pp. 583–593 (2010)

16. Silva, N.B., Tsang, I.-R., Cavalcanti, G.D.C., Tsang, I.-J.: A graph-based friend recommendation system using genetic algorithm. In IEEE Congress on Evolutionary Computation, pp. 1–7 (2010)

17. Bazelli, B., Hindle, A., Stroulia, E.: On the personality traits of stackoverflow users. Software Maintenance (ICSM) (2013)

18. Romero, D.M., Uzzi, B., Kleinberg, J.: Social networks under stress. In: Proceedings of the 25th International Conference on World Wide Web - WWW 2016, pp. 9–20 (2016)

19. Lika, B., Kolomvatsos, K., Hadjiefthymiades, S.: Facing the cold start problem in recommender systems. Expert Syst. Appl. **41**(4), 2065–2073 (2014)

Identifying Communities in Social Media
with Deep Learning

Pedro Barros[1], Isadora Cardoso-Pereira[1], Keila Barbosa[1], Alejandro C. Frery[1],
Héctor Allende-Cid[2], Ivan Martins[1], and Heitor S. Ramos[1(✉)]

[1] Instituto de Computação, Universidade Federal de Alagoas, Maceió, AL, Brazil
{pedro_h_nr,isadora.cardoso,keilabarbosa,acfrery,
ivan.martins,heitor}@laccan.ufal.br
[2] Escuela de Ingeniería Informática, Pontificia Universidad Católica de Valparaíso,
Valparaíso, Chile
hector.allende@pucv.cl

Abstract. This work aims at analyzing twitter data to identify communities of Brazilian Senators. To do so, we collected data from 76 Brazilian Senators and used autoencoder and bi-gram to the content of tweets to find similar subjects and hence cluster the senators into groups. Thereafter, we applied an unsupervised sentiment analysis to identify the communities of senators that share similar sentiments about a selected number of relevant topics. We find that is able to create meaningful clusters of tweets of similar contents. We found 13 topics all of them relevant to the current Brazilian political scenario. The unsupervised sentiment analysis shows that, as a result of the complex political system (with multiple parties), many senators were identified as independent (19) and only one (out of 11) community can be classified as a community of senators that support the current government. All other detected communities are not relevant.

Keywords: Community detection · Deep Learning
Text classification · Convolutional networks · Autoencoder

1 Introduction

The idea of community has changed with the growth of social media (such as forum, blogs, and micro-blogs), since it is possible to people, through the Internet, to connect and interact online based on shared interests and activities, even if they are not geographically close (Papadopoulos et al. 2012).

With 100 million daily active users, Twitter is one of the most popular social network nowadays. It enables the users to send short messages (called tweets) up to 280 characters and, has around 6000 tweets per second, which corresponds to over 500 million tweets per day[1]. Counting on this huge volume of data generated, Twitter can be seen as a valuable source of data that is useful for

---

[1] https://www.omnicoreagency.com/twitter-statistics/.

© Springer International Publishing AG, part of Springer Nature 2018
G. Meiselwitz (Ed.): SCSM 2018, LNCS 10914, pp. 171–182, 2018.
https://doi.org/10.1007/978-3-319-91485-5_13

monitoring several social aspects, such as detecting and analyzing communities. Jungherr (2016) reviews many studies that shows the power of Twitter to give meaningful insights about the American political scenario.

Brazil has a multi-party system, i.e., it admits the legal formation of several parties. This caused a highly fragmented party system, for instance, there are 81 senators currently in office, divided into 20 parties[2], which are: DEM (4 senators), PCdoB (1), PDT (3), PMDB (20), PODE (3), PP (3), PPS (1), PR (4), PRTB (1), PROS (1), PSB (4), PSC (1), PRB (1), PSD (4), PSDB (11), PT (9), PTB (2), PTC (1), REDE (1), and without parties (2). There are more parties in Brazil, but they do not have representative in Brazilian Senate.

Twitter data has been used to analyze political aspect in many different ways. For instance, Vaz de Melo (2015) showed that Brazil has and had many ideologically redundant parties, i.e., parties that are similar in the ideological space, being possible to reduce more than 20 to only 4 parties. Hadgu et al. (2013) analyzed tweets to show that some changes in political polarization of hashtags are caused by "hijackers" engaged in a particular type of hashtag war. Park (2013) have investigated the interrelationships between opinion leadership, Twitter use motivations, and political engagement.

Differently, our work aims to identify communities of Brazilian senators using Twitter posts. We analyze tweets of 76 senators (about 94% of the total) from the beginning of their mandate rather than specific events, so we can have a better understanding of their political vision and if they share it. Thereunto, we collect the data using a Twitter API. After cleaning, we apply autoencoder and bi-gram to the content of tweets to find similar subjects and hence cluster the senators into groups. Thereafter, we applied an unsupervised sentiment analysis to identify the communities of senators that share similar sentiments about a selected number of relevant topics.

This article is organized as follows: Sect. 2 describes the methodology used to collect and analyze data; Sect. 3 presents the main results; and Sect. 4 concludes this work.

2   Methodology

Figure 1 depicts a schematic view of the methodology applied to this work. First and second steps, Data collection and preparation, are described in Sect. 2.1 and consist of collecting and preprocessing the twitter data to rule out obvious cluster and meaningless data that hinder the community detection process. Processing and Sentiment Analysis are described in Sects. 2.2 and 2.3 and consist of the community detection technique used in this work. Finally, we described how we evaluated our approach in Sect. 2.4.

2.1   Data Collection and Preparation

We collected a real world dataset of tweets from Brazilian senators accounts. We obtained the list of senators official accounts from the Brazilian Senate official

---

[2] https://www25.senado.leg.br/web/senadores/em-exercicio/-/e/por-partido.

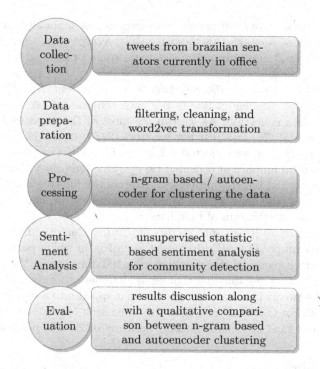

Fig. 1. Schematic view of the methodology used in this work

account. We found 76 accounts that represent a total of about 94% of the actual number of senators currently in office. All extracted data are in Brazilian Portuguese and also performed all processing in the original language. We translated the tweets and subjects only when showing the results, for the sake of clarity and readability.

To gather the tweets, we use the API provided by Twitter[3]. The API returns each tweet in JSON format, with the content of the tweet, metadata (e.g., timestamp, replied or not, retweeted or not, etc.), and information about the user (username, followers, etc.). In this work, we only consider the username (Senator's account), the date, and the content of the tweet, in which we applied some cleaning techniques.

To clean the data, we removed stopwords (such as prepositions and pronouns), words with less than 3 letters, punctuation marks, and URLs. We also lower all the letters and removed graphic accentuation, in order to normalize writing. Moreover, we used a time filtering to collect data from January 1st 2014 to November 8th 2017, which corresponds to the date of the beginning of the current mandate of this legislature up to the day we decide to stop the data collection and analyze the data. It is worth mentioning that we are not able to collect all tweets for all accounts, because the tweeter API provides only a

---

[3] http://www.tweepy.org/.

sample of the total amount of tweets. Furthermore, we manually removed some meaningless tweets, such as tweets with just greetings (e.g., "good morning", "hello friends", etc.) and automatic text (e.g., shared photo on different social networks).

We collected a total of 166, 893 tweets of 76 current Brazilian senators. Figure 2 shows the number of tweets per each senator twitter account during the period of data collection. As we can see most of the senators tweets more than 1000 times (59/76) and just few senators (9/76) tweets less than 5 times. This fact indicates that we collected a reasonable number of tweets. While red bars show the total number of collected tweets, blue bars show the final dataset after the cleaning and filtering processes. After all the filtering and the selection of relevant tweets (the ones related to a relevant topic), we have 33, 550 tweets in our final data set (the sum of blue bars).

We transformed all tweets into vectors of features using the Portuguese word embeddings set proposed by (Rodrigues et al. 2016), in which portuguese words are mapped into a vector of real numbers with 100 dimensions. We found in our sample of tweets that the maximum observed number of words is 29. Hence, we normalized the tweets in a way that they were mapped into a matrix of features of size 29×100. For tweets shorter than this dimension (with less than 29 words) we complete the gaps with zeros forming an array of $29 \cdot 10^2$ positions.

2.2   Processing

At this stage, the main goal is to find tweets of similar subject and cluster them into groups. In each group, we may find tweets that have positive or negative sentiment about the subject, and we will further find communities of Senators that have positive or negative sentiments about these subjects.

We applied two different strategies to cluster the data. The first is a simple strategy and was used as a baseline. We count the top most frequent bi-grams of the dataset and applied two filters: (i) we ruled out all bi-grams with frequency smaller than 50, (ii) with the remain dataset, we manually removed all bi-grams that are irrelevant, for instance, greetings messages. We chose the number 50 because we observed that all bi-grams less frequent than this threshold were irrelevant and disconnected to real political scenario and sometimes they represent popular language terms that have no connection to relevant subjects.

After the filtering, we identified 13 subjects we judge relevant and highly correlated to the current political scenario in Brazil. For each bi-gram, we identified and grouped all senators that have tweets about this subject and we disregarded the senators that tweet less than 5 times about a subject.

N-gram is a well known and simple technique to apply to such scenario, but it is difficult to choose the correct granularity. For instance, we observed that we have to discard some tweets that were related to the so-called "Car Wash Operation" because this approach formed two different clusters: (i) "Wash Operation", and (ii) "Car Wash".

To try to overcome such situation, we adopted another approach to cluster tweets into groups of similar subjects. We use autoencoder (Goodfellow et al. 2016)

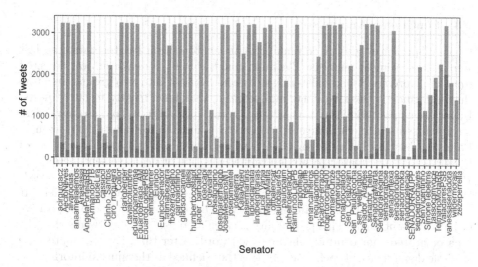

Fig. 2. Number of tweets per senator tweeter account during the period of data collection. Red bars show the total number of tweets, while blue bars show the number of tweets that were effectively used after cleaning and filtering the dataset. (Color figure online)

as an unsupervised clustering technique. Goodfellow et al. (2016) defines an autoencoder as a neural network trained with the goal of copy its input to its output. The network comprises of two parts: (i) an encoder function, (ii) a decoder. Autoencoders have been applied to dimensionality reduction and information retrieval.

An autoencoder should be able to learn in a restricted way that makes it to copy only useful properties of the data. Recently, autoencoder has been used to perform unsupervised clustering as in (Xie et al. 2016). Authors used a variation of (Maaten and Hinton 2008) and propose the Deep Embedded Clustering (DEC), which clusters a set of n points into a predetermined number of cluster (k). Instead of clustering directly in the data space, DEC transforms the data with a non-linear mapping $f_\theta: X \mapsto Z$. The new space Z has typically lower dimensionality than X, helping with the so-called curse-of-dimensionality. The function mapping f_θ is approximated with autoencoder, taking advantage of the theoretical function approximation property of neural networks.

DEC stands out from other clustering techniques due to the use of autoenconder to solve for feature space and cluster membership jointly. To do so, DEC uses a deep autoencoder network structure and finetunes it to minimize the reconstruction error. It then discards the decoder layers and use the encoder layers as its initial mapping between the data and the feature space. The second phase is the clustering itself that is done by computing an auxiliary target distribution and minimizing the Kullback-Leibler (KL) divergence to it. We chose DEC as our clustering method because we want to overcome the n-gram aforementioned problem.

2.3   Sentiment Analysis

In our approach to detect communities of senators from twitter data, the last step is the sentiment analysis. The main goal is to split, for each subject detected from the aforementioned clustering of tweets, senator into two groups: senators that have mostly positive posts and senators that have mostly negative posts. Hence, we can separate senators into two groups per subject. To do so, we use an unsupervised sentiment analysis proposed by Church and Hanks (1990), namely Pointwise Mutual Information (PMI), where words are statistically associated to words that are related to predetermined positive or negative sentiments.

PMI is an unsupervised technique that uses a list of predetermined words that are known in advance to be associated to positive or negative sentiment. It calculates the probability of an analyzed word be associated to a seed of positive or negative sentiment. This probability is calculated using the all tweets that the senator posted that contains the analyzed word. After that, it is calculated the statistic dependence of each word. PMI is thus defined as the mutual information between a given word c being positive (or negative) as follows

$$PMI(c, pos) = \log 2 \frac{\Pr(c, pos)}{\Pr(c)P(pos)}$$

and

$$PMI(c, neg) = \log 2 \frac{\Pr(c, neg)}{\Pr(c)P(neg)}$$

Informally, PMI is the probability of a given word c be positive with the probabilities of observing c and observing a positive pos word. Hence, for each word, it is calculated the difference between $PMI(c, pos)$ and $PMI(c, neg)$ as

$$\begin{aligned} PV_{PMI}(c) &= PMI(c, pos) - PMI(c, neg) \\ &= \log 2 \frac{\Pr(c, pos)/\Pr(pos)}{\Pr(c, neg)/\Pr(neg)} \\ &= \log 2 \frac{\Pr(c \mid pos)}{\Pr(c \mid neg)} \end{aligned}$$

Finally, we sum up the $PV_PMI(c)$ for all words of a tweet to calculate the sentiment about of that specific post.

It is worth mentioning that positive or negative sentiments are correlated to the opinion about a subject but it not necessarily means that a positive tweet agrees with a subject or, conversely, a negative tweet disagree with a subject. This happens because, sometimes, a text may use irony or some figure of speech that is difficult to be correctly recognized by the technique described herein.

2.4   Evaluation

We performed two analysis with the results of our proposal. In the first, we created a time series from the sentiment analysis of each senator per each subject we háve studied. Hence, we registered the PMI value for the time span of the analysis. The main goal of this analysis is to observe some non-usual behavior such as a steep change of a senator sentiment in a specific subject, suggesting that this senator changed the opinion due to some relevant event. The second analysis is related to the community detection itself. For this analysis, we created a vector of PMIs for each subject detected by the clustering analysis, in our case, a 13-dimension vector, with the goal of analyzing what senators have similar sentiments about the selected subjects. To do so, we associated for each dimension (subject) a sentiment value of -1 (negative), 0 (neutral) or 1 (positive).

We used Principal Component Analysis (PCA) to reduce the dimensionality to 4, where we used a DBSCAN clustering method with parameters $\epsilon = 0.4$ and a minimum of 3 senators per cluster. We chose the DBSCAN clustering due to the fact that a senator can be independent, i.e., not associated to any other cluster. This is a common situation where a senator does not necessarily follow the party orientation and have an independent opinion about different subjects.

3   Results and Discussion

Figure 3 shows the results of PMI changing in time for some selected senators, representing the changes of the sentiment of a given senator for a specific subject.

The Car-wash Operation started at March 2014 and consists of a set of investigations against corruption at the state-controlled oil company Petrobras, in progress by the Federal Police of Brazil. Many parties of the Brazilian political system have politicians investigated by the Car-wash Operation.

Randolfe Rodriguez was cited by an accused businessman to be a participant in corruption in early 2016. We can see in Fig. 3(a) that he expressed a very negative feeling about the operation that time, with PMI less than -20. After that, he showed less negative feelings about it, reaching almost a neutral sentiment by the middle of 2017. Figure 3(b) shows the PMI value for the tweets of Roberto Requião, senator for the same party (PMDB) as the current President of Brazil, Michel Temer. Senator Requião's tweets present a mild sentiment about the Car-wash Operation, varying from slightly positive to slightly negative and reaching a neutral sentiment by the end of 2017.

Michel Temer is the current Brazilian president, after the impeachment of Dilma Rousseff (August, 2016), which was received with mixed feelings by the Brazilian population. Humberto Costa is from the same party of the ex-president. We can see in Fig. 3(c) that he shows strong negative feelings about Michel Temer, but this behavior evolved to an almost neutral sentiment by the middle of 2017. On the other hand, Roberto Requião is from the same party of the current President, hence showing more positive feelings about him.

PSC and PMDB are both in favor of Labor-rights Reform. Eduardo Amoring and Romero Jucá are two party members of PSC and PMDB, respectively. We can see in Fig. 3(e) that Eduardo Amorim have neutral or positive feelings about the subject. Romero Jucá, the responsible for voting the reform, shows some negative feelings in the same time that the Labor-rights Reform was accused of

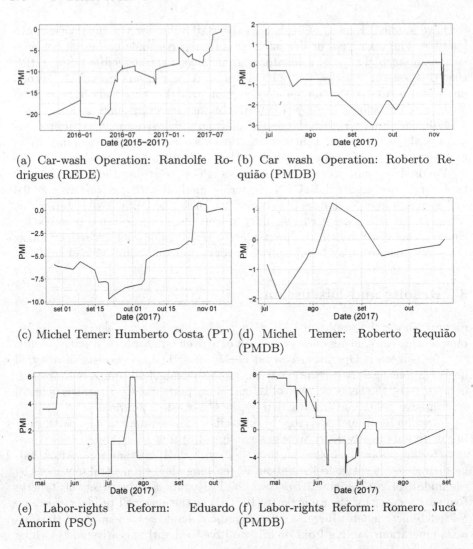

(a) Car-wash Operation: Randolfe Rodrigues (REDE)

(b) Car wash Operation: Roberto Requião (PMDB)

(c) Michel Temer: Humberto Costa (PT)

(d) Michel Temer: Roberto Requião
(PMDB)

(e) Labor-rights   Reform: Eduardo
Amorim (PSC)

(f) Labor-rights Reform: Romero Jucá
(PMDB)

Fig. 3. Variation of PMI for some selected senators

preventing the fight against slave labor, which forced him to change some reform
points. This such behavior is shown in Fig. 3(f).

The aforementioned results depicted how a sentiment analysis can be used in
the understanding of the political momentum. As we can observe in the example
of Romero Jucá about the Labor-rights reform, even tough he was in charge for
voting the reform, and hence, he supports the reform, he posted some negative
tweets showing his concern about the difficult he faced to conduct the process.
Therefore, we cannot state that a politician that presents positive (or negative)
sentiment about a topic necessarily supports or disapproves the matter.

Table 1. Number of senators per community per political party

Party	NC	C1	C2	C3	C4	C5	C6	C7	C8	C9	C10	C11	Total
DEM	0	1	1	0	0	0	0	0	0	0	1	0	3
PCdoB	1	0	0	0	0	0	0	0	0	0	0	0	1
PDT	0	1	0	0	0	1	0	0	0	0	0	0	1
PDT/NP	0	1	0	0	0	0	0	0	0	0	0	0	1
PMDB	5	9	0	0	1	0	0	1	1	0	0	1	18
PMDB/NP	0	0	1	0	0	0	0	0	0	0	0	0	1
PODE	0	1	0	0	0	0	0	0	0	1	0	0	2
PP	1	3	0	0	0	0	0	0	0	0	0	1	5
PPS	0	0	0	0	1	0	0	0	0	0	0	0	1
PR	1	3	0	0	0	0	0	0	0	0	0	0	4
PRB	0	0	0	0	0	0	1	0	0	0	0	0	1
PROS	0	1	0	0	0	0	0	0	0	0	0	0	1
PSB	1	0	0	1	0	0	1	0	0	1	0	0	4
PSC	0	1	0	0	0	0	0	0	0	0	0	0	1
PSD	0	2	0	0	0	1	0	0	0	0	0	0	3
PSDB	4	5	0	1	0	1	0	0	0	0	0	1	12
PT	6	1	0	0	0	0	0	0	0	1	1	1	10
PTB	0	1	0	0	0	1	0	1	0	0	0	0	2
PTC	0	1	0	0	0	0	0	0	0	0	0	0	1
PV	0	0	0	1	0	0	0	0	0	0	0	0	1
REDE	0	0	0	0	0	0	0	0	1	0	0	0	1
Total	19	31	2	3	2	4	2	2	2	3	2	4	76

In the following, we will present the results of the community detection method we have described and we will also cluster the politicians by the sentiment regarding a topic. Although some situations as the one described in Fig. 3(f) is possible to happen, we can say that the sentiment about a topic correlates with (but does not necessarily leads to) the opinion that a politician have about a specific matter.

Figure 4 shows the communities we were able to detect by using the method described in Sect. 2. To detect these communities, we created a vector of values -1, 0 or 1 corresponding to the sentiment about a given topic (each topic corresponds to a position in this vector) for each senator. The topics we have considered are: Pension Reform, Aécio Neves (a senator also cited in the Carwash Operation), Indigenous people, Michel Temer, President Dilma, Car-wash Operation, President Lula, Bolsa Família[4], Criminal responsibility age, Family agriculture, Eduardo Cunha, Labor-rigths Reform, and Political Reform.

---

[4] Assistance program for poor families.

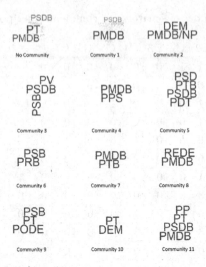

Fig. 4. Clusters of parties

All topics represent important aspects of the current Brazilian political momentum. In this plot, instead of showing individual politician names, we depicted how many politicians of each party showed up in each community. The font size is proportional to the number of senators of a given party that appears in that community. PMDB/NP and PDT/NP represents two senators that were elected by PMDB and PDT, but are not affiliated to these parties anymore.

The first group of parties (the top left group) depicts some senators that did not show any similarity with others. We can observe that PSDB, PT and PMDB present four, six and five senators, respectively, which have independent sentiment about the matters we have studied. These numbers correspond to about 33%, 60%, and 28%, respectively, of the total representation of these parties in the Brazilian Senate. The community C1 has the PMDB as principal party and presents other parties such as PSDB, and PP, all of them support the current president Michel Temer. Therefore, this community is likely to be the government base in the Brazilian Senate. As we can see in Table 1, there are 19 senators that cannot be associated to any community and 31 senators in the community that supports the government. These two groups amount about 66% of all senators considered in this study. All other communities are comprised by a small number of senators, typically 2−4 and represent groups that are most likely formed by chance. Observe in Table 1 that there is no community from C2 to C11 aggregating more than one senator of the same party. Showing that these communities are not too relevant from the point-of-view of the political analysis. Senators assigned to small communities like the ones from C2 to C11 are likely to have independent opinion, such as the ones assigned to NC. Hence, in this analysis, we conclude that there are only two relevant communities, C1, likely to be the community of senators that support the current government, and NC, formed by senators with independent opinion regarding the subjects studied herein.

4 Conclusion

This paper presents a study of a sample of 166, 893 tweets from Brazilian senators in office, in which a deep autoencoder was applied to cluster these tweets into groups of similar topics. We used autoencoder in contrast of a create clusters using the n-gram directly, because it is hard to find the granularity, n, which is more convenient to split the dataset into relevant topics. Hence, autoencoder is able to cluster the tweets without specify a specific granularity of important terms.

We have identified 13 relevant topics about the current Brazilian political scenario. The final dataset presents 33, 550 tweets of 76 different senators. We further performed a sentiment analysis to split the senators that tweeted about the same topic into three groups having: (i) positive, (ii) negative, and (iii) neutral sentiment about each selected topic. Furthermore, we created a vector of 13 positions for each senator and applied DBSCAN to detect communities of senators that have similar overall sentiment about the selected topics. We observed that, as a result of the complex political system (with multiple parties), many senators were identified as independent (19) and only one (out of 11) community can be classified as a community of senators that support the current government. All other detected communities are not relevant. The techniques used in this analysis have some limitations. For instance, although, we observed that the sentiment is correlated to the politician opinion about a topic, we cannot conclude the sentiment is sufficient to state the politician opinion about a topic. Moreover, despite some limitations, this correlation is useful to create the communities and we can use this technique to extract a big picture of the political scenario, although we found we cannot make individual conclusions. It is worth mentioning that this was an observational study, hence, we cannot state any causal relation. Moreover, the conclusions about this sample cannot be generalized, although they are useful to improve the understanding of the current Brazilian complex political scenario.

Acknowledge. This work was partially funded by Fapeal, CNPq, and SEFAZ-AL. The work of Héctor Allende-Cid was supported by the project FONDECYT Initiation into Research 11150248.

References

Church, K.W., Hanks, P.: Word association norms, mutual information, and lexicography. Comput. Linguist. **16**(1), 22–29 (1990)

Goodfellow, I., Bengio, Y., Courville, A.: Deep Learning. MIT Press (2016)

Hadgu, A.T., Garimella, K., Weber, I.: Political hashtag hijacking in the U.S. In: Proceedings of the 22nd International Conference on World Wide Web, WWW 2013 Companion, pp. 55–56. ACM, New York (2013). https://doi.org/10.1145/2487788. 2487809

Jungherr, A.: Twitter use in election campaigns: a systematic literature review. J. Inf. Technol. Polit. **13**(1), 72–91 (2016)

Maaten, L.V.D., Hinton, G.: Visualizing data using t-SNE. J. Mach. Learn. Res. **9**, 2579–2605 (2008)

Vaz de Melo, P.O.S.: How many political parties should brazil have? a data-driven method to assess and reduce fragmentation in multi-party political systems. PLOS ONE **10**(10), 1–24 (2015)

Papadopoulos, S., Kompatsiaris, Y., Vakali, A., Spyridonos, P.: Community detection in social media. Data Min. Knowl. Discov. **24**(3), 515–554 (2012)

Park, C.S.: Does twitter motivate involvement in politics? tweeting, opinion leadership, and political engagement. Comput. Hum. Behav. **29**(4), 1641–1648 (2013)

Rodrigues, J., Branco, A., Neale, S., Silva, J.: LX-DSemVectors: distributional semantics models for portuguese. In: Silva, J., Ribeiro, R., Quaresma, P., Adami, A., Branco, A. (eds.) PROPOR 2016. LNCS (LNAI), vol. 9727, pp. 259–270. Springer, Cham (2016). https://doi.org/10.1007/978-3-319-41552-9_27

Xie, J., Girshick, R., Farhadi, A.: Unsupervised deep embedding for clustering analysis. In: International Conference on Machine Learning, pp. 478–487 (2016)

Investigating the Generation- and Gender-Dependent Differences in Social Media Use: A Cross-Cultural Study in Germany, Poland and South Africa

Kaja J. Fietkiewicz[1]([⊠]), Elmar Lins[1], and Adheesh Budree[2]

[1] Heinrich Heine University Düsseldorf, Düsseldorf, Germany
Kaja.Fietkiewicz@uni-duesseldorf.de
[2] University of Cape Town, Cape Town, South Africa

Abstract. In social media research, there is an ongoing debate about whether and how much cultural and geographical differences impact social media interaction. There has not been reached a consensus yet, which is why we apply an extensive statistic model based on a unique and large dataset of German, Polish, and South African social media users. We aim to answer the following questions: How do different generation use social media? Are there any gender-dependent differences? How do these differences vary between three different countries?

Keywords: Social media use · Gender-dependent differences
Age-dependent differences · Cross-cultural study

1 Introduction

The modern world is shaped by new information and communication technologies, including the emergence of social media taking place in the context of Web 2.0 [1, 2]. With time, social media or "social networking sites (SNS) such as Facebook have become part and parcel of our daily lives" [3]. These new digital tools are slowly replacing the known, traditional means of communication [4]. They generate new ways of interaction not only between individuals, but also between firms and their clients [5]. However, not everyone applies the different SNSs in a same way. Different generations have different motivation for and manner of using the online media, and the same holds for male and female users [4, 6, 7].

Moreover, many social media researchers emphasize the need to examine demographic differences in the use of, for example, Facebook, among different age groups, cultures as well as genders [8–11]. This is important since most studies focus on Facebook users from the U.S. [8, 12–22] and students [9, 13–15, 18, 19, 21, 23–25]. The sole focus on Facebook instead of several social media platforms should also be broadened to further channels. Age and gender appear to be the key variable in understanding the user behavior on SNSs [26–29] as well as in the gratifications of internet use or the related accessibility [8, 30]. Therefore, this study will consider cross-cultural age- and

© Springer International Publishing AG, part of Springer Nature 2018
G. Meiselwitz (Ed.): SCSM 2018, LNCS 10914, pp. 183–200, 2018.
https://doi.org/10.1007/978-3-319-91485-5_14

gender-dependent differences in social media usage. The cross-cultural comparison includes three countries: Germany, Poland, and South Africa.

The current investigation of age-dependent differences is based on the motion of different generations. Here, the most distinct gap can be found between generations that were firstly confronted with the Internet and its applications in their late stages of life (Silver Surfers) [6], the ones that were raised without the Internet, however, had the possibility to adopt it from its beginning and in their adolescent or adult lives (Digital Immigrants) [4, 31], and, finally, generation that grew up in the omnipresence of the World Wide Web (the Digital Natives) [4, 31]. According to Prensky [31], the arrival and dissemination of digital technology at the end of the 20th century has "changed everything so fundamentally that there is no going back." Prensky calls the newest generation born and raised in this time the "Digital Natives". They spend their entire lives surrounded by computers, cell phones, and all other "toys and tools of the digital age" [31]. In this study, the investigation of age-dependent differences in social media usage is grounded on the notion of different generations and, therefore, the reference "generation-dependent" social media usage will be applied. Despite the differences in social media usage by different generations, there is a vast literature examining gender differences in online settings [27]. Even though there appear to be no gender differences in overall amount of Internet use, there could be gender-dependent divergences in motivations for it as well as the utilization of time spent online [6, 7, 27]. According to Muscanell and Guadagno [27], the gender differences in online behavior may apply especially to SNSs, since men and women use them for different reasons. In light of possible gender-dependent differences, this factor will be also regarded in the current study.

This study does not only focus on probability of use of a certain social media platform or the usage frequency, differentiated between generations, genders and cultures. Another relevant factor to investigate is the objective of using the service and the perspective of the uses and gratifications (U&G) of its users [29, 32–36]. More precisely, what exactly are the motives of users to create content in social media and how the gratifications of generating content affect the activities in social media [35]. The U&G theory (or UGT) [32, 33] is "a theoretical framework explaining how and why people actively seek out different media to fulfill their specific needs and wants. UGT posits that the gratifications users receive through the media they select, in turn, satisfy a variety of informational, social, and leisure needs" [37]. This framework enables an investigation of how individuals utilize media in a variety of contexts and settings, what desires motivate individuals to use specific technologies, and what are their social and psychological characteristics [38]. In this study, we incorporate two U&G factors possibly influencing the usage of diverse social media platforms, namely the "social interaction" and "approval seeking".

Another important aspect addressed especially in Facebook research, but also relevant in context of other social media, is the investigation of how users deal with privacy or how privacy concerns vary between different demographic groups and nationalities [9]. "Internet users, members of SNSs in particular, seem either not concerned about their privacy or not aware of the loss of privacy they suffer during their time online" [39]. Studies show that even though many users are concerned about the privacy issues, SNSs encouraging their users to reveal and exchange personal information are more and more popular [30]. Especially Facebook is not only known for its popularity, but also for the quality and quantity of personal information in it [30, 40] as well as reoccurring privacy issues related to acquisitions of other SNSs [41]. According to Acquisti and Gross [40], privacy concerns are a weak predictor of individual's membership in an SNS, and even if some have concerns about the privacy, they still join the networks and reveal personal information [30, 40]. Furthermore, there appear to be gender-dependent differences regarding the exposure of personal information on SNSs [30, 40].

2 Methods

2.1 Research Model

Even though the research on social media, especially Facebook, is very vast, there is need for more cross-cultural investigations, since the cultural background may have an influence on the adoption and usage of SNSs. Furthermore, many social media studies focus on certain age groups of the social media users, mostly teenagers and adolescents. To broaden this perspective, we conduct a more extensive, cross-generational investigation. Finally, we include gender as a factor. These three aspects—country, gender, and generation—are independent variables of our research model (Fig. 1). In this study, we investigate the differences in social media usage—the adoption of SNSs (probability of use), the frequency of continuous usage, the two U&G factors coming from the use of the SNSs—the social interaction and approval seeking, and, finally, the importance of privacy.

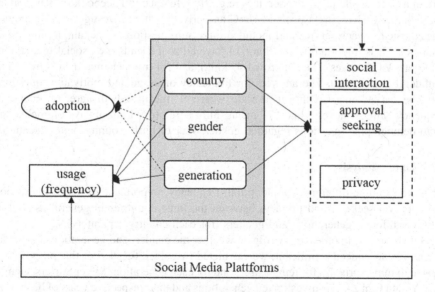

Fig. 1. Research model: cultural-, gender- and generation-dependent differences in adoption and usage of social media as well as in uses and gratification objectives and privacy concerns.

2.2   Online Questionnaire

The data presented in this paper was collected through an online survey which was distributed via social media, email and face-to-face. The survey was randomly distributed among males and females from different educational levels and ages within Germany, Poland, and South Africa. The survey focused on evaluating thirteen different social media platforms, and comprised of forty-two questions, which took respondents approximately 3–5 min to complete. The survey was designed in a way that respondents only had to select an option that closely related to the respondent. Furthermore, none of the question required an explanation for any given response.

Studies of online population, like the social media users, have led to an increase in the use of online surveys [6, 42]. There are many advantages of online surveys, including access to individuals from distant locations, automated data collection and analysis [42] as well as flexibility for the respondents to answer the question when and where they want to, question diversity, control of question order, and required completion of answers [43]. For this study the nonprobability sampling was applied, in form of purposive or judgment sampling (social media users), continued as snow-ball sampling (sharing on social media by participants). Judgment sampling is one of the most common sample techniques, where the researcher actively selects the most productive sample to answer the research question, whereas the subjects may recommend useful potential candidates for study [44].

The first question of the survey was a polar question about the use of a certain service, e.g., 'Do you use Facebook?' Dependent on the answer, two follow-up questions about the concerned service succeeded—about the frequency with which the service is used (e.g., 'How often do you use Facebook?') and about the importance of certain aspects while using the services (e.g., 'In reference to Facebook, it is important to me that...'). Here, the aspects of U&G and privacy where investigated. The inquiry was adjusted to each service and included three sub-questions, for example, in case of Facebook, 'It is important to me that (i) I have a lot of friends (i.e., social interaction) (ii) I get a lot of "likes" (i.e., approval seeking), and (iii) my personal data is treated as confidential' (privacy). The answers for frequency of usage and motivation questions could be marked on a 7-point Likert scale, where "1" meant fully disagree (or in case of frequency "almost never") and "7" meant fully agree (or "I am always online"). The socio-demographic questions regarded gender, year of birth, country, and education.

2.3   Data Analysis

First, we examine differences in adoption, frequency, importance of social interaction and approval seeking, and privacy, between the three countries in general as well as between different generations and genders (for each country separately).

Subsequently, in order to investigate whether the gender- and generation-dependent differences are indeed significant, we calculated two-sided t-tests. For this purpose, we created dummy variables for female users and for each generation (Silver Surfers, Gen X, Gen, Y, and Gen Z). The investigated generations and their respective years of birth were adopted from our previous research on inter-generational comparison of social media use [4, 6, 7]. The generations encompass the following years of birth: 1920–1959 for "Silver

Surfers", 1960–1979 for Gen X (or Digital Immigrants), 1980–1995 for Gen Y (or Digital Natives), and 1996–2010 for Gen Z [4, 6, 45–49]. Only significant outcomes ($p < 0.05$) of the t-test will be elaborated in more detail. Finally, a multivariate analysis of Variance (MANOVA) was conducted to investigate the influence of independent variables (country, gender, generation) on the dependent variables.

3   Results

From all 1,458 participants, 43% were male and 57% were female. As we can see in Table 1, the mostly represented generation group was the Gen Y (63.8%), followed by Gen X (22.6%). Our sample from South Africa was the biggest one (69.1%). Social media mostly applied by the respondents are Facebook (85.8%) and YouTube (80%). Around 45% of the respondents use Instagram, LinkedIn and Google+, the remaining

Table 1. Overall outcomes of the online survey.

General characteristics N = 1,458	
Gender	
Male	43%
Female	57%
Generations	
Silver Surfers	4.2%
Gen X	22.6%
Gen Y	63.8%
Gen Z	9.5%
Country	
Germany	25.5%
Poland	5.3%
South Africa	69.1%
Social media users	
Facebook	85.8%
YouTube	80.0%
Instagram	45.7%
LinkedIn	44.7%
Google+	44.0%
Twitter	36.3%
Pinterest	18.6%
9GAG	14.0%
Xing	6.4%
Tumblr	5.8%
Flickr	2.5%
Foursquare	1.7%
YouNow	0.6%

platforms are applied by less than 40%. The least popular social media platforms, Flickr, Foursquare and YouNow, with usage probability of less than 3%, will not be included in further analysis.

3.1   General Cross-Cultural Differences

Table 2 shows the probability of use (or the percentage of participants using the platform) and the mean frequency of usage (between 1 "seldom" and 7 "always online") of the ten investigated social media platforms in Germany, Poland and South Africa. It appears that the use of Facebook is similar in all three countries (it is the mostly and most frequently applied platform), however, slightly less popular in South Africa (84% adoption probability as compared to 90% in Germany and 95% in Poland). Instagram is most popular in South Africa, but if applied, then most frequently in Poland (5.64). YouTube is mostly applied in Germany (86%) and most frequently in Poland (5.31).

Table 2. Cross-cultural differences in social media usage probability (P) and mean usage frequency (F).

F(P)	Germany	Poland	South Africa
	N = 372	N = 78	N = 1008
Facebook	5.69 (90%)	5.95 (95%)	5.31 (84%)
Instagram	4.93 (41%)	5.64 (46%)	4.88 (48%)
YouTube	4.87 (86%)	5.31 (83%)	4.41 (78%)
Tumblr	4.77 (8%)	4.13 (21%)	3.67 (4%)
9GAG	4.47 (29%)	4.00 (3%)	3.71 (9%)
Twitter	3.82 (35%)	3.00 (9%)	3.61 (39%)
LinkedIn	3.52 (13%)	2.20 (6%)	3.51 (59%)
Pinterest	3.48 (13%)	3.00 (4%)	3.44 (22%)
Xing	3.23 (25%)	–	3.00 (0.3%)
Google+	2.64 (21%)	4.00 (51%)	4.10 (52%)

The platform Tumblr shows the biggest divergence between the countries, as for Germany and South Africa the adoption probability lies under 10%, whereas in Poland it is 21%. However, the continuous usage is most frequent in Germany (4.77). The platform 9GAG shows the next biggest divergence. It is mostly and most frequently applied in Germany (29%), as compared to under 10% in Poland and South Africa. Twitter is least popular in Poland (9%), but very popular in Germany in South Africa (over 30%). The business platform LinkedIn is most popular in South Africa (59%), less popular in Germany (13%) and even less in Poland (6%). A similar difference is given for Pinterest, most popular in South Africa (22%), followed by almost half as many users in Germany (13%), and even less in Poland (4%). The business network Xing is only popular in its origin country Germany (25%). Google+ is most popular in South Africa and Poland (slightly over 50%), and half as much popular in Germany (21%).

Table 3 includes the mean importance values for the factor "social interaction" while using the investigated platforms. The social interaction was most important on Instagram (especially in Poland, 3.778) and LinkedIn (especially in South Africa, 3.72, and Germany, 3.574). As for German users, this factor is most important on LinkedIn and Xing (mean over 3.0), which are both business networks, and least important on YouTube (1.356) and 9GAG (1.455), but also Google+ (1.608). As for the Polish sample, social interaction is most important on Instagram, followed by Facebook (3.365) and Tumblr (3.188), and not at all important for Pinterest (1) and 9GAG (1). Regarding the participants from South Africa, social interaction is most important on LinkedIn (3.72), Xing (3.33; again, two business networks) and Instagram (3.26). It is least important for Pinterest (1.913), YouTube (1.81) and 9GAG (1.85). These three platforms also got the lowest values in the context of social interaction in all three countries.

Table 3. Cross-cultural differences in average importance of social interaction while using social media (scale 1–7, where 1 indicates the lowest and 7 the highest importance level).

Social interaction	Germany	Poland	South Africa
Facebook	2.054	3.365	2.802
Twitter	2.192	2.714	2.594
Instagram	2.947	3.778	3.263
LinkedIn	3.574	2.800	3.721
Xing	3.451	–	3.333
Google+	1.608	2.050	2.683
Pinterest	1.900	1.000	1.913
YouTube	1.356	2.000	1.810
Tumblr	2.767	3.188	2.103
9GAG	1.455	1.000	1.851

Table 4 shows the mean importance values for the factor "approval seeking" while using the investigated platforms. Here, almost all values are under the mean of 4, except for LinkedIn usage in South Africa (4.235). For all three countries, the highest values of approval seeking are given for the networks LinkedIn, Instagram and Xing (except for Poland, where this platform is not broadly adopted). As for the German users, approval seeking is most applied on Instagram, LinkedIn, Tumblr and Xing. The least approval seeking values are given for Google+ and 9GAG. Regarding the participants from Poland, the highest values of approval seeking are given for Instagram, Tumblr, Facebook, and LinkedIn, whereas the lowest ones for Pinterest and Google+. For the South African participants, the highest approval seeking values are given for LinkedIn, Instagram and Xing, and the lowest one for YouTube.

The mean importance values of privacy are shown in Table 5. They are overall much higher than the values for social interaction or approval seeking. Especially for German users, where all the values are above 5. The highest ones are given for Tumblr, Facebook and Xing, whereas the lowest one (5.044) is given for YouTube. As for

Table 4. Cross-cultural differences in average importance of approval seeking while using social media.

Approval seeking	Germany	Poland	South Africa
Facebook	2.275	3.135	2.855
Twitter	2.177	2.286	2.398
Instagram	3.322	3.883	3.361
LinkedIn	3.447	3.000	4.235
Xing	3.121	–	3.333
Google+	1.633	1.900	2.310
Pinterest	2.020	1.000	2.320
YouTube	1.433	1.848	1.773
Tumblr	3.433	3.438	2.282
9GAG	1.450	2.000	2.011

Poland, there are bigger divergences between the privacy needs on different platforms. The highest values are given for 9GAG and Pinterest (7), and the lowest one for YouTube (3.923). As for South African users, the values oscillate between 4.333 (Xing) and 5.664 (Facebook). Other high values are given for Pinterest and Instagram, whereas lower ones for YouTube and Tumblr.

Table 5. Cross-cultural differences in average importance of privacy while using social media.

Privacy	Germany	Poland	South Africa
Facebook	5.585	4.541	5.664
Twitter	5.138	4.000	5.176
Instagram	5.210	4.472	5.541
LinkedIn	5.191	4.000	5.396
Xing	5.593	–	4.333
Google+	5.494	3.775	5.275
Pinterest	5.260	7.000	5.601
YouTube	5.044	3.923	4.973
Tumblr	5.767	4.000	4.667
9GAG	5.241	7.000	5.362

3.2 Generation-Dependent Differences

Another factor possibly influencing the usage of social media is the age of the user. The following tables show the usage probability and the mean use frequency of social media by different generations, separately for each country. In Table 6 presented are the outcomes for German users. The most popular platforms among all generations are Facebook and YouTube. Facebook is most popular among Gen X and Gen Y (above 90%), whereas the usage probability of YouTube is very similar for all four generational groups (80%–87%). Twitter is also equally popular among all generations (between 30% and 39%). When considering Instagram, there is a noticeable decrease in popularity with the age of the users, starting from 59% for the youngest generation (Gen Z), through 35%–40% for Gen X and Gen Y, to only 8% for the Silver Surfers. The business network Xing and LinkedIn are most popular among the Gen X. Google+ is more popular among older generations (Silver Surfers and Gen X; 33%–32%) rather than the younger ones (19%–24%). As for the mean usage frequency, the most frequently used services are Facebook and YouTube. Despite Facebook, Silver Surfers apply Instagram, Twitter and Google+ most frequently. Gen Y applies 9GAG, Twitter and YouTube more frequently than other platforms. Most frequently used services by Gen Y are Instagram, followed by YouTube and 9GAG. Finally, Gen Z uses YouTube (5.29) and Tumblr (5.17) even more frequently than Facebook (4.97).

Table 6. Cross-generational differences in social media usage probability (P) and mean usage frequency (F) in Germany.

Germany F(P)	Silver	Gen X	Gen Y	Gen Z
Facebook	5.66 (75%)	5.79 (92%)	5.71 (95%)	4.97 (67%)
Twitter	4.50 (33%)	4.36 (30%)	3.58 (35%)	4.55 (39%)
Instagram	6.00 (8%)	3.61 (35%)	5.13 (40%)	4.77 (59%)
LinkedIn	2.67 (25%)	3.93 (38%)	3.40 (11%)	–
Xing	2.33 (25%)	3.47 (57%)	3.26 (24%)	1.00 (4%)
Google+	4.50 (33%)	3.23 (32%)	2.30 (19%)	2.92 (24%)
Pinterest	–	3.60 (14%)	3.40 (15%)	4.00 (9.8%)
YouTube	4.20 (83%)	4.25 (86%)	4.90 (87%)	5.29 (80%)
Tumblr	–	1.50 (5%)	4.95 (8%)	5.17 (12%)
9GAG	1.00 (8%)	5.00 (3%)	4.52 (34%)	4.28 (25%)

Table 7 presents the social media usage probability and mean frequency of Polish participants. The most popular platform is Facebook, applied mostly by Gen Z (98%). It is followed by YouTube, which is applied mostly by Gen Y (100%). As for the oldest generation, they only apply Facebook, Google+ and YouTube, however, quite regularly (6.0). Gen X prefers, despite Facebook, YouTube and Google+, also Instagram and LinkedIn (22%). They use Facebook and Instagram most frequently (6.0). Gen Y is most probable to use YouTube and Facebook, but also Pinterest and LinkedIn (25%). The most frequently applied networks are Facebook (6.0) and YouTube (5.13). Finally, Gen Z applies all social networks except for Xing. The most popular ones are Facebook (98%), YouTube (81%), Instagram (56%), and Google+ (56%). The most frequently used services are Facebook, Instagram and YouTube, whereas least frequently used are LinkedIn and Twitter.

In Table 8 presented are the social media usage outcomes of South African participants. Here, the oldest generation, Silver Surfers, prefers Facebook (45%) and YouTube (43%), followed by LinkedIn (38%) and Google+ (30%). The remaining platforms are applied by under 10% of the Silver Surfers. Interestingly, they apply Xing most frequently, followed by Instagram and Facebook. Gen X has similar preferences

Table 7. Cross-generational differences in social media usage probability (P) and mean usage frequency (F) in Poland.

Poland F(P)	Silver	Gen X	Gen Y	Gen Z
Facebook	6.00 (50%)	6.00 (89%)	6.00 (88%)	5.93 (98%)
Twitter	–	5.00 (11%)	–	2.67 (10%)
Instagram	–	6.00 (22%)	4.00 (13%)	5.67 (56%)
LinkedIn	–	2.50 (22%)	2.00 (25%)	2.00 (2%)
Xing	–	–	–	–
Google+	6.00 (50%)	4.80 (56%)	4.00 (13%)	3.83 (56%)
Pinterest	–	0.00 (0%)	3.00 (25%)	3.00 (17%)
YouTube	6.00 (50%)	4.33 (89%)	5.13 (100%)	5.51 (81%)
Tumblr	0%	2.00 (11%)	–	4.27 (25%)
9GAG	–	5.00 (11%)	–	3.00 (2%)

for SNS adoption, whit general higher probability of Facebook, LinkedIn and YouTube adoption. For them, the least popular platforms (under 5%) are Xing, Tumblr and 9GAG. However, they use Facebook and Google+ most frequently. Gen Y applies most of the investigated services, except Xing (0.3%) and Tumblr (4%). They apply Facebook, Instagram and YouTube most frequently. Finally, Gen Z is most probable to apply Facebook (96%), followed by Instagram (93%), and YouTube (86%). The least applied networks are Xing (none) and 9GAG (7%). Similar to Gen Y, they use Facebook, Instagram and YouTube most frequently.

Table 8. Cross-generational differences in social media usage probability (P) and mean usage frequency (F) in South Africa.

South Africa F(P)	Silver	Gen X	Gen Y	Gen Z
Facebook	4.24 (45%)	5.05 (77%)	5.45 (88%)	5.15 (96%)
Twitter	4.00 (9%)	3.62 (31%)	3.59 (44%)	3.90 (36%)
Instagram	4.75 (9%)	4.00 (26%)	5.04 (58%)	5.19 (93%)
LinkedIn	2.72 (38%)	3.33 (58%)	3.65 (63%)	2.22 (32%)
Xing	6.00 (2%)	–	1.50 (0.3%)	–
Google+	3.82 (30%)	4.52 (55%)	3.94 (52%)	3.78 (64%)
Pinterest	1.67 (6%)	3.67 (20%)	3.41 (23%)	2.83 (21%)
YouTube	3.91 (43%)	4.05 (72%)	4.56 (82%)	4.72 (86%)
Tumblr	3.00 (2%)	4.17 (2%)	3.55 (4%)	4.00 (11%)
9GAG	4.00 (2%)	4.13 (3%)	3.65 (13%)	4.00 (7%)

T-test Results for Silver Surfers. In the following only significant outcomes of the t-test between the different generations in U&G and privacy factors will be elaborated. Regarding the differences between Silver Surfers and other generations, as for the German sample they care less about social influence on Instagram (−0.9), approval (−0.97), or even privacy (−2.03) on Instagram. They are also slightly less interested in social interaction (−0.27), approval (−0.28), or privacy (−0.73) on Pinterest. The mean differences for Tumblr are also negative, for social interaction (−0.23), approval seeking (−0.29), and privacy (−0.48). Finally, they care less about social interaction (−0.36), approval (−0.35), and privacy (−1.49) on 9GAG.

As for Polish Silver Surfers, they are less interested in social interaction (−0.25), approval (−0.21), or privacy (−0.37) on Twitter as well as on Instagram (−1.79, −1.82 and −2.12 respectively). As for the business network LinkedIn, they care less about social interaction (−0.18), approval (−0.2), or privacy (−0.26). Finally, the results indicate significant negative differences in the importance of social interaction (−0.67), approval seeking (−0.72), or privacy (−0.84) on Tumblr.

Regarding our South African sample almost all differences were significant on at least 5%-level. Most of the differences were negative, like for the other two countries. As for Facebook, there are negative differences for social interaction (−1.14), approval seeking (−1.32), and even privacy (−2.55). The outcomes for Twitter are also negative: social interaction (−0.75), approval seeking (−0.67), and privacy (−1.64). As for Instagram, there is less interest in social interaction (−1.36), approval (−1.39), and privacy (−2.25). Furthermore, the Silver Surfers have less interest in social interaction (−1.1), approval (−1.1), or privacy (−1.35) on LinkedIn. Interestingly, there is slightly more interest in social interaction (+0.08), approval seeking (+0.08), and privacy (+0.14) on another business network, Xing. This is the only platform with positive mean differences. As for Google+, there is slightly less interest in social interaction (−0.7), approval (−0.63), or privacy (−1.09). For Pinterest, the mean differences are also very small: social interaction with −0.34, approval seeking with 0.37, and privacy with −0.96. The mean differences for YouTube usage by Silver Surfers are also negative, social interaction with −0.7, approval seeking with −0.73, and privacy with total −2.17. There is one significant, however, small difference in usage of Tumblr − the importance of privacy (−0.14). Finally, Silver Surfers care less about social interaction (−0.16) or privacy (−0.44) on 9GAG.

T-test Results for Gen X. When compared to Silver Surfers, there were less significant outcomes for mean differences between social media usage by Gen X and other generations. As for Germany, there were few positive differences, like the importance of privacy on Facebook (+0.87), LinkedIn (+2) and Xing (+2.65). Furthermore, this generation cares more about social capital on LinkedIn (+1.24) and Xing (+1.25) as well as approval (+1.11 and +1.07 respectively). Also, Gen X'ers appear to be more interested in privacy on Google+ (+1.08). However, they care less about social interaction (−0.39), approval (−0.41), or privacy (−1.66) on 9GAG. Finally, they care less about social interaction (−0.19) and approval (−0.25) on Tumblr. As for Polish Gen X participants, they are less interested in social capital (−1.47) and approval (−1.37) on Instagram. The same holds for Tumblr (−0.61 and −0.67, respectively). The remaining differences in mean values were not statistically significant. Regarding the Gen X participants from South Africa, they are less interested in social interaction (−0.56) or approval (−0.56) on Facebook. They are also less interested in social capital (−0.5), approval (−0.4), or privacy (−0.56) on Twitter, as well as social capital (−1.37), approval (−1.39) and privacy (−1.57) on Instagram. Furthermore, the approval on LinkedIn is less important (−0.37). As for YouTube, they are less interested in social capital (−0.34) and approval (−0.33). Finally, there are less interested in social capital (−0.18) and approval (−0.2) on 9GAG as well as on Tumblr (−0.07 and −0.08 respectively).

T-test Results for Gen Y. There are only few significant differences between Gen Y and other generations for German and Polish sample. As for German Gen Y participants, they care more about privacy on Facebook (+0.79), however, less about privacy on LinkedIn (−0.47). They are also less interested in social interaction (−0.27), approval (−0.24), and privacy (−0.57) on Google+. In turn, they care slightly more about social interaction (+0.25), approval (+0.24), and privacy (+1.06) on 9GAG. As for Polish Gen Y users, they care less about approval on Facebook (−1.5) and social interaction (−0.27), approval (−0.23) and privacy (−0.4) on Twitter. Also, they are less interested in social interaction (−1.8), approval (−1.83) and privacy (−2.02) on Instagram. As for the business network LinkedIn, they care more about social capital (+0.91), approval (+0.9), and privacy (+1.25). In turn, they are less interested in social interaction (−1.03), approval (−0.95), and privacy (−2.02) on Google+ and on Tumblr (−0.73, −0.79 and −0.91 respectively). Finally, they are more interested in social interaction (+0.24), approval (+0.24) and privacy (+1.65) on Pinterest. As for the South African Gen Y participants, there are many significant positive differences. As for Facebook, they are more interested in social interaction (+0.54), approval (+0.59) and privacy (+0.65). The same holds for Twitter (+0.54, +0.45, and +0.83 respectively) and Instagram (+1.19, +1.23, and +1.53 respectively). This Gen Y is also more interested in social capital (+0.56) and approval (+0.66) on LinkedIn. They are also slightly more interested in social interaction on Pinterest (+0.14) than other generations. Finally, they care more about social interaction (+0.41), approval (+0.38), and privacy (+0.41) on YouTube. The same holds for 9GAG (+0.2, +0.19, and +0.52 respectively).

T-test Results for Gen Z. Finally, the significant t-test results for the generation Z are elaborated. As for Germany, Gen Z is less interested in social interaction (−0.85), approval (−0.7), or privacy (−1.72) on Facebook. However, they care more about these aspects on Instagram (+0.67, +0.65, +1.05 respectively). As for the business network LinkedIn, they care less about social interaction (−0.52), approval (−0.5) or privacy (−0.76) than other generations. The same holds for Xing (−0.93, −0.84 and −1.3 respectively). As for the Gen Z from Poland, they are way more interested in approval on Facebook (+1.71) as well as social capital (+2), approval (+1.92), and privacy (+1.96) on Instagram. Finally, they are less interested in social interaction (−0.67), approval (−0.65), and privacy (−0.91) on LinkedIn. Finally, the Gen Z users from South Africa are more interested in social capital (+1.45) and approval (+1.37) on Facebook as well as social capital (+2.41), approval (+2.21) and privacy (+2.47) on Instagram. In turn, they care less about social capital (−1.54) and approval (−1.05) on LinkedIn.

3.3   Gender-Dependent Differences

The following analysis concerns the differences in social media usage between male and female users for each country separately (Tables 9 and 10). As for German users, women are more likely than men to apply Facebook, Instagram, Google+, Pinterest and

Tumblr. The mean usage frequencies are partially comparable. Men use Facebook and YouTube most frequently, whereas Google+, Pinterest and Xing least frequently. As for women, they use Facebook, Tumblr and Instagram most frequently, whereas, Google+, Xing and LinkedIn least frequently.

Table 9. Cross-cultural differences in social media usage probability (P) and mean usage frequency (F) of male users.

Men F(P)	Germany	Poland	South Africa
Facebook	5.74 (87%)	5.71 (91%)	5.09 (80%)
Twitter	3.50 (42%)	2.00 (4%)	3.85 (43%)
Instagram	4.67 (31%)	5.13 (35%)	4.79 (44%)
LinkedIn	3.45 (16%)	3.00 (4%)	3.59 (68%)
Xing	3.18 (27%)	–	4.00 (0.4%)
Google+	2.46 (19%)	3.90 (43%)	3.85 (50%)
Pinterest	2.86 (6%)	–	3.26 (12%)
YouTube	5.35 (92%)	5.29 (87%)	4.69 (85%)
Tumblr	3.29 (6%)	4.00 (4%)	4.00 (5%)
9GAG	4.33 (40%)	–	3.84 (13%)

As for the Polish participants, both gender are most likely to use Facebook, YouTube and Google+. Women also prefer Instagram (51%). Regarding the mean usage frequencies, Polish men use Facebook, Instagram and YouTube most frequently, whereas Twitter and LinkedIn least frequently. As for women, they apply Facebook, Instagram, and YouTube most frequently, whereas LinkedIn and Pinterest least frequently.

Finally, as for participants from South Africa, both, men and women, are most likely to apply Facebook and YouTube. The male participants apply Facebook, Instagram and YouTube most frequently, whereas Pinterest and LinkedIn least frequently. As for the female participants from South Africa, they apply Facebook most frequently, followed by Instagram and Google+, whereas Xing and Tumblr least frequently.

Like for the different generations, we conducted a two-sided t-test for the U&G and privacy values and elaborate the significant differences between male and female users. Regarding German participants, female users are slightly less concerned about privacy on Facebook (−0.95), however, care more about approval (+0.52) and privacy (+0.93) on Instagram than men. Furthermore, female German users care more about approval (+0.21) and privacy (+0.7) on Pinterest, however, less about social interaction (−0.35) or approval (−0.41) on YouTube. As for Tumblr, women care slightly more about social interaction (+0.25) and approval (+0.28) than men. Finally, female users care less about social interaction (−0.23), approval (−0.27), or privacy (−0.69) on 9GAG.

As for Polish participants, the only significant difference between male and female users is given for the platform Tumblr. Apparently, women care a little bit more about social interaction (+0.87), approval (+0.94), and privacy (+1.04).

Table 10. Cross-cultural differences in social media usage probability (P) and mean usage frequency (F) of female users.

Women F(P)	Germany	Poland	South Africa
Facebook	5.66 (92%)	6.04 (96%)	5.49 (87%)
Twitter	4.04 (31%)	3.17(11%)	3.34 (35%)
Instagram	5.03 (46%)	5.79(51%)	4.96 (51%)
LinkedIn	3.57 (11%)	2.00 (7%)	3.42 (51%)
Xing	3.26 (23%)	–	1.00 (0.2%)
Google+	2.71 (23%)	4.03 (55%)	4.31 (54%)
Pinterest	3.58 (17%)	3.00 (6%)	3.51(30%)
YouTube	4.62 (83%)	5.33 (82%)	4.09 (71%)
Tumblr	5.22 (9%)	4.13 (27%)	3.13 (3%)
9GAG	4.59 (24%)	4.00 (4%)	3.46 (6%)

Finally, regarding the participants from South Africa, women appear to care more about approval (+0.36) and privacy (+078) on Facebook as well as privacy on Instagram (+0.71). In turn, they are less interested in social interaction (−0.29) and approval (−0.31) on Twitter, as well as social interaction (−0.8), approval (−0.67) and privacy (−0.6) on LinkedIn. Furthermore, women care slightly more about approval (+0.22) and privacy (+0.44) on Google+ as well as social interaction (+0.36), approval (+0.35) and privacy (+1.08) on Pinterest. In turn, they are less interested in social interaction (−0.4) and approval (−0.4) on YouTube, as well as social interaction (−0.1) or privacy (−0.44) on 9GAG, than men.

3.4 Multivariate Analysis of Variance Between Countries, Genders and Generations

First, a three-way MANOVA test was conducted to compare differences in social media usage frequency as well as the two uses and gratification scales and importance of privacy, dependent on country of origin, generation and gender. There was no statistically significant three-way interaction between country, generation and gender regarding usage frequency. However, there was a statistically significant country* generation interaction for Facebook, Instagram, Xing and 9GAG.

Regarding the "social interaction" factor, there was statistically significant three-way interaction between country, generation and gender $F(4, 1436) = 3.751$, $p = .005$ for Facebook. Furthermore, there was a statistically significant country* generation interaction for Facebook, Instagram, LinkedIn and Xing, as well as generation*gender interaction for Facebook.

Concerning the "approval seeking" factor, there was no statistically significant three-way interaction between country, generation and gender. There was a statistically significant country*generation interaction for Facebook, Instagram, LinkedIn and Xing.

Finally, regarding the importance of privacy, there was no statistically significant three-way interaction between country, generation and gender. There was a statistically significant generation*country interaction for Facebook, Instagram, LinkedIn, Xing, and 9GAG.

4 Discussion and Conclusion

The results offer broad insights into the research field of geographical and cultural differences in social media interaction. Major implications can be drawn by our findings. First, when considering solely geographical issues, we find striking differences for Poland with regard of consuming videos via social media, i.e. YouTube, and mobile content, i.e. Instagram. Polish social media users are more receptive to those features than those from Germany and South Africa. Interestingly, we can rule out that spatial proximity in general can serve as an explanatory approach for the observed differences, since Germany and Poland are sharing borders, whereas South Africa is located on a different continent. We rather argue that a bundle of socio-economic factors, such as the political orientation, demography and the availability of social media products serve to explain cultural differences in social media use.

Furthermore, when considering age structures of the users, certain patterns in the use of social media are cross-culturally consistent. Younger generations tend to discover and occupy new media forms, such as Instagram, and simultaneously exhibit an increasing tendency to move away from established platforms, such as Facebook and Twitter. This finding serves moreover as evidence that generational shifts towards more mobile oriented social media interaction are taking place.

This study also shows striking differences in the perception of privacy concerns with regard to gender. In South Africa, females tend to care substantially more about privacy issues on social media platforms, similarly to Polish female users, when compared to their German counterparts. Again, our results show that pure partial proximity issues cannot explain those geographical differences. Possible reasons are rather bound by historical and socio-economic issues, particularly with regard to the status quo of gender equality.

Overall, our study emphasizes that cross-cultural differences in social media cannot be explained by generally-valid patterns. Studies on social media have to be scrutinized with regard to a bundle of specific socio-economic factors. Thus, social media can also be seen as a reflection of the current state of a certain society. This implicates that future studies have to refrain from generalizing empirical findings in social media research for geographical contexts.

References

1. Kilian, T., Hennigs, N., Langner, S.: Do millennials read books or blogs? Introducing a media usage typology of the internet generation. J. Consum. Mark. **29**, 114–124 (2012)
2. Shuen, A.: Web 2.0: A Strategy Guide. O'Reilly, Sebastopol (2008)
3. Dhir, A., Pallesen, S., Torsheim, T., Andreassen, C.S.: Do age and gender differences exist in selfie-related behaviours? Comput. Hum. Behav. **63**, 549–555 (2016)

4. Fietkiewicz, K.J., Lins, E., Baran, K.S., Stock, W.G.: Inter-generational comparison of social media use: investigating the online behavior of different generational cohorts. In: 49th Hawaii International Conference on System Sciences 2016, HICSS, pp. 3829–3838. IEEE, Washington, D.C. (2016)

5. Fietkiewicz, K.J., Hoffmann, C., Lins, E.: Find the perfect match: the interplay among Facebook, YouTube and LinkedIn on crowdfunding success. Int. J. Entrep. Small Bus. **33**(4), 472–493 (2018)

6. Fietkiewicz, K.J.: Jumping the digital divide: how do "silver surfers" and "digital immigrants" use social media? Netw. Knowl. **10**, 5–26 (2017)

7. Fietkiewicz, K.J., Lins, E., Baran, K.S., Stock, W.G.: Other times, other manners: how do different generations use social media? In: Arts, Humanities, Social Science & Education Conference 2016, AHSE, pp. 1–17. Hawaii University, Honolulu (2016)

8. Dhir, A., Torsheim, T.: Age and gender differences in photo tagging gratifications. Comput. Hum. Behav. **63**, 630–638 (2016)

9. Caers, R., De Feyter, T., De Couck, M., Stough, T., Vigna, C., Du Bois, C.: Facebook: a literature review. New Media Soc. **15**, 982–1002 (2013)

10. Dhir, A., Kaur, P., Chen, S., Lonka, K.: Understanding online regret experience in Facebook use – effects of brand participation, accessibility and problematic use. Comput. Hum. Behav. **59**, 420–430 (2016)

11. Joinson, A.N.: Looking at, looking up or keeping up with people? In: 26th Annual Chi Conference On Human Factors in Computing Systems 2008, pp. 1027–1036. ACM Press, New York (2008)

12. Dhir, A., Tsai, C.: Telematics and informatics understanding the relationship between intensity and gratifications of Facebook use among adolescents and young adults. Telemat. Inf. **34**, 350–364 (2017)

13. Ellison, N.B., Steinfield, C., Lampe, C.: The benefits of facebook "friends:" social capital and college students' use of online social network sites. J. Comput. Commun. **12**, 1143–1168 (2007)

14. Steinfield, C., Ellison, N.B., Lampe, C.: Social capital, self-esteem, and use of online social network sites: A longitudinal analysis. J. Appl. Dev. Psychol. **29**, 434–445 (2008)

15. Valenzuela, S., Park, N., Kee, K.F.: Is there social capital in a social network site? Facebook use and college students' life satisfaction, trust, and participation. J. Comput. Commun. **14**, 875–901 (2009)

16. Phua, J., Jin, S.-A.A.: "Finding a home away from home": the use of social networking sites by Asia-Pacific students in the United States for bridging and bonding social capital. Asian J. Commun. **21**, 504–519 (2011)

17. Lampe, C., Wohn, D.Y., Vitak, J., Ellison, N.B., Wash, R.: Student use of Facebook for organizing collaborative classroom activities. Int. J. Comput. Collab. Learn. **6**, 329–347 (2011)

18. Sheldon, P.: The relationship between unwillingness-to-communicate and students' Facebook use. J. Media Psychol. **20**, 67–75 (2008)

19. Kwon, M.-W., D'Angelo, J., McLeod, D.M.: Facebook use and social capital. Bull. Sci. Technol. Soc. **33**(1–2), 35–43 (2013)

20. Alhabash, S., Chiang, Y., Huang, K.: MAM & U&G in Taiwan: differences in the uses and gratifications of Facebook as a function of motivational reactivity. Comput. Hum. Behav. **35**, 423–430 (2014)

21. Alhabash, S., McAlister, A.R.: Redefining virality in less broad strokes: predicting viral behavioral intentions from motivations and uses of Facebook and Twitter. New Media Soc. **17**, 1317–1339 (2015)

22. Park, N., Lee, S.: College students' motivations for Facebook use and psychological outcomes. J. Broadcast. Electron. Media. **58**, 601–620 (2014)
23. Ross, C., Orr, E.S., Sisic, M., Arseneault, J.M., Simmering, M.G., Orr, R.R.: Personality and motivations associated with Facebook use. Comput. Hum. Behav. **25**, 578–586 (2009)
24. Kalpidou, M., Costin, D., Morris, J.: The relationship between Facebook and the well-being of undergraduate college students. Cyberpsychol. Behav. Soc. Netw. **14**, 183–189 (2011)
25. Labrague, L.J.: Facebook use and adolescents' emotional states of depression, anxiety, and stress. Heal. Sci. J. **8**, 80–89 (2014)
26. Pfeil, U., Arjan, R., Zaphiris, P.: Age differences in online social networking – A study of user profiles and the social capital divide among teenagers and older users in MySpace. Comput. Hum. Behav. **25**, 643–654 (2009)
27. Muscanell, N.L., Guadagno, R.E.: Make new friends or keep the old: gender and personality differences in social networking use. Comput. Hum. Behav. **28**, 107–112 (2012)
28. Sheldon, P.: Student favorite: Facebook and motives for its uses. Southwest. Mass Commun. J. **23**, 39–53 (2008)
29. Yuan, Y.: A survey study on uses and gratification of social networking sites in China (2011). https://etd.ohiolink.edu
30. Taraszow, T., Aristodemou, E., Shitta, G., Laouris, Y., Arsoy, A.: Disclosure of personal and contact information by young people in social networking sites: an analysis using Facebook profiles as an example. Int. J. Media Cult. Polit. **6**, 81–101 (2010)
31. Prensky, M.: Digital natives, digital immigrants – part 1. MCB Univ. Press **9**(5), 1–6 (2001)
32. Katz, E.: Mass communications research and the study of popular culture: an editorial note in a possible future for this journal. Stud. Public Commun. **2**, 1–6 (1959)
33. Katz, E., Blumer, J.G., Gurevitch, M.: Utilization of mass communication by the individual. In: Blumer, J.G., Katz, E. (eds.) The Uses of Mass Communications: Current Perspectives on Gratifications Research, pp. 19–32. Sage Publications, Beverly Hills (1974)
34. Whiting, A., Williams, D., Whiting, A., Williams, D.: Why people use social media: a uses and gratifications approach. Qual. Mark. Res. Int. J. **16**(4), 362–369 (2013)
35. Leung, L.: Generational differences in content generation in social media: the roles of the gratifications sought and of narcissism. Comput. Hum. Behav. **29**, 997–1006 (2013)
36. Papacharissi, Z., Rubin, A.M.: Predictors of internet use. J. Broadcast. Electron. Media. **44**, 175–196 (2000)
37. Phua, J., Jin, S.V., Kim, J.: Gratifications of using Facebook, Twitter, Instagram, or Snapchat to follow brands: the moderating effect of social comparison, trust, tie strength, and network homophily on brand identification, brand engagement, brand commitment, and membership intention. Telemat. Inf. **34**, 412–424 (2017)
38. Magsamen-Conrad, K., Dowd, J., Abuljadail, M., Alsulaiman, S., Shareefi, A.: Life-span differences in the uses and gratifications of tablets: implications for older adults. Comput. Hum. Behav. **52**, 96–106 (2015)
39. Wildemuth, B.M.: The illusion of online privacy (2006). https://ils.unc.edu/~wildem/Publications/CHI2006-Privacy.pdf
40. Acquisti, A., Gross, R.: Imagined communities: awareness, information sharing, and privacy on the Facebook. In: Danezis, G., Golle, P. (eds.) PET 2006. LNCS, vol. 4258, pp. 36–58. Springer, Heidelberg (2006). https://doi.org/10.1007/11957454_3
41. Fietkiewicz, K.J., Lins, E.: New media and new territories for European law: competition in the market for social networking services. In: Knautz, K., Baran, K.S. (eds.) Facets of Facebook: Use and Users, pp. 285–324. De Gruyter, Berlin/Boston (2016)
42. Wright, K.B.: Researching internet-based populations: advantages and disadvantages of online survey research, online questionnaire authoring software packages, and web survey services. J. Comput. Commun. **10**, 00 (2006)

43. Evans, J.R., Mathur, A.: The value of online surveys. Internet Res. **15**, 195–219 (2005)
44. Marshall, M.N.: Sampling for qualitative research. Fam. Pract. **13**, 522–525 (1996)
45. Bolton, R.N., et al.: Understanding Generation Y and their use of social media: a review and research agenda. J. Serv. Manag. **24**, 245–267 (2013)
46. Brosdahl, D.J.C., Carpenter, J.M.: Shopping orientations of US males: a generational cohort comparison. J. Retail. Consum. Serv. **18**, 548–554 (2011)
47. Freestone, O., Mitchell, V.: Generation Y attitudes towards e-ethics and internet-related misbehaviours. J. Bus. Ethics **54**, 121–128 (2004)
48. McIntosh-Elkins, J., McRitchie, K., Scoones, M.: From the silent generation to Generation X, Y and Z: strategies for managing the generation mix. In: Proceedings of 35th Annual ACM SIGUCCS Fall Conference, pp. 240–246 (2007)
49. Tapscott, D.: Grown Up Digital: How the Net Generation is Changing Your World. McGraw-Hill, New York (2009)

Interactions of Twitch Users and Their Usage Behavior

Daniel Gros[✉], Anna Hackenholt, Piotr Zawadzki,
and Brigitta Wanner

Department of Information Science, Heinrich Heine University Düsseldorf,
Düsseldorf, Germany
{daniel.gros, anna.hackenholt, piotr.zawadzki,
brigitta.wanner}@hhu.de

Abstract. In recent years social live streaming services (SLSSs) like YouNow
or Periscope are becoming more and more popular. Within the gaming and
E-sports industry, Twitch became one of the biggest live streaming platforms for
video games in late 2014. This paper analyzes the usage behavior and interactions of Twitch users, especially their wish to get involved in a live-stream.
The study is a quantitative analysis based on an online questionnaire, which was
completed by 603 Twitch users. Both user groups, 'viewers' and 'streamers',
participated in the study. Additionally, streamers with different popularity
(follower size) were observed to get an insight of the interactions between
Twitch users.

Keywords: Streaming · Social live streaming services · Twitch
Media usage · Usage behavior · User interaction

1 Introduction

In recent years social live streaming services (SLSSs) like YouNow or Periscope are
becoming more and more popular. Being able to broadcast your own content in
real-time, interacting with your audience and the possibility of a gratification system
are the main characteristics of an SLSS [1]. SLSSs are represented for different purposes, but particularly the streaming of video game content enjoys an enormous
viewership [2].

In the gaming and E-sports industry, Twitch became one of the biggest live
streaming platforms for video games in late 2014. Apparently, a substantial number of
gamers prefer to watch video games on a live stream than playing the game themselves
[3]. On Twitch, viewers can communicate with the streamer or other viewers via chat,
while streamers are broadcasting a game or starting an "IRL" (In Real Life) stream to
simply talk to the community. By talking with viewers on stream or raffle (mostly
gaming related) prizes, the streamer creates an interactive community.

In 2016, Twitch had collectively 550,000 years-worth of video content, created by
2.2 million unique streamers [4]. Another feature of Twitch are donations and subscriptions, which are mostly used to support the streamer financially. However, some
streamers are collecting money for different charities. More than $25,300,000 were

raised for different charities in 2016 [4]. The popularity of Twitch has risen not only in the United States but also in Germany, as it is ranked on place 29 of the most visited websites in Germany[1].

With growing usage of streaming platforms like Twitch, the motivation of this paper is to look closer at the reasons why and how Twitch is used in Germany and why some of the users spend money on it even though it is free of charge in general. Although Twitch is used so heavily by a huge number of users, research is still sparsely conducted. This paper will explore the motivation for using Twitch regarding the different application types to "watch" a stream.

In the following, related work regarding SLSSs and Twitch are presented (Sect. 2). Afterwards, an overview of the applied methods is given by presenting the developed research questions, the use of a questionnaire and the calculated statistical tests to substantiate the results (Sect. 3). Subsequently, the research questions are answered by analyzing the obtained data (Sect. 4). In conclusion, an evaluation of the findings and limitations are given (Sect. 5).

2   Related Work

Even before the SLSSs, social media services allowed users to both consume and produce content, creating a new kind of user. The so-called "prosumers" [5] form virtual communities, which usually pursue a common goal [6]. Unlike social media services such as Facebook, SLSSs are not asynchronous, but take place in real time [1]. SLSSs come in many different forms, but especially in the video game sector, thousands of spectators are cast in the spell [2]. There exist various subcategories in this subject area. Smith, Obrist and Wright [7] define three popular sub-categories: "eSports" (electronic sports), "speed running" and "Let's Plays".

An "ethnographic investigation of the live streaming of video games on Twitch", is conducted by Hamilton et al. [8], finding that "Twitch streams act as virtual third places, in which informal communities emerge, socialize, and participate". These virtual worlds are an example where "social and commercial realizations of an emergent streaming culture" are combined [9]. Even the gameplay of a single player game can be a social activity while streaming on Twitch [11].

Lessel et al. [10] investigated the influence of the audience on the gameplay in a stream by two different case studies and came to the conclusion, that "more influence options [for the audience] are appreciated and considered as important". Another research about the live-streaming community shows the possibility of "predicting the number of chat messages based on the number of spectators" [12]. Also, research about the motivations of Twitch users exists. Sjöblom and Hamari [13] identify "tension release, social integrative and affective motivations" to have an increasing impact on the usage time on Twitch and that "social integrative motivations are the primary predictor of subscription behavior".

---

[1] www.alexa.com/siteinfo/twitch.tv [Retrieved 02-05-2018].

Gros et al. [14] investigated the motivational factors to use Twitch based on the uses and gratifications theory by surveying German Twitch viewers in 2016. While test items regarding the entertainment factor as a motive are valued high, e.g. *"I use Twitch to follow tournaments and events"*, test items regarding socialization aspects become less important to most of the viewers, e.g. *"... to communicate with other viewers through the chat"*. However, the socialization factor becomes more important for viewers who spend more time or money on Twitch, reasoning its usage. This finding leads to the main motivational factor for this study, as one of the main motives *"... to get in touch with a streamer"* seem to be insignificant compared to other motives. How is the desire to get involved in a live-stream and the interaction between the streamer and its audience perceived?

3 Methods

To investigate the perceived involvement (desire) of Twitch users, the following research questions were developed:

RQ1a: How strong is the desire for involvement and the actual involvement of Twitch users in a stream?

RQ1b: How do socio-demographic factors (gender, age, education, profession), the time spent or the money spent on Twitch influence the desire for involvement in a stream?

RQ1c: How does the usage behavior influence the desire for involvement in a stream?

RQ2: How are the interactions between streamer and viewers perceived?

Based on these research questions, a research model was created to investigate the involvement desire of Twitch users on two different dimensions: interaction and usage behavior (Fig. 1).

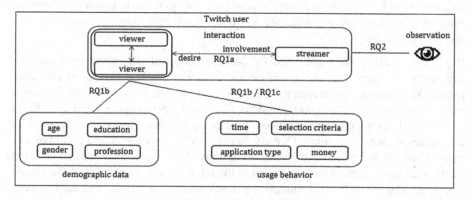

Fig. 1. Research model

Test subjects were surveyed through an online questionnaire, which contained in total 23 major items, in early 2016 (from December 30th, 2015 to February 15th, 2016). The target group were German-speaking Twitch users. This way an unequal distribution of participants from different countries was prevented. The questionnaire was spread via different social media networks, e.g. in Facebook groups, under gaming related hashtags on Twitter or on Reddit. Moreover, Twitch streamers were contacted and asked to spread the link to the questionnaire through the chat. A pretest with ten participants was performed. The time to fill in the questionnaire takes 10 to 15 min. In total, the survey consists of 23 items.

RQ1a. A five-point Likert scale is used to measure the test item regarding the involvement desire, "*It is important to me, that I get involved by the streamer.*". The following answer options were given: "Strongly Disagree" (1), "Disagree" (2), "Undecided" (3), "Agree" (4) and "Strongly Agree" (5), while every option has the same distance on a scale of sentiments. In addition, participants were asked whether and how they got involved by a streamer, to test the actual involvement.

RQ1b. Do the socio-demographic factors (gender, age, education, profession) influence the desire for involvement in a stream? The age of the participants was divided into four different groups: teenagers (≤ 18 years), younger adulthood (19–25 years), middle adulthood (26–35 years) and older adulthood (≥ 36 years). Also, the usage time on Twitch and the willingness to financially support a streamer may be influential factors to the involvement desire. Therefore, the data is analyzed by the participants weekly average usage time on Twitch and whether a donation or subscription was made. Regarding the factor time, the participants were grouped into five groups depending on their average usage time per week: ≤ 2 h, 3–6 h, 7–12 h, 13–20 h and ≥ 21 h.

RQ1c. As a stream can be used in different ways, e.g. a viewer can watch or just listen to a stream, different application types could be possible influential factors. Furthermore, criteria regarding the selection of a stream may be relevant. Under what criteria do Twitch users choose a stream? Participants of the conducted survey had to choose from a list of possible ways to use Twitch and decide whether the game, the streamer (person) or the channel is relevant for the selection to watch a broadcast. The test items were rated on a five-point Likert scale and compared with the rating of the desire for involvement. Moreover, a regression is calculated to show which test items are reasoning the involvement desire.

RQ2. The findings are highlighted by an observation of the interactions between streamer and viewers. The selection of the streamers was based on a randomly generated list of 100 streamers. As the number of followers of the streamers diverge, three groups were classified: small (<3,000), middle (3,000–14,999), large ($\geq$15,000). Eight streamers of each group were randomly selected for this observation, to identify differences among these groups. Overall, 24 streamers were observed in a period of four months in early 2016 in a total broadcasting time of ten hours. Each session lasted at least for an hour. A checklist was prepared to take notes for the concept of the stream (e.g. streaming schedule or advantages for subscriptions), the overlay and the observed interactions between streamer and viewers. Each broadcast was observed by two individuals.

The results of RQ1(a–c) were proven by various statistically applicable tests. As the data is not normally distributed, non-parametric tests have been used: Mann-Whitney U test [15], Kruskal-Wallis test [16], Pearson's chi-squared test (χ^2), Dunn-Bonferroni test [17, 18]. To calculate the strength of a relation, the symmetric mass Phi (φ), Cramer's V (CV) and the contingency coefficient (CC) are used. To measure the effect size of relations between variables in case of a significant difference, Cohen's d (r_d) was calculated and classified as weak ($r_d \geq 0,1$), medium ($r_d \geq 0,3$) or strong ($r_d \geq 0,5$) [19].

4 Results

In total, 791 people were surveyed. The introductory questions intended to filter out participants who are acquainted with and use Twitch. While 695 (89.9%) answer to be familiar with the live-streaming platform, 603 (86.6%) are actively using it, out of which 470 (77.9%) use Twitch only as a 'viewer' and are not broadcasting content. The following results are focusing on Twitch spectators. As 133 (22.1%) participants consider themselves as 'viewer and streamer' (further labeled as 'streamer'), the outcomes are analyzed separately and compared.

4.1 Involvement Desire and Actual Involvement

To answer RQ1a, "How is the involvement desire and the actual involvement of Twitch users perceived?", the results of the corresponding test items "*It is important to me, that I get involved by the streamer.*", "*Did you get involved by a streamer before?*" and "*How did you get involved by a streamer?*" are presented.

On the one hand, 50.4% (n = 67) of the 'streamers' and 37% (n = 174) of the 'viewers' agree with the statement, that it is important to them to get involved by a streamer. On the other hand, about one fifth of the 'streamers' (19.5%, n = 26) and about a third of the 'viewers' (34%, n = 160) disagree with the statement. The average ratings are 3.03 ('viewer') and 3.46 ('streamer').

To identify a significant difference between the two user groups 'viewer' (n = 470) and 'streamer' (n = 133), the middle ranks by means of the Mann-Whitney U test are calculated. 'Viewers' reach a middle rank of 288.67, whereas 'streamer' reach a value of 349.11. The higher value for 'streamer' indicate a stronger involvement desire. This outcome is significant (z = −3.628; p < 0.001). However, the effect size for this difference is classified as weak ($r_d = 0,15$).

Regarding the actual involvement, 64% (n = 301) of the 'viewers' and 89.5% (n = 119) of the 'streamers' stated they got involved in a broadcast by a streamer before, which is a significant difference ($\chi^2 = 32.152$; p < 0.001). It seems 'streamers' are involved more often in a stream – or at least have the feeling of being involved in a stream. However, the relation is low ($\varphi = 0.231$, p < 0.001; CV = 0.231; p < 0.001).

At last, the different types of involvement are presented (Fig. 2).

The most common type of getting involved seems to get mentioned in a stream by the streamer, as 80.4% (n = 82) of the 'streamers' and 52.1% (n = 245) of the 'viewers' answered this option. Furthermore, being able to make a stream-related decision, e.g. choosing a hero, and participating in raffles are popular ways to involve

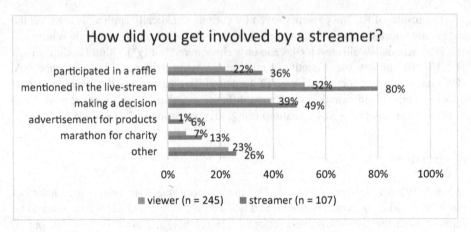

Fig. 2. Types of involvement.

the viewers, too. It is striking, that the group 'streamer' are more often involved by the given answer options. Participants who chose 'other' as an option ('viewer' = 23.2%, n = 57; 'streamer' = 26.1%, n = 28), stated viewer games as another type of involvement.

4.2 Demographic Data

The gathered data is analyzed under different aspects of the demographic data and usage behavior, to answer RQ1b: "How do socio-demographic factors (gender, age, education, profession), the time spent, or the money spent on Twitch influence the desire for involvement in a stream?". The following table gives an overview of the middle rank regarding the involvement desire (Table 1).

Gender. Regarding the gender, female 'viewers' have a stronger involvement desire (middle rank = 242.58) than males (middle rank = 234.78) and those who did not provide any information (middle rank = 154.13). However, the statistics of the Kruskal-Wallis test for the 'viewer' indicate no significant difference regarding the gender ($\chi^2 = 1.769, p = 0.413$), as well as for 'streamers' ($\chi^2 = 0.021, p = 0.883$).

Age. The biggest share of 'viewers' and 'streamers' are young adults between 19 and 25 years. In both cases the middle rank is highest for teenagers. Generally, the middle rank decreases by their age group. In conclusion, teenagers have a stronger involvement desire. Though, the results are not significant for the group of 'streamers' ($p = 0.570$). As there exists a significant difference for 'viewers' ($p < 0.001$), a post hoc analysis (Dunn-Bonferroni test) is conducted to identify which groups differ significantly from teenagers. It reveals that all three groups differentiate from teenagers, while the effect size shows partly a strong relevance: young adulthood ($r_d = 0.27$, $p = 0.002$), middle adulthood ($r_d = 0.53$, $p < 0.001$) and older adulthood ($r_d = 0.64$, $p = 0.049$).

Table 1. Involvement desire on different aspects

	Demographic data	viewer		streamer	
		n = 470	Middle rank	n = 133	Middle rank
Gender	female	85	242.58	17	68.24
	male	381	234.78	116	66.82
	no information	4	154.13	–	–
Age	≤ 18 years	113	277.31	29	74.74
	19–25 years	235	237.25	65	66.78
	26–35 years	105	200.02	36	61.65
	≥ 36 years	17	152.47	3	61.00
Education	still at school	57	284.94	15	90.57
	Hauptschulabschluss (basic school education)	11	286.45	9	63.06
	Realschulabschluss (high school diploma)	66	232.56	25	78.74
	advanced technical college certificate	59	258.63	23	62.61
	European Baccalaureate	175	229.27	41	62.80
	Bachelor's degree	68	214.18	16	49.06
	Master's degree	25	164.54	3	33.00
Profession	unemployed	20	290.50	13	56.88
	pupil	93	265.41	18	89.22
	vocational training	39	272.13	14	80.00
	student	187	227.15	41	61.50
	part-time working	17	214.44	4	55.50
	full-time working	103	203.02	40	62.98
Time	0–2 h	152	210.69	21	55.40
	3–6 h	125	235.64	33	64.61
	7–12 h	81	245.98	30	70.52
	13–20 h	53	259.25	23	61.76
	≥ 21 h	59	263.41	26	79.98
Money	did spend money on Twitch	125	290.85	65	73.25
	did not spend money on Twitch	345	215.44	68	61.02

Education. Most of the participants have an European Baccalaureate ('viewer': 37.2%, n = 175; 'streamer': 30.8%, n = 41) as the highest educational attainment. 'Viewers' who have a high school diploma and those who are still in school have a higher middle rank than the other groups. Furthermore, it is noticeable that in general participants with educational qualifications attainable at a university (Bachelor, Master), have low middle ranks. This outcome is significant ($\chi^2 = 27.812, p = 0.883$). In case of the 'streamers' similar results are given ($\chi^2 = 17.426, p = 0.015$). The subsequent post-hoc analysis for 'viewers' shows that only the education groups 'Master' and 'still in school education' differ significantly (z = 3.803, p = 0.009) with a medium effect size (r_d = 0.42).

Profession. In this study, most of the participants of both groups are students, full-time working or pupils. The middle ranks for 'viewers' are between 203.02 (full-time working) and 290.50 (unemployed). A significant difference regarding the desire to be involved could be found for 'viewers' in different occupational situations ($\chi^2 = 19.536, p = 0.012$). However, the subsequent post hoc analysis reveals only a significant difference between pupils and full-time employees (p = 0.035), with a weak effect size (r_d = 0.24). In case of the 'streamers', where pupils have the highest middle rank (= 89.22), no significant difference could be found ($\chi^2 = 12.715, p = 0.079$).

Time. While the biggest share of 'viewers' use Twitch less than 3 h (n = 152, 32.3%) and between 3 and 6 h (n = 125, 26.6%) per week, the share of the 'streamers' is distributed evenly on the groups. For 'viewers', the middle rank increases as more time is spent on the platform ($\chi^2 = 10,195$, p = 0.037). This suggests that viewers who spend more time on Twitch most likely agree to the desire for involvement than those who spend less time on Twitch. The followed-up post hoc analysis could not find significant differences between the individual groups. Again, in case of the 'streamers' no significant difference could be found ($\chi^2 = 6.034$, p = 0.197).

Money. For both groups, the middle ranks are higher if a donation or subscription to a streamer has been made in the past. The difference of the values for the 'viewers' is 75.4 (z = −5.462, p < 0.001), while for the 'streamers' it is 12.23 (z = −1.890, p = 0.059). The identified significant difference for the viewers has a weak effect size (r_d = 0.25).

4.3   Usage Behavior

To answer RQ1c, "How does the usage behavior influence the involvement (desire) of a Twitch user?", Spearman's rank correlation coefficient for different test items regarding the usage behavior and the involvement desire is calculated (Table 2).

On the one hand, the highest correlation for 'viewers' exists between the involvement desire and test item 5, to watch the broadcast and use the chat simultaneously (r_s = 0.476, p < 0.001). On the other hand, 'streamers' have a stronger involvement desire as they are using the chat while playing a game (r_s = 0.306, p < 0.001). These test items represent the anticipated use of Twitch for a regular viewer and streamer. Regarding the selection criteria of a stream (items 9, 10 and 11), for 'streamers' a significant correlation for test item 9 (r_s = 0.202, p = 0.020) and for 'viewers' for test item 11 (r_s = −0.094, p = 0.042) exists.

For the conduction of a regression, the test items 3, 5, 6, 7 and 10 were chosen as predictors in consideration of the correlations as well as the corresponding collinearity. The multiple correlation of the criterion with all predictors is 0.493 for 'viewers' and 0.392 for the 'streamers'. Furthermore, the 'viewers' have a higher explanatory variance (R^2) with 0.243 than the 'streamers' (R^2 = 0.154), meaning 24.3% of the variance of the involvement desire of the 'viewers' can be explained by

Table 2. Correlations between involvement desire and usage behavior.

Test item	viewer r_s	streamer r_s
1: *I watch broadcasts of games that I personally play, too*	0.076	0.119
2: *I watch broadcasts of games that I personally do not play*	0.067	0.074
3: *I watch the broadcast and play games concurrent*	0.116**	−0.069
4: *I listen to the broadcast and play games concurrent*	0.107*	0.030
5: *I watch the broadcast and use the chat concurrent*	0.476***	0. 264***
6: *I just use the chat*	0.189***	0.074
7: *I play games and use the chat concurrent*	0.253***	0.306***
8: *I just listen to the broadcast*	−0.018	0.051
9: *I choose the broadcast depending on the channel*	0.068	0.202*
10: *I choose the broadcast depending on the streamer*	0.083	0.137
11: *I choose the broadcast depending on the game*	−0.094*	−0.011

***$p \leq 0,001$; **$p \leq 0,01$; *$p \leq 0,05$

these five predicators. In contrast, these predictors account for 15.4% of the variance of the 'streamers' involvement desire. The values of the Durbin-Watson statistics are 1.956 and 1.599 respectively, which speaks for no autocorrelation and independent error values.

4.4   Interactions on Twitch

Supplementary to the conducted survey, the second research question (RQ2) "How are the interactions between streamer and viewers perceived?", is answered by summarizing the results of the observation. While streamers with a different follower size were monitored, the results are restricted to streamers with a small (<3,000 followers), middle (3,000–14,999 followers) and large follower base ($\geq$ 15,000 followers). The observation put emphasis on the actions of a streamer to interact with his or her audience. In addition, the actions of viewers to interact with the streamer and other viewers are investigated. Besides, advantages of donations and subscriptions on Twitch will be analyzed.

It appears that all the observed streamers are offering their viewers the option for donations, which can be considered as interactions. A form of gratitude is generally expressed by mentioning the username, the amount of money and thanking the user for the financial support. Most streamers implemented a donation alert to fade-in an animation, insert text messages which are read aloud or play a short video on screen. It was noted, that none of the streamers offer an influence on the game play by a donation.

Another way to support a streamer financially, in case of a Twitch partner, is a subscription. Like donation alerts, most streamer implemented a subscription alert to highlight a new subscriber. It stands out that most of the streamers with a large (7 out of 8) and middle (6 out of 8) follower base offer benefits for subscribers. These benefits are in most instances channel-exclusive emoticons and to be able to have no restrictions regarding the chat, e.g. no slow-motion mode or a subscriber-only mode. Moreover, one streamer (middle follower base) setup a TeamSpeak server to create a better network exclusively for subscribers. Less popular streamers (<3,000 followers) in this study were too unknown and therefore most of the test subjects of this group were not a Twitch partner.

Besides, it was striking that most of the streamers with a middle or large follower base, in contrast to smaller ones, advertised gaming-related products or channel-related merchandise in their stream.

During the observation, only a few of the streamers offered their audience to actively participate on the game, e.g. by organizing 'viewer games' and playing with or against a viewer.

Moreover, half of the streamers gave the audience a decision-making power. In most cases, the audience could decide which character should be picked. In terms of the chat, almost every streamer was actively referring to the content of the chat – regardless of his or her follower size.

The interactions between viewers through the chat were also observed. The content of the conversations was mostly referring to the broadcasted gameplay (e.g. questions about a specific gameplay) or the streamer (e.g. about his or her appearance). Even if not directed to the audience, viewers were mostly trying to help each other and answering questions of other viewers. This community feeling could be a reason for many users to greet and to say good bye to others. Furthermore, especially in the chat of streamers with a small follower base, some users were talking about personal experiences and plans, which, again, shows the community and socialization aspect of Twitch.

5   Discussion

In recent years, Twitch has become a huge platform for streaming video game related content with a high number of users. The community aspect could be a reason for the regular use and reasoning its consumption. As stated in Sect. 2, the findings of a previous study are the main motivational factor to investigate the involvement desire and interactions on Twitch.

While more than a fifth of the participants stated to be a 'streamer and viewer' (22.1%), the results for each research questions are compared with the data of the participants who are using Twitch only as a viewer (77.9%). The comparison shows that 'streamers' in general have a stronger desire for involvement than 'viewers'.

What are possible indicators for the involvement desire of 'viewers'? While the gender does not seem to play an important role, the age of the participants does. The ratings of teenagers (≤ 18 years) regarding the involvement desire are significantly higher. Regarding the education, the results indicate that Twitch users who are still at school or have a high school diploma as the highest educational attainment are rating the involvement higher than users with a higher educational attainment (e.g. bachelor's or master's degree). Under consideration of the average weekly time spent on Twitch, a higher usage time leads to a stronger desire for involvement. Furthermore, participants who donated or subscribed to a streamer have a higher desire for involvement.

The analysis of the data given the socio-demographic factors as well as the factors time or money spent on Twitch could not reveal significant outcomes for 'streamers', except for the aspect 'education'. A possible reason for the higher ratings could be the experience of a streamer. They know exactly the difficulties of starting a stream and reaching a worthwhile viewer- and followership. Broadcasting for a non-existent audience is not considered as entertaining and fun. Therefore, the involvement of the viewers by the streamer is important to create a community.

Another possible influence could be the different application types of how Twitch is used. While for 'viewers' the desire for involvement rises due to the agreement to the use of the chat while watching a live-stream, the 'streamers' desire increases while they play games and use the chat.

But, what types of interactions do exist? Donations and subscriptions are ways to support the streamer financially, which are highlighted by alert systems. While donations are gratified by the streamer, benefits for subscriptions are e.g. channel-specific emoticons or no limitations while a subscriber-only mode is active. A characteristic for most streamers with a middle (3,000–14,999) and large follower base ($\geq15,000$) was the advertisement for gaming related products. However, especially this feature is not considered as an involvement by participants of this study (Fig. 2). Moreover, the chat of streamers with a small follower base (<3,000) sometimes includes personal experiences and plans, which is probably not possible in chats with thousands of viewers. This speaks for the community and socialization aspect of streams.

Limitations. The developed questionnaire was only used on German-speaking Twitch users to prevent an uneven distribution among different countries. Though, it would be interesting to survey English-speaking Twitch users to compare the results and get a deeper insight on the usage behavior of Twitch users.

With a low participation rate of female Twitch users (16.9%, n = 102), which appears unbalanced at first sight, this represents a realistic distribution of the gender in online game cultures [20, 21]. Moreover, the numbers of participants of some educational attainments and professions – e.g. state examination, PhD, federal voluntary service or retired – were too low to draw specific conclusions and thus are excluded in Table 1.

While there exist more application types of Twitch, which may differ from time to time, RQ1c does not represent a complete analysis of the usage behavior of Twitch users. However, an insight of the usage behavior is given for the used test items.

Although each test subject was observed for ten hours, the observation is only limited to 24 streamers with different follower sizes. Especially streamers with a slight followership were difficult to monitor for ten hours, as some had no streaming schedule. Furthermore, the classification of these groups may be suboptimal, as the aggregation of the popularity given by the follower size of a streamer is most likely not representing a list of all Twitch users. Thus, the results may not be sufficient to draw conclusions for 'small', 'middle' and 'large' sized streamers. Nevertheless, the results give an interesting insight of the interactions between Twitch users in general.

Acknowledgement. We would like to thank Kathrin Knautz for her guidance throughout this study.

References

1. Scheibe, K., Fietkiewicz, K., Stock, W.G.: Information behavior on social live streaming services. J. Inf. Sci. Theory Pract. **4**, 6–20 (2016)
2. Kaytoue, M., Silva, A., Cerf, L., Meira, W., Raïssi, C.: Watch me playing, i am a professional: a first study on video game live streaming. In: Proceedings of the 21st international conference companion on World Wide Web - WWW 2012 Companion, p. 1181. ACM Press, New York (2012)
3. Cheung, G., Huang, J.: Starcraft from the stands: understanding the game spectator. In: Proceedings of the 2011 annual conference on Human factors in computing systems - CHI 2011, p. 763. ACM Press, New York (2011)
4. Twitch: Welcome Home: The 2016 Retrospective. https://www.twitch.tv/year/2016
5. Toffler, A.: The Third Wave. Bantam books, New York (1981)
6. Linde, F., Stock, W.G.: Information Markets: A Strategic Guideline for the I-commerce. De Gruyter Saur, New York (2011)
7. Smith, T., Obrist, M., Wright, P.: Live-streaming changes the (video) game. In: Proceedings of the 11th european conference on Interactive TV and video - EuroITV 2013, p. 131. ACM Press, New York (2013)
8. Hamilton, W.A., Garretson, O., Kerne, A.: Streaming on twitch: fostering participatory communities of play within live mixed media. In: Proceedings of the 32nd annual ACM conference on Human factors in computing systems - CHI 2014, pp. 1315–1324. ACM Press, New York (2014)
9. Burroughs, B., Rama, P.: The eSports trojan horse: Twitch and streaming futures. J. Virtual Worlds Res. **8**, 1–5 (2015)
10. Lessel, P., Mauderer, M., Wolff, C., Krüger, A.: Let's play my way: Investigating audience influence in user-generated gaming live-streams. In: Proceedings of the 2017 ACM International Conference on Interactive Experiences for TV and Online Video. ACM Press, New York (2017)
11. Consalvo, M.: Player one, playing with others virtually: what's next in game and player studies. Crit. Stud. Media Commun. **34**, 84–87 (2017)
12. Nascimento, G., Ribeiro, M., Cerf, L., Cesario, N., Kaytoue, M., Raissi, C., Vasconcelos, T., Meira, W.: Modeling and analyzing the video game live-streaming community. In: Proceedings - 9th Latin American Web Congress, LA-WEB 2014, pp. 1–9 (2014)
13. Sjöblom, M., Hamari, J.: Why do people watch others play video games? an empirical study on the motivations of Twitch users. Comput. Human Behav. **75**, 985–996 (2016)
14. Gros, D., Wanner, B., Hackenholt, A., Zawadzki, P., Knautz, K.: World of Streaming. Motivation and gratification on Twitch. In: Meiselwitz G. (ed.) Social Computing and Social Media. Human Behavior, pp. 44–57 (2017)
15. Mann, H.B., Whitney, D.R.: On a test of whether one of two random variables is stochastically larger than the other. Ann. Math. Stat. **18**, 50–60 (1947)
16. Kruskal, W.H., Wallis, W.A.: Use of ranks in one-criterion variance analysis. J. Am. Stat. Assoc. **47**, 583 (1952)
17. Dunn, O.J.: Estimation of the medians for dependent variables. Ann. Math. Stat. **30**, 192–197 (1959)
18. Dunn, O.J.: Multiple comparisons among means. J. Am. Stat. Assoc. **56**, 52–64 (1961)
19. Cohen, J.: A power primer. Psychol. Bull. **112**, 155–159 (1992)

20. Sundén, J., Malin, S.: Gender and Sexuality in Online Game Cultures: Passionate Play. Routledge Chapman & Hall, New York (2011)
21. Williams, D., Martins, N., Consalvo, M., Ivory, J.D.: The virtual census: representations of gender, race and age in video games. New Media Soc. **11**, 815–834 (2009)

Does Age Influence the Way People Interact with Social Live Streaming Services?

Thomas Kasakowskij[✉]

Department of Information Science, Heinrich Heine University Düsseldorf,
Düsseldorf, Germany
thomas.kasakowskij@hhu.de

Abstract. The interest in social media and in particular the social live streaming services (SLSSs) is increasing, as can be observed by the growing number of users in different age groups. However, the social live streaming services have not been satisfactory investigated yet. Therefore, knowledge gaps in this subject area are still present. This study focuses on the use of SLSSs in terms of content, motivation, and gender depending on the age of the streamer. A research team has been assembled for this purpose. 4,937 streams were analyzed for content, motivation, age and gender on three different platforms in three different countries. Dependencies of content and motivation regarding the streamers age could be determined. The results indicate that older age groups are more likely to share information and therefore broadcast content related to information. It could be observed that younger users are more likely to film their lifestyle on this medium. The genders differ for the age groups significantly. So, it can be assumed that a correlation between age and the content, motivation, and genders on social live streaming services is given.

Keywords: Social live streaming service · Age · Content
Motivation · Gender · Social network service

1 Introduction: Information Behavior on SLSSs

Today, almost everyone having an internet connection uses at least one social medium to communicate with others or share their opinion with the public. This is a possible reason why social networking services (SNSs) are becoming more and more popular. One can be connected with others regardless of time and place [1]. In recent years social live streaming services (SLSSs), which are part of SNSs, are gaining popularity [2]. Yet, the user behavior of the streamers on these platforms has hardly been scientifically investigated. This is exactly what should be done to find out why user numbers in this area of social media are rising and which clientele is addressed by this medium, which content they produce, and what their motivations to do so are. Since user groups on SNSs cover a large age rang, it would be especially interesting to investigate the content and the motivation related to different age groups.

There are already some studies dealing with the use of social live streaming services [3–6]. However, there is still no study that investigates the usage behavior of SLSSs in relation to the streamers' age. Moreover, this should be done since there are

© Springer International Publishing AG, part of Springer Nature 2018
G. Meiselwitz (Ed.): SCSM 2018, LNCS 10914, pp. 214–228, 2018.
https://doi.org/10.1007/978-3-319-91485-5_16

already studies that show the influence of age on motivation and content in the use of SNSs. For example, Brell et al. [7] have already discovered that the gender has an influence on the motivation to use a social medium. A dependency on the motivation to use social networks and the users' age can be observed [8, 9]. In addition, age is considered a key variable in the studies. Perhaps a similar dependence could be found between the age of the user and the motivation to use SLSSs. Pfeil et al. [10] were able to detect a difference in the use of a social medium depending on the age of the user.

An SLSS is an application on which users can generate content by filming themselves or others and broadcast the stream directly online to allow other users to participate. Depending on the platform, streamers can use their mobile devices or a webcam for filming. It is open to the streamer which content he or she generates, whether it is a city tour or an excerpt of his or her everyday life, everything can be streamed. However, one should try not to break the law [5, 11]. For our research we investigated the produced content and motivations of the streamers on the platforms Ustream, YouNow, and Periscope.

1.1 Investigated Social Live Streaming Services

Periscope[1] was developed by Kayvon Beypour and Joe Bernstein and launched by Twitter in March 2015 [12]. As a live streaming application for Android and iOS, Periscope enjoys a high level of mobility. Periscope itself presents the service as a tool by which the world can be seen through the eyes of someone else. Periscope was developed with the idea of creating something that comes close to teleportation [13]. Just like Twitter itself, Periscope is an information sharing platform that can be used for any purposes [14].

Ustream[2] was developed by Brad Hunstable, John Ham and Gyula Feher in 2007 with offices in San Francisco as well as in Budapest. In January 2016, Ustream was acquired by IBM and is now part of IBM Cloud Video [15]. Unlike YouNow and Periscope, Ustream tends to target companies rather than users. To use a professional streamer account, a monthly fee of $99 up to $999 is required. There are also free accounts offered, but these are provided with advertising. As a leading provider of cloud-based, end-to-end video solutions for media and enterprises, Ustream offers 80 million viewers per month the chance to watch live or on-demand streams from internal meetings to press conferences up to worldwide entertainment [15].

With the mission to create an interactive platform where anyone can participate and express themselves live [16], YouNow[3] was founded by Adi Sideman in 2011. YouNow is a live streaming service which is mostly used as a web application, but it is also offered for Android and iOS. According to YouNow [16], the service hosts more than 100 million user sessions a month and about 50,000 h of live videos every day. The most appreciated target group are teenagers [3].

---

[1] https://www.pscp.tv/.

[2] http://www.ustream.tv/.

[3] https://www.younow.com/.

There are many more than the three streaming providers we studied. These were not included in this study because of their specializations on subject areas or locations. These include, for example, Twitch[4], with focus on gaming, or niconico[5], addressing Japanese-speaking users or YY[6] in China.

1.2   Research Model

A streamer broadcasts a stream on a platform (Ustream, Periscope, and YouNow) and has an age and a gender (Fig. 1). The produced content could possibly depend on the age of the streamer. The content is divided into five categories: food & lifestyle; information; entertainment; nature & spirituality; sports & arts. Probably dependent on the age, a streamer has several motivations to stream. The motivations were divided into four categories based on the Uses & Gratification Theory: entertainment; social interaction; social realization; information [6, 17]. To be investigated is the frequency distribution of the content as well as the motivation depending on the age of a streamer. Also, the age-dependent change in the gender-distribution will be examined.

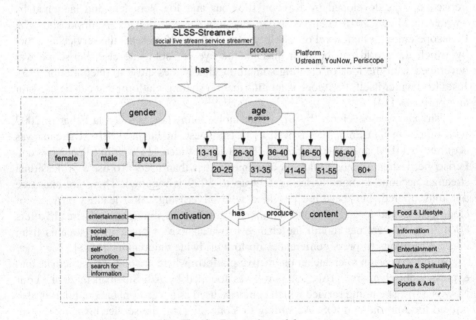

Fig. 1. Research model

[4] https://www.twitch.tv/.

[5] http://www.nicovideo.jp/.

[6] https://www.yy.com/.

2 Method

To calculate meaningful statistics, it was necessary to create standardized data sets. For this purpose, a codebook [18] based on literature concerning the usage of social media was made. Two different approaches were applied to ensure a qualitative content analysis with a great dependability. The directed approach was used with assorted literature to get guidance for the research variables and categories. Additionally, the conventional approach via observation was implemented [19] to get a general idea of the streams' content in each country (U.S., Germany, and Japan) and on each platform (Periscope, Ustream, and YouNow) that were chosen for this examination.

In addition to the literature review, streams on SLSSs were analyzed to determine appropriate categories for the content of the streamers' and the users' motives. The motives can be classified according to the Uses & Gratifications Theory: entertainment; social interaction; information; self-realization [20]. The Uses & Gratifications Theory asserts that users of a (social) medium use media to fulfill specific gratifications. Therefore, they are looking for a medium that best satisfies their needs [21]. Due to a high number of different content categories, related ones were aggregated into main categories. The chosen categories were influenced by commonly used topics. The topics are: entertainment {entertainment media, comedy}; nature & spirituality {nature, animal, spirituality}; information {share information, news, STM (science, technology and medicine), politics, advertising, business information}; sports & arts {make music, draw/paint a picture, gaming, fitness, sports}; food & lifestyle {to chat, 24/7, slice of life, food}.

For example, *entertainment media* and *comedy* are in the *entertainment* category, as they both provide entertainment for the viewers [22]. *Nature & spirituality* has been summarized as a top category since the idea of spirituality is closely related to nature and offers different approaches and forms to reconnect with nature [23]. Even according to the Bible, a connection between faith, nature, and animals can be observed [Job 12: 7 New Living Translation]. The category *lifestyle* describes a certain way of life. This includes content such as social interactions (to chat) as well as everyday tasks (slice of life, 24/7), which can be regarded as subfields. Food is also part of our everyday life and can reflect our lifestyle, for example through food culture. There are even fairs evolving around "food & lifestyle", for example the "Chester Food, Drink & Lifestyle Festival" or the "Ingolstädter Food & Lifestyle Messe". Sports can be considered as a form of art, for example through special movements [24]. Due to these points of contact, the contents of the sports activities (gaming, fitness, sport) were accommodated with those of the artistic actions (draw/paint a picture, make music) for this research in the *sports & arts* category. For the category *information*, the contents that serve to convey professional, educational, or business-related information were brought together (share information, news, STM, politics, advertising, business information).

A spread sheet was generated for the content and motivation categories as well as socio-demographic data. Norm entries were used for the formalities. Those were: gender {male, female, group} and the age of the streamer in groups {13–19, 20–25, 26–30, 31–35, 36–40, 41–45, 46–50, 51–55, 56–60, 60+}. Similar to the study of

Pfeil et al. [10] this study compares the user behavior of teenagers (13–19 years old) with those of older people (60+ years old). However, as not only these two age groups are relevant, other age groups were added for our study, each limited to a 5-year period. The investigation only applies to streamers over the age of 13, as younger children are not allowed to use social live streaming services according to the terms and conditions of the streaming platforms.

For each streamer, the top categories of content and motivation were marked (binary coded) once one of their respective subcategories were fulfilled. This prevents weightings of individuals on the content category. The data of the streams was collected from three different countries, namely Germany, Japan, and the United States of America, to ensure representativeness. To ensure that the streams originated from those countries the declaration of the country for a broadcast on each platform was checked for every stream. Additionally, the collectors of the data had the required language skills for those countries.

It was necessary to train the coders to ensure a good quality of their coding skills is given [25]. To guarantee this, coders need to work in teams with a minimum of two coders [26]. Twelve teams, each consisting of two persons, were formed. The coder gathered the data in two phases. In phase one they watched the streams and extracted the content [18]. In the next phase, they communicated with the streamers to find out their motivation to do a broadcast. To support the uniform analysis of the content, the 'four eyes principle' [27] was used. Each stream was observed simultaneously but independently by two people for two to a maximum of ten minutes. Communication always happened between the two observers to guarantee a 100% intercoder reliability.

The streams were not recorded, as for this, a consent of the streamer would be needed but could not be always obtained, possibly resulting in violation of personality rights.

This way, a data set of a total of 7,667 streams in a time span of four weeks, from the 26th of April to the 24th of May 2016, was collected. Of these, 4,937 streams were broadcasted from streamers who stated their age and therefore could be evaluated for this study. The analysis of the streams focused only on the producers (streamer); information on user-behavior of the participants and consumers (viewer) was not collected. The results of the observations were statistically analyzed and compared regarding different aspects of our research model.

3   Results

3.1   Gender

Of the 4,937 participants we evaluated, 33.42% were female, 50.53% were male and 16.05% were active in a group (Fig. 2). The age- and gender-dependent distribution of users shows surprising results. It can be observed that in younger age groups (13–19) a much higher proportion of female streamers (51.34%) prevails. This decreases with advancing age (8.33%). The proportion of male streamers is the opposite of female streamers. Percentages are lower in younger age groups (32.70%) and rise with increasing age (75%). The proportion of streamers who are active in a group remains between 15% and 20% in all age groups and is therefore hardly dependent on age.

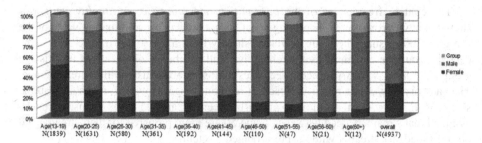

Fig. 2. Age division of gender on SLSSs (N = 4,937)

3.2   Age and Content

There also appear to be age-dependent differences regarding the streamed content (Fig. 3). By looking at the percentages, a ranking of the content categories can be determined. Streams that are related to the *food & lifestyle* category are broadcasted the most often for each age group, which ranges between 42% and 81%. In the second place are streams in the *information* category. These make up between 19% and 57% of the streams. Third rank the streams of the *sports & arts* section. These account for a percentage of 13% to 29% of all streams. The fourth place is occupied by the *nature & spirituality* category with values ranging between 2% and 34%. Lastly, there are the *entertainment* and *nothing* categories with values between 8% and 16% for streams related to entertainment and 5% to 18% for no content at all.

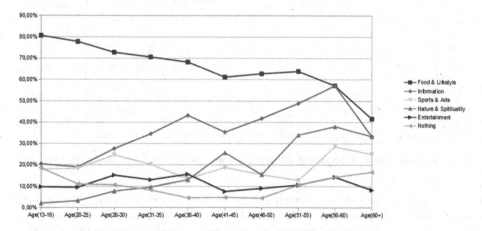

Fig. 3. Content distribution on SLSSs (N = 4,937)

It can be generalized that the content of the streams changes with age. In the groups *nature & spirituality*, *food & lifestyle*, and *information* a change of content frequency is clearly recognizable. The groups *entertainment* and *sports & arts* show only small fluctuations. Therefore, no dependency of content frequency on age can be observed for these groups.

The content frequencies of the *information* and *nature & spirituality* categories show potential growth with increasing age. Conspicuous is the information value of the age group 60+ , since the value decreases sharply from 57.14% down to 33.33%. This decrease could be explained by the relatively small number of observations for this age group. The *food & lifestyle* sector decreases almost continuously with increasing age. The category *Nothing* is rather unsteady in relation to age. However, it has a parabolic shape, with a low point in the age group between 41 and 45 years. Thus, it can be assumed that streamers aged between 36 and 50 are more prepared for their streams and do not broadcast streams without pre-arranged content.

The reasons for these changes can be clarified by looking at the individual contents of the groups. Beginning with the *food & lifestyle* sector (Fig. 4), it can be seen that the sector is heavily influenced by the content *to chat*, as it accounts for the largest percentage. It is easy to recognize that chatting with the viewer decreases with increasing age of the streamer. Streams broadcasting 24/7 are the only ones of the lifestyle group that is becoming more prevalent with age. *Slice of life* and *food* are rather unsteady, whereby slice of life has a tendency to decrease, whereas food has a tendency to increase with raising age.

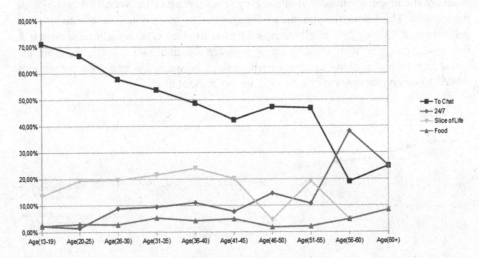

Fig. 4. Distribution of the *food & lifestyle* category split into the respective subcategories

The *information* domain is dominated by the content category *share information*, with values ranging between 17.04% and 44.68% (Fig. 5). The *news, STM, business information, advertising,* and *politics* categories range in values between 0% and 15%. Between the age of 41–50 there is a decrease in frequency that runs through all the content categories, except for advertising. The same can be observed for streamers who are 55 years old, except with a much sharper decline in frequencies.

A particularly surprising result is the continuously strong increase in spiritual related content in the *nature & spirituality* category (Fig. 6). The occurrence of this content starts at 0.44% of the streamers in the age group 13–19 and increases with

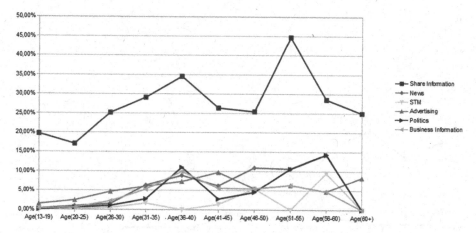

Fig. 5. Distribution of the *information* category split into the respective subcategories

rising age to 33.33%. Especially broadcasted Holy Masses increase the frequency of the spirituality category in the mature age range. The frequency of streamers broadcasting nature increases from the age of 13 to 45 and has a steady decline after that but rises again with the age groups between 50 and 60+.

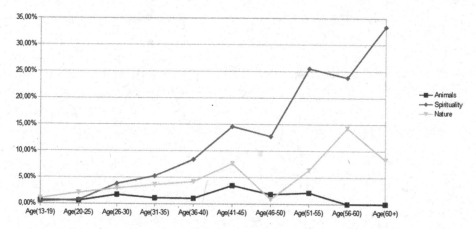

Fig. 6. Distribution of the *nature & spirituality* category split into the respective subcategories

The self-made music category of the *sports & art* sector varies greatly between the age groups (Fig. 7). This increase could also be possibly explained by the streaming of church music, as explained earlier for spirituality. The other content categories (sports, fitness, draw/paint a picture, and gaming) show slightly fluctuating frequencies ranging between 0% and 6%, with no discrepancies within the age groups.

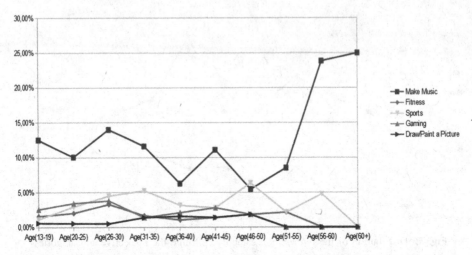

Fig. 7. Distribution of the *sports & art* category split into the respective subcategories

The group *entertainment* is, with a maximum value of 14%, seldom represented in streams (Fig. 8). There are no clear tendencies in terms of age groups. The values show high fluctuations in this small frequency range for both content categories. However, entertainment media is much more common on SLSSs than comedy related streams.

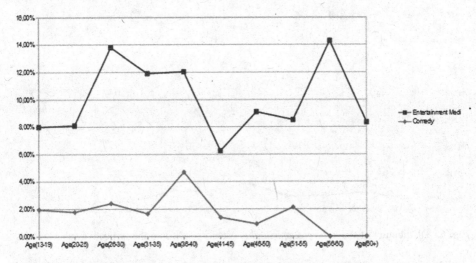

Fig. 8. Distribution of the *entertainment* category split into the respective subcategories

Thus, it can be said that older streamers are less prone to talk nonsense and to devote to other, more concrete topics like spirituality, making music, or news. This could be explained by the high number of professional channels or the increasing life experience of older streamer [28]. They tend to share information and approach specific topics. In contrast, young streamers tend to stream without a precise plan.

3.3   Age and Motivation

An equivalent occurrence of the topic *information* between motivation and content becomes apparent (Fig. 9).

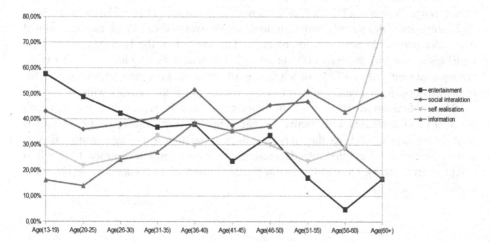

Fig. 9. Distribution of the motivations among SLSSs

The streamers' willingness to entertain decreases with increasing age from 57.64% to 4.67%. In contrast, the will to inform increases from 13.92% to 51.06%. These two motives behave almost in opposite directions. The motivations *social interaction* and *self-realization* hardly show dependencies regarding the age of the streamer. Noticeable is the sharp rising of the need for self-realization in the age group 60+. When the motivation category *social interaction* is broken down into its individual components, it becomes clear that the need to communicate and desire to socialize have the greatest importance for the streamers (Fig. 10).

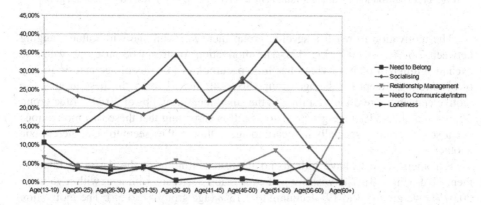

Fig. 10. Distribution of the *social interaction* motive split into the respective subcategories

The motive *need to communicate* behaves similarly to the content *share information*, both show a significant increase from age 13 to 40 and decrease in the years between 41–50. In the group of 51–55 year old streamers, this increases again and then decreases slightly in the age group 56+ .

Streamers aged between 13 and 50 years have a need for socializing for which the values range between 17% and 28%. However, from the age of 51, the interest in socializing decreases sharply and even drops to 0% eventually. Thus, it can be assumed that older people do not want to socialize. The interest in the management of relationships is rising by the age of 60+ (16.67%). The need to belong has its highest value among teenagers with 10.77%, which falls to 0% with maturing age. This could be due to the self-discovery phase (puberty) of the youth, which also passes with age.

The motivation to stream out of boredom is mostly represented by teenagers (41.87%) and decreases with increasing age (Fig. 11). Boredom could be associated with the content *to chat*, as they have parallels in relation to the age groups. Broadcasting for fun is represented by a decreasing line from 19.36% (age group 13 to 19) to 0% (age group 60+). Only streamers in the age group 46 to 50 seem to really enjoy streaming (24.55%).

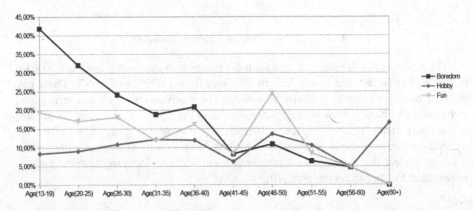

Fig. 11. Distribution of the *entertainment* motive split into the respective subcategories

The motivation to reach a specific group increases with age with values ranging between 7.67% and 50% (Fig. 12). The motivation to stream because one wants to exchange different point of views has little peculiarities in terms of age and frequency of occurrence, except for the age groups 36 to 40 and 51 to 55. In these groups, the motive appears more often, resulting in the small measuring tips of 20.31% (age group 36 to 40) and 25.53% (age group 51 to 55). It is surprising that these two motivations occur so differently, especially in regard to age, although they seem to be related to one another.

Streamers seem to have little to no interest in trolling or the desire to improve themselves (Fig. 13). The wish for self-expression decreases with age, with a value of 20.07% (age group 13 to 19) declining to 2.13% (age group 51 to 55). The motivation to make money, and a sense of mission show a tendency to rise with increasing age.

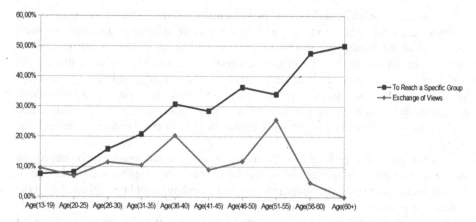

Fig. 12. Distribution of the *information* motive split into the respective subcategories

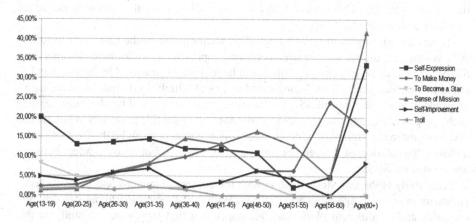

Fig. 13. Distribution of the social-realization motive split into the respective subcategories

Noticeable are the high percentages of the streamers being 60+ and their sense of mission as well as their desire for self-expression when using SLSSs. Apparently, in this age group, the need for self-realization through streaming is high.

4 Discussion

In this study, we examined whether age has an influence on the content of streams or the motivation of the streamer on SLSSs. We conducted a broad analysis to compare social media usage of SLSSs for different age groups. The results indicate that there were differences in the generated content and the driving motivations of the users in relation to the respective age groups. Also, some connections between content and motivations could be observed.

Some age-dependent differences in streamed content could be recognized, for example, the *information* and *nature & spirituality* content categories increase with rising age of the streamer. By contrast, streams with *food & lifestyle* related content decline with advancing age. From this observation it could be concluded that people with increasing life experience are more likely to share their knowledge [28] through SLSSs. A need for spirituality becomes more and more apparent as we grow older, often associated with the recognition of mortality and failure [30, 31]. This could explain the sharp increase of spiritually related content in context with increasing age on SLSSs.

Similar results can be found for the motivations of the streamers. According to our study, the motivation to gather and search for information is influenced by the age of the streamer, it increases with maturing age. Those findings could be explained with the appearance of professional services, such as news or radio broadcasts, which are more likely to be represented by older age groups [32], which could also apply to streams on SLSSs.

The motivation to entertain others or be entertained through SLSSs is strongly decreasing with age. This observation could be explained by the declining desire for attention with rising age [33].

It was astonishing to observe an age-dependent change in the gender-distribution. At a young age, mostly female streamers were represented, while male streamers use SLSSs even at an advanced age. This could have different causes. On the one hand, it could be assumed that a more pronounced technical affinity [34] tends to make males more likely to venture into SLSSs. On the other hand, there could be also other reasons such as parenting and the associated shortage of time [35], or job relationships involving more masculine individuals [32]. This phenomenon can be the foundation for a closer examination of the relationship between the distribution of genders between age groups on SLSSs.

This study is the first study dealing with the content of streams and motivations of streamers in terms of age on general SLSSs. With 4,937 examined streams in different countries and on different platforms, this paper is a first representative study in this area. However, this number comparatively small to the monthly streams broadcasted on each platform. So, it is just a drop in the ocean and should be further investigated.

For possible further research, the connection between the streamers' motivation and the produced content on SLSSs could be investigated in more detail, since correlations between these variables seem to exist. Another possible aspect should be the examination of the streamers' gender and his or her motivations and produced content on SLSSs. This promises to gain further insights into the users' behavior on SLSSs.

Acknowledgements. The authors want to thank Wolfgang G. Stock for his valuable feedback and insights for this study, the help is much appreciated.

References

1. Shah, C.: Social Information Seeking: Leveraging the Wisdom of the Crowd. Springer, Cham (2017). https://doi.org/10.1007/978-3-319-56756-3
2. Friedländer, M.B.: And action! Live in front of the camera: an evaluation of the social live streaming service YouNow. Int. J. Inf. Commun. Technol. Hum. Dev. (IJICTHD) 9(1), 15–33 (2017)
3. Scheibe, K., Fietkiewicz, K.J., Stock, W.G.: Information behavior on social live streaming services. JISTaP 4(2), 6–20 (2016)
4. Friedländer, M.B.: Streamer motives and user-generated content on social live-streaming services. JISTaP 5(1), 65–84 (2017)
5. Zimmer, F., Fietkiewicz, K.J., Stock, W.G.: Law infringements in social live streaming services. In: Tryfonas, T. (ed.) HAS 2017. LNCS, vol. 10292, pp. 567–585. Springer, Cham (2017). https://doi.org/10.1007/978-3-319-58460-7_40
6. Scheibe, K., Zimmer, F., Fietkiewicz, K.: Das Informationsverhalten von Streamern und Zuschauern bei Social Live-Streaming Diensten am Fallbeispiel YouNow. Information - Wissenschaft & Praxis 68(5–6), 352–364 (2017)
7. Brell, J., Calero Valdez, A., Schaar, A.K., Ziefle, M.: Gender differences in usage motivation for social networks at work. In: Zaphiris, P., Ioannou, A. (eds.) LCT 2016. LNCS, vol. 9753, pp. 663–674. Springer, Cham (2016). https://doi.org/10.1007/978-3-319-39483-1_60
8. Fietkiewicz, K.J.: Jumping the digital divide: how do "silver surfers" and "digital immigrants" use social media? Netw. Knowl. 10(1), 5–26 (2017)
9. Fietkiewicz, K.J., Baran, K.S., Lins, E., Stock, W.G.: Other times, other manners: how do different generations use social media? In: 2016 Hawaii University International Conferences. Arts, Humanities, Social Sciences & Education, January 8–11, 2016, Honolulu, Hawaii, Proceedings, pp. 1–17. Hawaii University, Honolulu (2016)
10. Pfeil, U., Arjan, R., Zaphiris, P.: Age differences in online social networking–A study of user profiles and the social capital divide among teenagers and older users in MySpace. Comput. Hum. Behav. 25(3), 643–654 (2009)
11. Honka, A., Frommelius, N., Mehlem, A., Tolles, J.N., Fietkiewicz, K.J.: How safe is younow? An empirical study on possible law infringements in Germany and the United States. J. MacroTrends Soc. Sci. 1(1), 1–17 (2015)
12. Edelmann, M.: From Meerkat to Periscope: Does intellectual property law prohibit the live streaming of commercial sporting events? Columbia J. Law Arts 39(4), 469–495 (2015)
13. Periscope. https://www.periscope.tv/about (2018)
14. Fiecht, E.S., Robinson, J.J., Dailey, D., Starbird, K.: Eyes on the ground: emerging practices in Periscope use during crisis events. In: ISCRAM 2016 Conference Proceedings – 13th International Conference on Information Systems for Crisis Response and Management, pp. 1–10. Federal University of Rio de Janeiro, Rio de Janeiro (2016)
15. Ustream. https://www.ustream.tv/our-company?itm_source=footer&itm_medium=onsite&itm_content=About_Ustream&itm_campaign=about_us_link. Accessed 2018
16. YouNow (2018). https://www.younow.com/press
17. McQuail, D.: Sociology of Mass Communications: Selected Readings. Penguin Books, Harmondsworth (1972)
18. MacQueen, K.M., McLellan, E., Kay, K., Milstein, B.: Codebook development for team-based qualitative analysis. CAM J. 10(2), 31–36 (1998)
19. Hsieh, H.F., Shannon, S.E.: Three approaches to qualitative content analysis. Qual. Health Res. 15(9), 1277–1288 (2005)

20. Hsu, M.-H., Chang, C.-M., Lin, H.-C., Lin, Y.-W.: Determinants of continued use of social media: the perspectives of uses and gratifications theory and perceived interactivity. Inf. Res., **20**(2), paper 671 (2015)

21. Sangwan, S.: Virtual community success: A uses and gratifications perspective. In: Proceedings of the 38th Annual Hawaii International Conference on System Sciences. IEEE, Washington, DC (2005)

22. Zillmann, D., Vorderer, P. (eds.): Media Entertainment: The Psychology of its Appeal. Lawrence Erlbaum Associates, Mahwah (2000)

23. Taylor, B.R.: Dark Green Religion: Nature Spirituality and the Planetary Future. University of California Press, Berkeley, Los Angeles, London (2010)

24. Kovich, M.: Sport as an art form. J. Health Phys. Educ. Recreation **42**(8), 42 (1971)

25. McMillan, S.J.: The challenge of applying content analysis for the world wide web. In: Krippendorff, K., Bock, M. A. (eds.) Content Analysis Reader, pp. 60–67. Sage, Thousand Oaks (2009)

26. Krippendorff, K.: Content Analysis: An Introduction to its Methodology, 2nd edn. Sage, Thousand Oaks (2004)

27. Winter, S., Kreuzinger, H.: The Bad Reichenhall ice-arena collapse and the necessary consequences for wide span timber structures. In: Proceedings World Conference on Timber Engineering (WCTE) 2008 Conference, Miyazaki, Japan (2008)

28. Glück, Judith, Bluck, Susan: The MORE life experience model: a theory of the development of personal wisdom. In: Ferrari, Michel, Weststrate, Nic M. (eds.) The Scientific Study of Personal Wisdom, pp. 75–97. Springer, Dordrecht (2013). https://doi.org/10.1007/978-94-007-7987-7_4

29. Katz, E., Blumler, J.G., Gurevitch, M.: Uses and gratifications research. Public Opin. Q. **37**(4), 509–523 (1973)

30. Mulholland Jr., M.R.: Invitation to a Journey: A Road Map for Spiritual Formation. InterVarsity Press, Westmont (2016)

31. Ironson, G., Stuetzle, R., Fletcher, M.A.: An increase in religiousness/spirituality occurs after HIV diagnosis and predicts slower disease progression over 4 years in people with HIV. J. Gen. Intern. Med. **21**(S5), S62–S68 (2006)

32. Stewart, P., Alexander, R.: Broadcast Journalism: Techniques of Radio and Television News, 5th edn. Focal Press, Oxford (2016)

33. Welford, A.T.: Desire for attention. Aust. N. Z. J. Psychiatry **11**(3), 157–161 (1977)

34. Baumann, E., Czerwinski, F., Reifegerste, D.: Gender-specific determinants and patterns of online health information seeking: results from a representative German health survey. J. Med. Internet Res. **19**(4), e92 (2017)

35. McStay, R.L., Dissanayake, C., Scheeren, A., Koot, H.M., Begeer, S.: Parenting stress and autism: the role of age, autism severity, quality of life and problem behaviour of children and adolescents with autism. Autism **18**(5), 502–510 (2014)

A Text Analysis Based Method for Obtaining Credibility Assessment of Chinese Microblog Users

Zhaoyi Ma[✉] and Qin Gao[✉]

Tsinghua University, Peking 100084, China
mazhaoyi25@126.com, gaoqin@tsinghua.edu.cn

Abstract. Nowadays, social network platforms, such as Weibo have become important ways for people to get information. However, considering their ability to disseminate information quickly and widely, they also become breeding ground of rumors. Thus, the information credibility on Weibo has become a meaningful issue to be paid attention to. This research investigated on the process of rumors spreading and refuting, using a new method which can quantify users' credibility assessment. The method was based on text analysis, designing weighted sum of words frequency as microblog's credibility score. By this method, two significant processes related to rumor spreading, the credibility dissemination process of a microblog and the change of people's credibility assessment towards a rumor, were investigated and assessed. Shown from the results some enlightenments about refuting rumors could be concluded. The findings are: (1) the designed method is reliable in some situation. Although it is a simple, rough classification method, it can find the credibility attitude the microblog expressed and help us finding essential users. (2) Popular individuals is important in the dissemination of credibility assessment in Weibo, they can easily convey their judgement. (3) The key to rumor refutation is not only timely reaction, but also wide and sustaining spread to form consensus. If the disputation is not continuous, people may forget it soon.

Keywords: Language and culture in social computing and social media
User generated content (wikis, blogs, etc.) · Information credibility

1 Introduction

1.1 Background

With the development and popularization of Internet, social media has become an important way for people to get information. According data from CINI (China Internet Network Information Center), there were approximately 700 million Internet users in China, and 34% were using a popular kind of social media called "Weibo". For Weibo users, this social media has been an important information source so far, derived from the report about China new media. Actually, Weibo has become a major source of information, especially in young people.

© Springer International Publishing AG, part of Springer Nature 2018
G. Meiselwitz (Ed.): SCSM 2018, LNCS 10914, pp. 229–235, 2018.
https://doi.org/10.1007/978-3-319-91485-5_17

However, the openness and flatness of Weibo has made information disseminate quickly and widely via it. Consequently, Weibo provides a convenient entrance for rumors to arise and spread. Every year, rumors emerge on Weibo and some of them spread to whole Internet widely. There even exited some rumors that have misled many people and caused chaos. In this case, information credibility on Weibo needs to be studied, focusing more on information about dissemination of rumors on this social media.

1.2   Literature Review

In recent years, many researchers tried to explain the process of information credibility assessment by different models. There were several kinds of models, based on different assumptions. The first is The elaboration likelihood model proposed by Petty and Cacioppo (1986, 1996). In this model, ability, peripheral cues are taken as essential factors in credibility assessment. The second one is Dual-process model (Chaiken and Trope 1999). In this model, motivation and ability play important roles in the process of credibility assessment. The third model is presented by Rieh (2002). It is found that people will make prejudge based on experience before visit the website, and the new credibility assessment will be cues to do further assessment. The fourth model is "Three-Stage" model, in which the process of website evaluation is considered as three stages. Other models, such as "MAIN" model, translate the process in different points of view. Synthesizing previous research, Liao (2015) proposed a model to describe information assessment process on Sina Microblog. The model combine socialized and nonsocialized factors, in which users use different heuristic evaluation rules aiming at each kind of cues. In most cases, evaluation process accompanied with many circular assessment process and each of them end with personal interpretation. After several rounds, people will produce credibility judgement to the microblog.

Assessment of information credibility is highly correlated with information source, but the topic of information is found not significantly correlated with it. Westerman et al. (2012) find curvilinear effects for number of followers, meaning that having too many or too few connections results in lower judgements of expertise and trustworthiness. The ratio of followers/followees significantly correlates with the competence. In addition, the appearance characteristics of information source will influence the credibility assessment of information source.

Carlos Castillo et al. (2013) focused on automatic methods for assessing the credibility of a given set of tweets by machine learning. They find active users and new users who had many followers and followees tend to publish credible information, and characristics based on emotion is related to credibility evaluation. Similarly, machine learning is used by Krzysztof et al. (2015) to assess information credibility automatically. Combining content and feature, the research proposed a method and greatly improved the prediction accuracy.

We proposed a text analysis based method to obtain credibility assessment of Weibo content. In the method, "perceived credibility" is what we want. And in fact, what people take as "credible" rather than whether the information is "credible" make sense. What's more, this method can provide advice on rumor control and marketing.

2 Methodology

2.1 Corpus

The research based on Sina Weibo, an online social network used by 240 million Chinese. According to study of Bollen et al. (2009), corpus containing different credibility assessments is necessary to automatic acquisition of credibility assessment. Words in the corpus were from previous studies and crawled microblogs. Former studies proved that some words related to certain kind of credibility assessment, but in real language environment, those were insufficient. Thus, we crawled microblogs and manually labelled these microblogs with different credibility assessment. For labelling accuracy, two other people participated to get the Kappa index, and result showed our manual labelling was reliable. Then we did word frequency analysis to them, and sorted words that often appeared in certain kind of microblogs. So far, corpus was set up. However, it was obviously different words had different weight on the expression of credibility assessment. For example, "fake news" expressed incredible attitude intensively while "verify" just showed tendency.

Thus, we grouped those words in five groups (Table 1). Among them, first three groups (G1–G3) related to "disbelieve". G4 showed tendency to "believe" while G5 correlated with "questionable" attitude. To verify the difference, we selected 232 tagged microblogs, counted the number of words of each group in those microblogs and recorded their corresponding credibility assessments. Using binary regression for the two variables, we found that different groups of words indeed had different weight to assessment

Table 1. The grouping of words in the corpus

G1	rumors (谣言, 谣传) reputation, refute a rumor, dispel a rumor, contradict a rumor, deny a rumor (辟谣) start a rumor (造谣) an old rumor reappears recently (旧谣新传) charge, accuse (举报) fake news (假消息) nonsense (瞎扯) hollow, false, fake, untruthful, inauthentic, untrue, unreal, inveracious (不实) adverse, negative (负面) jackal, accomplice, accessary(帮凶) ignorant, fatuity, stupid, benighted, foolish, blockhead, simpleton, fool, ass, idiot, flathead, dummy (有病, 弱智, 愚昧, 无知, 脑残, 白痴, 脑子, 智障, 傻, 猪脑) mistake, error, fault, slip, inaccuracy (错误) absurd, ridiculous, fallacious, preposterous (荒谬) misunderstand, misconception, misinterpretation (误解) mislead, misdirect, misguide, misadvise (误导) pop science, dissemination of science (科普) clarify (澄清)
G2	bug, loophole, leak (漏洞) incomplete, partial (不全) one-sided, unilateral (片面)
G3	indeed, groundless (纯属) query, doubt (质疑) panic, scare, trepidation, horror(恐慌) stand hand(a emoticon), truth (真相) falsify, cheat, forge, counterfeit (作假, 造假) false, sham, fake (虚假) unreliable, uncertain, unsoundness (不靠谱,不可靠) doubt, disbelieve, discredit (不信, 别信) verify (核实) cheat (欺骗, 忽悠) expert (专家) verify, affirm (证实) examine, inspect (检查, 核查) test, check (检验) treat unjustly, grievance (冤枉) rubbish, waste, garbage (垃圾)
G4	black heart, evil mind (黑心) have no conscience (无良)
G5	I don't know true or false, I couldn't tell the real from the unreal (不知真假, 不知道真假, 求真假) seek confirmation, make clear, figure out, get it right (求证) Seriously? Really? (真的假的)

deduction and designed their weight in the classification method. In addition, we found people who said nothing special just believed and commented the news.

2.2 Weighted Scoring Process

Based on above work, we designed the weighted sum of words frequency as microblog's credibility score. Detailed classification process was as Fig. 1.

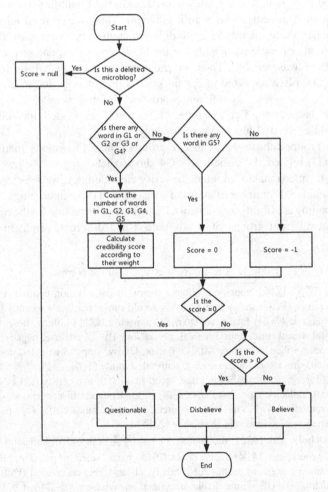

Fig. 1. The weighted scoring process

First, check if there are words related to "incredible" in the microblog, if so, then check if there are words related to "questionable", If so,then calculate the credibility mark according to the credibility weight, decide which assessment the microblog can be labelled. If no word about "incredible" is found, then words about "questionable" is checked. If found, the microblog is labelled as "questionable". If no words about "incredible" or "questionable" is found in the microblog, then it is labelled as "credible".

2.3   Validation

Two other people are recruited to label the microblogs to guarantee the accuracy. The kappa coefficient of them is larger than 0.8, meaning that two labelers are reliable. Then we apply the method to the same material, and find the agreement is 92.07%, which is acceptable.

3   Result

3.1   The Dissemination Process of Credibility Assessment in Weibo

First, we chose news "a baby who burn with an intrauterine ring" to study the dissemination process of credibility assessment in Weibo. We crawled the reposts of source microblog from May 9th to June 2nd, 2017 and applied the designed method to them. Then we obtained the credibility assessments the reposts expressed and observed the change. Figure 2 showed that the proportion of "believe" went up and down for a while, and came up again at the end time.

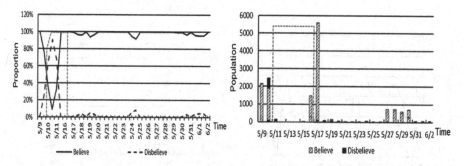

Fig. 2. The change of repost users' credibility assessment

To figure out what happened, we chose the change time as a point to collect data and painted repost path network. Result was in Fig. 3. Nodes represented users, and were colored according to their credibility assessments.

Comparing two figures, we found influential users played significant roles in repost users' credibility assessments and the decline of "believe" proportion was due to refutation of several popular users. News source and those key users led most users and conveyed their credibility assessments successfully. However, combined with Fig. 2, we knew people forgot disputation soon. With the method, we can find the essential users in credibility dissemination intuitively and easily.

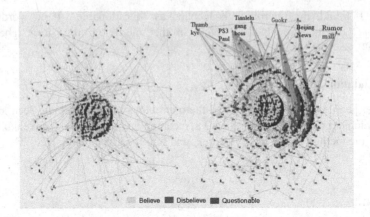

Fig. 3. Repost path network with credibility assessment

3.2   The Change of People's Credibility Assessment of to a News

We chose "plastic rice" for the topic to observe the change of people's credibility assessment of to a news. We crawled related microblogs from May 14th to May 30th, 2017, which is the dissemination period of the rumor. Via the method, we got the credibility attitude they expressed. The change was as follow (Fig. 4):

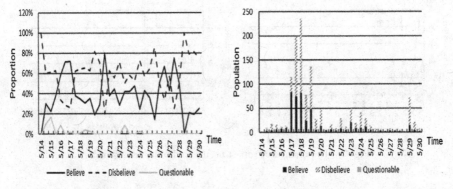

Fig. 4. The change of users' credibility assessment to a news

Apparently, there exited a period, during which the "believe" proportion declined and more people participated the topic. Back to the raw data, we found the phenomenon was caused by many official users' refutation. Those professional and reliable disputations made the topic hotter and "believe" proportion drop sharply. After May 19th, few people participated in the topic and most users took it as fake news. The rumor was controlled.

4 Conclusion

From the above two cases, we find (1) the designed method is reliable in some situation. Although it is a simple, rough classification method, it can find the credibility attitude the microblog expressed and help us finding essential users. (2) Popular individuals is important in the dissemination of credibility assessment in Weibo, they can convey their judgement easily. (3) The key to rumor refutation is not only timely reaction, but also wide and sustaining spread to form consensus. If the disputation is not continuous, people may forget it soon.

References

Bollen, J., Pepe, A., Mao, H.: Modeling public mood and emotion: twitter sentiment and socioeconomic phenomena. Comput. Sci. **44**(12), 2365–2370 (2009)

Carlos, C., Marcelo, M., Barbara, P.: Predicting information credibility in time-sensitive social media. Internet Res. **23**(5), 560–588 (2013)

Chaiken, S., Trope, Y.: Dual-process Theories in Social Psychology. Guilford Press, New York (1999)

Krzysztof, L., Jacek, S.W., Michał, J.L., Amit, G.: Automated credibility assessment on twitter. Comput. Sci. **16**(2), 157–168 (2015)

Petty, R.E., Cacioppo, J.T.: The elaboration likelihood model of persuasion. Adv. Exp. Soc. Psychol. **19**(4), 123–205 (1986)

Petty, R.E., Cacioppo, J.T.: Attitudes and Persuasion: Classic and Contemporary Approaches. Westview Press, Boulder (1996)

Rieh, S.Y.: Judgment of information quality and cognitive authority in the Web. J. Am. Soc. Inform. Sci. Technol. **53**(2), 145–161 (2002)

Westerman, D., Spence, P.R., Van Der Heide, B.: Social network as information: the effect of system generated reports of connectedness on credibility on twitter. Comput. Hum. Behav. **28**(1), 199–206 (2012)

廖晨: 微博信息可信度的评判模型和可视化工具研究 (硕士论文). 清华大学, 清华大学 (2015)

An Experience of Textual Evaluation Using the MALTU Methodology

Marilia S. Mendes[1(✉)] and Elizabeth Furtado[2]

[1] Federal University of Ceará (UFC), Russas, CE, Brazil
marilia.mendes@ufc.br
[2] University of Fortaleza (UNIFOR), Fortaleza, CE, Brazil
elizabet@unifor.br

Abstract. This paper presents an experience of textual evaluation in Usability and User eXperience in an academic system. The methodology used for the textual evaluation was MALTU. In this study, we analyzed 650 postings from an academic system and these posts have gone through a process of textual evaluation whose results will be presented in this paper.

Keywords: Textual evaluation · Usability · User eXperience (UX)
Postings related to the use (PRUs)

1 Introduction

Methods of collecting user opinion about a system, such as: interviews, questionnaires, schedules and Experience Sampling Method (ESM) [18] have been used to obtain User eXperience (UX) data from a product. Although such methods provide valuable data, they do not provide rich UX descriptions of users' daily life, primarily because they are applied at predefined times by researchers (for example developers and evaluators) of systems [9].

In [13–15], the authors of this paper investigated post messages of the users of Social Systems (SS): Facebook and Twitter. Postings that revealed reports of users' experiences will be called herein as Postings Related to the Use (PRUs). Unlike the other textual evaluation works [5, 9, 13, 21, 26], in which users are asked to write about their experience, these posts are spontaneous and report the user's perceptions about the system during its use. A PRU is a post in which the user refers to the system in use, for example: "*I can't change the Twitter profile photo*". A non-PRU is any post that does not refer to the use of the system, such as: "*Let's go to the show on Friday?*". The capture of spontaneous posts is obtained because we collect posts exchanged by the users in the system itself, when it has a forum or space to exchange messages.

In [12] we proposed the Maltu methodology and since then we have been experimenting with textual evaluation in different systems [4, 25]. The purpose of this paper is to present a detailed textual evaluation and discuss interesting points of this new form of systems evaluation. In this work, we analyzed 650 postings of an academic system with social characteristics (e.g., communities, forums, chats, etc.).

G. Meiselwitz (Ed.): SCSM 2018, LNCS 10914, pp. 236–246, 2018.
https://doi.org/10.1007/978-3-319-91485-5_18

This paper is organized as follows: in the next section, we present a background on textual evaluation of systems and the Maltu Methodology. In Sect. 3, we present some researches related to ours. In Sect. 4, we describe the Textual evaluation with Maltu Methodology, followed by results, conclusion and future works.

2 Background

2.1 Textual Evaluation of Systems

The textual evaluation of systems consists of using user narratives in order to evaluate or obtain some perception about the system to be evaluated [12]. It is possible to evaluate one or more criteria of quality of use with textual evaluation, such as usability, UX and/or its facets (satisfaction, memorability, learnability, efficiency, effectiveness, comfort, support, etc.) [6, 9, 12, 13, 21]. Other criteria can be evaluated, such as privacy [11], credibility [3, 11] and security [23]. Evaluation forms vary from identifying the context of use to identifying the facets of Usability or UX. Some papers have analyzed specifically the most satisfactory and unsatisfactory user experiences with interactive systems [5, 20, 21, 26].

The textual evaluation can be manual, through questionnaires with questions about the use of the system or experience reports, in which the users are requested to describe their perceptions or sentiments about the use of the system. The other way is automatic: evaluators can collect product evaluations on rating sites [6] or extract PRUs from Social Systems (SS) [10, 12–15, 17, 19]. The automatic form allows more spontaneous reports, including doubts when using the system, but, on the other hand, may also contain many texts that are not related to the use of the system, and these must be discarded.

Textual evaluation has its advantages and disadvantages, similar to other types of HCI assessment, such as user testing, heuristic evaluation, among others. The main advantage is to consider users' spontaneous opinions about the system, including their doubts. The main disadvantage is the long time of texts analysis. However, there are few initiatives of automatic textual evaluations [16], since it is an new evaluation type.

2.2 The Maltu Methodology

The MALTU methodology [12] for the Usability and UX (UUX) textual evaluation, mentioned in the introduction, consists in using user-generated narratives (postings) done in the own system, usually a SS, where spontaneous comments about the system are reported by users while using it; or from the extraction of postings on product/service evaluation websites [4, 25]. A user's posting can have more than one sentence, which in turn has multiple terms (words, symbols, scores), and those can help investigate what motivated (the cause of the problem) the user to write their posting, as well as what their reaction (behavior) was to the system in use, for example.

The methodology uses five steps for evaluation: (1) definition of the evaluation context; (2) extraction of PRUs; (3) classification of PRUs; (4) results and (5) report of results. In step 1, we define the system under evaluation; the users whose opinion matters to the evaluators; and the purpose of the evaluation. In step 2, the extraction of PRUs can be carried out either manually or automatically, by using the patterns of extraction proposed by the methodology described in [12]. When the extraction is manually done, the evaluators should use the search fields of the system under evaluation by informing the extraction patterns for the recovery of PRUs. When extraction is done automatically, the evaluators should use a posting extraction tool [16]. In step 3, we apply a process of classification of PRUs. This step can also be performed either manually or automatically (by using a tool [16]). When this step is performed manually, the sentences are analyzed by specialists for classification. The methodology proposes the minimum number of two specialists for classification. In addition to the previously mentioned criteria (classification by UUX facets, type of posting: complaint, doubt, praise), it is possible to analyze the user's feelings and intentions regarding the system in use and identify the functionality that may be the cause of the problem. In step 4, we interpret the results, and in step 5 we report them. In the next section, these steps will be more detailed in the evaluation of the academic system.

3   Related Works

Some studies that have focused on user narratives in order to study or evaluate usability or UX. In [5], the authors, focusing on studying UX from positive experiences of users, collected 500 texts written by users of interactive products (cell phones, computers etc.) and presented studies about positive experiences with interactive products. In [9], the authors collected 116 reports of users' experiences about their personal products (smartphones and MP3 players) in order to evaluate the UX of these products. Users had to report their personal feelings, values and interests related to the moment at which they used those. In [20], the authors collected 90 written reports of beginners in mobile applications of augmented reality. The focus was also evaluating the UX of these products, and the analysis consisted in determining the subject of each text and classifying them, by focusing attention on the most satisfactory and most unsatisfactory experiences. Following this line, in [26], the authors studied 691 narratives generated by users with positive and negative experiences in technologies in order to study the UX from them.

In the four studies mentioned above, the information was manually extracted from texts generated by users. The users were specifically asked to write texts or answer a questionnaire, unlike the spontaneous gathering of what they post on the system.

In [6], the authors extracted reviews of products from a reviews website and did a study in order to find relevant information regarding UUX in texts classified by specialists. However, they did not investigate SS, but other products used by users. In this case, the texts were written by products reviewers. It is believed that the posture of users in a product review website is different from that when they are using a system and face a problem, then deciding to report this problem just to unburden or even to suggest a solution. In addition, in none of these studies was a methodology used to present system evaluation results. In this work, we focused on considering the opinions of users about the system in use from their postings on the system being evaluated. We intend thereby to capture the user spontaneously at the moment they are using the system and evaluate the system.

4 Textual Evaluation Using the MALTU Methodology

The evaluation will be described, following the steps of the Maltu methodology.

(1) Definition of the evaluation context

The investigations were carried out in PRUs written in Brazilian Portuguese, collected from the database of an academic system with social characteristics (communities, discussion forums, chats, etc.) called SIGAA [24], which is the academic control system of the Federal Universities in Brazil. In this system, students can have access to several functionalities, such as: proof of enrollment, academic report, enrollment process, etc. The system allows the exchange of messages from a discussion forum. Its users are students and employees from the university. The system can be accessed by a browser on computers and mobile phones.

(2) Extraction of PRUs

For this work, 650 PRUs were selected from a part of the database coming from a previous work [12]. In this previous work, from a total of 295,797 posts, this sample of posts was collected by IHC specialists. The selection criteria was to collect postings in which users were talking about the system. An example of a PRUs collected was: "*I cannot stand this SIGAA anymore!*". Postings from students asking questions about their graduation courses, grades, location, etc. were not selected, for example: "*Will the coordination work during vacation?*" and "*Professors did not post the grades yet*".

(3) Classification of PRUs

The PRUs contained between one and six sentences each. That is why many times the post starts praising the system and ends up criticizing it, for example: "*I think this new system has a lot to improve*" (Negative Feeling)…"*However, it is already much better than the previous one*" (Positive Feeling). In this way, we divided the PRUs into sentences. After this division, we performed another analysis in order to verify the related and unrelated sentences to the use of the system, because there were sentences such as: "*Good morning*", "*Thank you*", "*Sincerely...*", which were not related to the use of the system. In this way, we discarded such sentences.

The rating process consists of categorizing a post into an evaluation category. There are seven types of classification categories for evaluation: (i) type of message to be investigated; (ii) intention of the user; (iii) polarity of Sentiment; (iv) intensity os sentiment; (v) quality in use criterion; (vi) functionality; and (vii) platform.

(i) **Type of message:** this type of classification refers to investigating what type of message the user is sending over the system in use, which can be: **(a) critical**: containing complaint, error, problem or negative comment regarding to the system; **(b) praise** or positive comment about the system; **(c) help** (giving of) to carry out an activity in the system; **(d) doubt** or question about the system or its functionalities; **(e) comparison** with another system; and **(f) suggestion** about a change in the system;

(ii) **Intention of the user:** the intention classification aims to classify the PRUs according to the user's intention with the system. In [17], a classification of PRUs was made in the categories: visceral, behavioral and reflexive. The definitions that emerged from the PRU were as follows:

(a) **Visceral PRU:** has greater intensity of user's sentiment, usually to criticize or praise the system. It is mainly related to attraction and first impressions. It does not contain details of use or system features. These are two examples: *"I'm grateful to SIGAA which has errors all the time: ("* and *"This System does not work!!! < bad language > !!"*;

(b) **Behavioral PRU:** has lower intensity of user's sentiment and is also characterized by objective sentences, which contain details of use, actions performed, functionalities, etc. Two examples are the following: *"I would like to know how you can add disciplines to SIGAA"*; and *"It's so cool to be able to enter here"*;

(c) **Reflective PRU:** is characterized by being subjective, presenting affection or a situation of reflection on the system. One example: *"The system looks much better now than it did last semester, when it was installed"*.

Information of Sentiment: in this category, two forms of classification are presented to analyze the sentiment in the PRUs: **(iii) polarity:** a PRU can demonstrate positive sentiment, neutral sentiment and negative sentiment; and **(iv) intensity:** allows us to classify how much of sentiment (positive or negative) is expressed in a PRU. In the examples: *"I like this system..."* and *"I really love using this system"*. The positive sentiment observed is more intense in the second PRU. This type of classification is only performed automatically [12].

(v) **Quality in use criterion:** this category involves determining the criterion of quality in use. The Maltu uses the following criteria: **(a) usability** and/or **(b) UX.** This category involves relating a facet of each criterion to a PRU. Maltu uses the following facets for Usability: efficacy [7], efficiency [7], satisfaction [7], security [22], usefulness [22], memorability [22] and learning [22]. For UX, the facets used are: satisfaction [7], affection [1], confidence [1], aesthetics [1], frustration [1], motivation [2], support [8], impact [8], anticipation [8] and enchantment [8];

(vi) **Functionality:** there are PRUs that detail the use of the system, making it possible to classify the functionality of the system and is referred to by the user or the cause of the problem to which the user refers. In the exemplo: *"I can not exclude disciplines. Can someone help me?"*, the functionality is "exclude disciplines"; and

(vii) **Platform:** this category consists of identifying the operating system and device that the user was using at the time of the relative posting. There are systems, like Twitter and Facebook, for example, where the PRUs extracted from the system can come from different devices. On SIGAA, as access is by browser, it can also be accessed from different devices.

We illustrate (Fig. 1) the following some examples of classification of postings.

Type of message	Intention of the user	Sentiment polarity	Usability	UX	Functionality	Platform
Critical	Visceral	Negative	-	Frustration	-	-

"Really, think of a bad system. Now raise to the square, multiply by a thousand, and so it still gets very bad, the worst system!!!!!"
Sentiment

Type of message	Intention of the user	Sentiment polarity	Usability	UX	Functionality	Platform
Critical	Behavioral	Negative	Security	Frustration	Update the screen	-

Functionality
"Every time I update the screen, I put my finger to the left and goes to the 'discover' tab grrr!!!!!"
Sentiment

Type of message	Intention of the user	Sentiment polarity	Usability	UX	Functionality	Platform
Doubt	Behavioral	Neutral	Efficacy	-	Disciplines choices	-

Functionality
"How to make the disciplines choices if the system is not showing the menu?"
Problem

Type of message	Intention of the user	Sentiment polarity	Usability	UX	Functionality	Platform
Praise	Reflective	Positive	satisfaction	satisfaction, affection	-	-

Reflection
"It seems that the system is much better now than last semester, when it was adopted for the first time. I am optimistic about SIGAA. It is very interactive and has everything to be a good tool for the
Praise   *entire academic community."*
Sentiment

Fig. 1. Examples of classification of postings

According to the examples presented, it is not always possible to categorize a post in all proposed classification forms. The classification form took place as follows: 500 PRUs were classified by 10 undergraduate students and 150 by IHC specialists, totaling 650 PRUs, corrected by two IHC specialists.

(4) Results and (5) Report results

The graphs and tables presented below, in this section, present the relationship between the classifications obtained, providing an overview of the evaluated system. Graph 1 illustrates the percentages obtained in each usability facets related to PRUs of the critical type. The efficacy facet, for example, obtained a higher percentage (48%). Graph 2 shows the percentages obtained in each UX facet related to PRUs of the critical type. The frustration facet, for example, obtained higher percentage (84%).

Graph 3 illustrates the percentages obtained in each UX facet related to praise PRUs. The satisfaction facet, for example, obtained a higher percentage (43%).

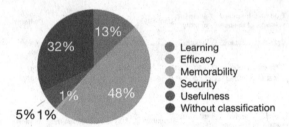

Graph 1. Quality of use criteria = usability x type of PRU = critical

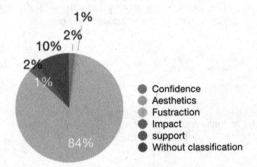

Graph 2. Quality of use criteria = UX x type of PRU = critical

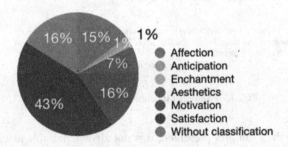

Graph 3. Quality of use criteria = UX x type of PRU = praise

Table 1 presents the functionalities collected from the critical-type PRUs in each usability facet. In the memorization facet, the criticisms were referring to: "*a lot of information*", "*how to register*", "*visual*". Table 2 presents the percentages and functionalities collected from PRUs of praise type in each usability facet. The highest percentage, satisfaction facet, indicates that users are satisfied with SIGAA for the following reasons: "communication", "interaction", "beauty", "new features", "practicality" and "sociable".

Table 3 presents the functionalities collected from the critical-type PRUs in each UX facet. The frustration facet, for example, presents a greater number of causes cited in the PRUs. The others have few functionalities, because, through the analysis

Table 1. Quality of use criteria = usability x type of PRU = criticism x cause.

Facets of usability	Functionalities
Learning	View, download or insert file; view or error in the disciplines; edit information; view or history error; error in calculating the media; perform, display or error in the registration; view notes, classes, frequency or faults; lock registration
Efficacy	View, download, open or insert file; credits less; view or history error; Perform, display or error in registration; view notes or times; error in calculating the media; blocking in the system, system in general
Security	Perform, display or error in registration; view or error in the disciplines
Usefulness	Browser; room location
Memorability	A lot of information; how to registration; visual

Table 2. Quality of use criteria = usability x type of PRU = praise x cause.

Percentage	Usability facet	Functionalities
2%	Learning	General system
5%	Efficacy	Make registration; General system
63%	Satisfaction	Communication; interaction; Beauty; new features; practical; sociable
15%	Usefulness	Warnings; Communication; interactivity; Discussion forums; system in general
15%	Without classification	–

Table 3. Quality of use criteria = UX x type of PRU = critical x cause

UX Facet	Functionalities
Fustraction	Menu unavailable; View, download or insert file; Calendar; accounting for claims; view or history error; Make registration; view or understand the disciplines' schedules; access only by the Firefox browser; error in calculating the media; view groups
Support	Make registration
Impact	Previous system
Confidence	Grades; Registration
Esthetics	Visual
Without classification	–

performed PRUs – UX classifications, the users did not present details of the system. Table 4 presents the main functionalities that the users had doubts and Table 5 presents suggestions of functionalities for the system.

Table 4. Main features that users had doubts

Type of PRU = doubt x functionalities
Edit information; View, download or insert file; Lock in system; View or error in the disciplines; how to hide board registration numbers; View or error in history; To visualize or to understand the class schedule, notes, amendment of the disciplines; media calculation error; How to make a lock

Table 5. Main features suggestions

Type of PRU = suggestion x features
option to "enjoy", to do tests at home; Location map of the room allied to the disciplines; improvement of the system, explanation of the time code;

The Fig. 2 illustrates the system usage context obtained from the evaluation of a set of PRUs.

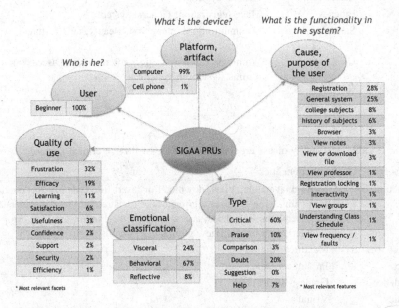

Fig. 2. Context of use of the SIGAA system evaluation

5 Final Considerations and Future Work

The results obtained using the methodology pointed to UUX problems, the main functionalities in which the users have doubts, criticisms and suggestions about SIGAA. As for the evaluation experience using Maltu, the classification stage was sometimes not simple, since the extracted PRUs were characterized by an average of 3 lines each, being at least 1 and at most 10 lines. In this way, the classification has become, at times, a slow and tiring process for the evaluators.

This paper reported a textual evaluation experience of UUX of SIGAA. The results have shown that the application receives many criticisms from various causes, mainly being support and efficacy problems that cause frustration to users of the application. Maltu is a recent methodology. Its use in this work consisted in the validation of the methodology by the application in different contexts. Future work will seek new ways to improve the classification process of PRUs with Maltu, in order to simplify and automate the extraction, classification and interpretation of results. Other suggested forms of classification will also be used. Another activity to be carried out is the expansion of the database, since only a specific source of complaints was used.

Acknowledgment. This project is funded by the Fundação Cearense de Apoio a Desenvolvimento Científico e Tecnológico (FUNCAP) and supported by LINCE Laboratories (Laboratory of Computer Science and Software Engineering) and LUQS (Laboratory of user studies and Quality of use of the Systems).

References

1. Bargas-avila, J.A., Hornbæk, K.: Old wine in new bottles or novel challenges: a critical analysis of empirical studies of user experience. In: CHI 2011, Vancouver, pp. 2689–2698 (2011)
2. Bevan, N.: What is the difference between the purpose of usability and user experience evaluation methods? In: UXEM 2009, INTERACT 2009, Uppsala, Sweden, pp. 24–28 (2009)
3. Castillo, C., Mendoza, M., Poblete, B.: Information credibility on Twitter. In: Proceedings of the 20th international Conference on World Wide Web, Hyderabad, India, pp. 675–684 (2011)
4. Freitas, L., Silva, T., Mendes, M.: Avaliação do Spotify – uma experiência de avaliação textual utilizando a metodologia MALTU. In: IHC 2016, São Paulo, Brazil (2016)
5. Hassenzahl, M., Diefenbach, S., Goritz, A.: Needs, affect, and interactive products. J. Interact. Comput. **22**(5), 353–362 (2010)
6. Hedegaard, S., Simonsen, J. G.. Extracting usability and user experience information from online user reviews. In: Proceedings of CHI 2013, Paris, France, pp. 2089–2098 (2013)
7. ISO DIS 9241 - 210:2008: Ergonomics of human system interaction - Part 210: Human - centred design for interactive systems (formerly known as 13407)
8. Ketola, P., Roto, V.: Exploring user experience measurement needs. In: 5th COST294-MAUSE Open Workshop on Valid Useful User Experience Measurement (2008)
9. Korhonen, H., Arrasvuori, J., Väänänen-vainio-mattila, K.: Let users tell the story. In: Proceedings of CHI 2010, pp. 4051–4056 (2010)
10. Lima, A., Silva, P., Cruz, L., Mendes, M.: Investigating the polarity of user postings in a social system. In: 19th International Conference on Human-Computer Interaction, HCII 2017 (2017)
11. Mao, H., Shuai, X., Kapadia, A.: Loose tweets: an analysis of privacy leaks on Twitter. In: Proceedings of WPES 2011, Chicago, IL, USA, pp. 1–12 (2011)

12. Mendes, M.S.: MALTU - model for evaluation of interaction in social systems from the users textual language, 200 f. Thesis (Ph.D. in Computer Science) – Federal University of Ceará (UFC), Fortaleza, CE – Brazil (2015)

13. Mendes, M.S., Furtado, E.S., Furtado, V., Castro, M.F.: Investigating usability and user experience from the user postings in social systems. In: HCI International (2015)

14. Mendes, M.S., Furtado, E.S., Militao, G., Castro, M.F.: Hey, I have a problem in the system. Who can help me? An investigation of Facebook users interaction when facing privacy problems. In: HCI International, pp. 391–403 (2015)

15. Mendes, M.S., Furtado, E., Castro, M.F.: Do users write about the system in use? An investigation from messages in natural language on Twitter. In: 7th Euro American Association on Telematics and Information Systems, Valparaiso, Chile (2014)

16. Mendes, M., Furtado, E.: UUX-Posts: a tool for extracting and classifying postings related to the use of a system. In: VIII Latin American Conference on Human-Computer Interaction, CLIHC 2017 (2017)

17. Mendes, M., Furtado, E., Furtado, V., Castro, M.: How do users express their emotions regarding the social system in use? A classification of their postings by using the emotional analysis of Norman. HCI Int. **2014**, 229–241 (2014)

18. Obrist, M., Roto, V., Väänänen-vainio-mattila, K.: User experience evaluation – do you know which method to use? In: CHI 2009, pp. 2763–2766 (2009)

19. Oliveira, D., Furtado, E., Mendes, M.: Do users express values during use of social systems? A classification of their postings in personal, social and technical values. In: HCI International, Los Angeles, CA, USA (2016)

20. Olsson, T., Salo, M.: Narratives of satisfying and unsatisfying experiences of current mobile augmented reality applications. In: CHI 2012, pp. 2779–2788 (2012)

21. Partala, T., Kallinen, A.: Understanding the most satisfying and unsatisfying user experiences: emotions, psychological needs, and context. Proc. Interact. Comput. **24**(1), 25–34 (2012)

22. Preece, J., Rogers, Y., Sharp, H.: Interaction Design: Beyond Human-Computer Interaction. Wiley, New York (2002)

23. Reynolds, B., Venkatanathan, J., Gonçalves, J., Kostakos, V.: Sharing ephemeral information in online social networks: privacy perceptions and behaviours. In: Campos, P., Graham, N., Jorge, J., Nunes, N., Palanque, P., Winckler, M. (eds.) INTERACT 2011. LNCS, vol. 6948, pp. 204–215. Springer, Heidelberg (2011). https://doi.org/10.1007/978-3-642-23765-2_14

24. SIGAA: Integrated Management System for Academic Activities. https://si3.ufc.br/sigaa/verTelaLogin.do. Accessed 10 Mar 2018

25. Silva, T., Freitas, L., Mendes, M.: Beyond traditional evaluations - users view in app stores. In: IHC 2017, Joinville, Brazil (2017)

26. Tuch, A.N., Trusell, R.N., Hornbæk, K.: Analyzing users' narratives to understand experience with interactive products. In: Proceedings of CHI 2013, pp. 2079–2088 (2013)

Early Tracking of People's Reaction in Twitter for Fast Reporting of Damages in the Mercalli Scale

Marcelo Mendoza[1](✉), Bárbara Poblete[2], and Ignacio Valderrama[2]

[1] Centro Científico y Tecnológico de Valparaíso,
Universidad Técnica Federico Santa María, Valparaíso, Chile
marcelo.mendoza@usm.cl
[2] Department of Computer Science, Universidad de Chile, Santiago, Chile
{bpoblete,ivalderr}@dcc.uchile.cl

Abstract. The Modified Mercalli Intensity Scale is a measure of the severity of an earthquake for a nonscientist. Since the Mercalli scale is based on perceived effects, it has a strong dependence on observers. Typically, these reports take time to be prepared and, as a consequence, Mercalli intensities are published hours after the occurrence of an earthquake. The National Seismological Center of Chile needs to provide a preliminary overview of the observed effects of an earthquake. This has motivated us to create a system for early tracking of people's reaction in social networks to infer Mercalli intensities. By tracking people's comments about the effects of an earthquake, a collection of Mercalli point estimates is retrieved at county level of granularity. We introduce the concept of Reinforced Mercalli support that combines Mercalli point estimates with social support, allowing to discard social unsupported estimates. Experimental results show that our proposal is accurate providing early Mercalli reports 30 min after an earthquake, detecting the maximum Mercalli intensity of an event with high accuracy in terms of mean absolute error (MAE).

Keywords: Social networks · Disaster management
Mercalli intensity · Social media during emergencies

1 Introduction

The Modified Mercalli intensity scale (from now on Mercalli) is a commonly used measure that summarizes the effects of an earthquake in public infrastructure and damages perceived by people. Unlike the *moment magnitude* scale, which quantifies the released energy during an earthquake. Energy and damages may differ due to a number of physical variables, as the depth of the seismic movement and the geological composition of the ground. In addition, energy and damages may differ due to the standard used to certify the quality of buildings.

Mercalli reports are prepared by observers providing ratings for earthquakes in a given location. Seismological centers keep groups of observers distributed

© Springer International Publishing AG, part of Springer Nature 2018
G. Meiselwitz (Ed.): SCSM 2018, LNCS 10914, pp. 247–257, 2018.
https://doi.org/10.1007/978-3-319-91485-5_19

along territories providing these reports. Usually, Mercalli reports are released hours after an earthquake, as the strong dependence on local observers makes difficult to provide fast and fresh information. Many factors obstruct fast reporting, among them the quality of communications during a disaster or the observer availability. To provide a first fast report of damages, the National Seismological Center have pay attention to the information propagated through social networks.

In this paper we study how social networks can be used to infer damages in the Mercalli scale after an earthquake. The state of the art show some efforts in this direction with promising results in the problem of earthquake detection. We extend the state of the art providing fast Mercalli intensity reports, focusing our efforts in the estimation of the maximum intensity in the Mercalli scale of damages. The spatial dimension of the problem, namely how people distribute along a territory and how this piece of information is involved in the Mercalli inference process, is the key building block of our fast Mercalli intensity report method.

Main contribution of the paper: In this paper we address the problem of Mercalli intensity inference using the spatial dimension of the data. From our point of view, social tracking is naturally related to the description of the effects associated to a given earthquake, as the Mercalli scale is a scale of perceived damages. However, the spatial dimension of the problem, namely how people is distributed along a territory and how this information is included in the inference process, is a key aspect of the problem not addressed in the state of the art. We claim that this aspect can not be discarded. Experimental results will show that our method outperforms the state of the art in the specific task of maximum Mercalli intensity detection.

This paper is organized as follows. Related work is discussed in Sect. 2. Our method is introduced in Sect. 3. Experiments are discussed in Sect. 4. Finally, we conclude in Sect. 5 giving conclusions and outlining future work.

2   Related Work

Twitter has been a social network of much research along time, as it is a huge source of user-generated content, which currently reaches more than 300 million active users per month. This is the reason why several scientists have researched Twitter with the aim of exploiting the information available, such as finding correlations between Twitter and physical events [6]. These efforts have shown interesting results. For instance, during the Tohoku earthquake in 2011, a research highlighted high correlations between the amount of tweets and the intensity of the disaster in some locations [5].

The elaboration of early reports of seismic event based on Twitter has been of growing interest during the last five years. Quake alert systems have been developed in different places of the world such as Australia [7] or Italy [1]. These systems use a burst detection algorithm to report an earthquake, where a burst is defined as a large number of occurrences of tweets within a short time window [9].

Despite the fact that these systems just report that an earthquake happened in a given location, they have shown that it is possible to infer more information. Maybe the most salient result on seismic-event report relies on the estimation of the epicenter of an earthquake event using only information recovered from Twitter [8] as tweets counts and tweets rates.

Burks *et al.* [2] proposed the first approach to estimate the Mercalli intensity of an earthquake using Twitter. Conditioned on a set of reports retrieved from seismological recording stations provided by the Japanese seismological center, the area around each recording station is segmented into 9 radial areas, mapping tweets that mention the word 'earthquake' to these areas. Lexical features in each areal disc are calculated to study the correlation of these features with the Mercalli intensity. To do this, the authors explored a number of linear regression models, showing good results in terms of accuracy. We take some inspiration from the ideas explored by Burks *et al.* to design our method, but discarding the use of data recovered from seismological stations. The point here is that Japan has a huge network of seismological recording stations distributed along its territory, providing valuable information to the method. The aim of our study is to explore the predictive power of the social network itself, in specific Twitter, in absence of seismological recording stations reports, to provide a first fast Mercalli intensity report that does not depend on the quality and coverage of the seismological sensor network.

An on-line system named TwiFelt [4] has exploited the Twitter stream to provide an estimation of the extension area where an earthquake was felt in Italy. The system only use geo-located tweets to infer the area showing promising results for high intensity earthquakes. One limitation of this system relies on the availability of geo-located tweets, as a great proportion of the tweets recovered in our country from the stream does not include the tweet location.

Maybe the closest work to our proposal is the one authored by Cresci *et al.* [3]. In that paper the authors studied how to use Twitter to estimate the maximum intensity of an earthquake using only Twitter features. Using linear regression models over a huge collection of aggregated features (45 features were tested in that proposal), the authors showed that Twitter has enough predictive power to infer the maximum intensity of an earthquake in the Mercalli scale. The set of features tested by the authors comprises features extracted from user profiles, tweets contents and time-based features of the tweet stream (e.g. tweet interval rates). Our proposal can be considered as an extension of this work but focusing only on tweets contents. We will use a linear regression model for a first estimation of the Mercalli intensity. The main difference in the maximum intensity task between our method and the method proposed by Cresci *et al.* is that our proposal works over a reduced set of features (only 12 lexical features) in comparison with the 45 features used in [3]. We will show in our experiments that out method performs well in this specific task, taking advantage of the spatial dimension of the data boosting the results achieved by Cresci *et al.* [3].

3 Early Inference of Mercalli Intensities

We use information gathered from social networks, in specific from Twitter, to infer damages in the Mercalli scale. As this information can be collected and summarized in short periods of time, it is possible to infer at the early stages of an emergency the Mercalli intensities of a given earthquake. We divide the inference process in 3 stages: (1) Social tracking of earthquake effects, (2) Estimation of a region of interest, and (3) Inference of the maximum Mercalli intensity.

The first stage corresponds to the social tracking of an earthquake's effects. Each event of interest is characterized at county level, the finer level of geolocation considered in our method. The second stage of the process starts with a regression process applied to infer the region of interest of a given earthquake. The last stage of our method takes the collection of point estimates to infer the maximum intensity in the Mercalli scale for a given seism. We introduce a Reinforced Mercalli variable that is used to adjust the Mercalli estimate according to the level of support of the point estimate. Finally, we look for the maximum intensity at the area of interest, fixing this value as the maximum intensity of the earthquake.

3.1 Social Tracking of Effects

Posts are collected to extract features of the event that characterize the social perception of the earthquake. In our study we use Twitter to collect the data. Each perceived event is characterized at a level of aggregation that describes the perception of the earthquake in a county. For each county batch, a set of features is calculated to describe the earthquake.

County batches are built as follows. After each earthquake, a set of tweets that matches the keywords "quake", "earthquake" or "seismic" are retrieved from Twitter. The time considered to collect the data is a parameter of our system, with a window length of 30 min by default. Shorter windows can be considered but at the cost of less accurate Mercalli predictions. Tweets that are mapped to counties are aggregated into county batches. This piece of data is the basic unit of earthquake characterization used for feature extraction.

We map tweets to counties using the user location field. We were forced to use this field as only a very small fraction of the tweets in our country is geo-located. The user location field is retrieved from the user profile and then, using a fuzzy string matching procedure, it is mapped to a specific county. We understand that many tweets will be effectively posted from a location matching the user location, giving us a trace of the tweet spatial distribution. More accurate methods for tweet geolocation can improve this aspect of our method but to the best of our knowledge, this task is challenging and it is still open.

Twelve features are considered at this level of aggregation, as is shown in Table 1. These features are calculated in each county data batch, characterizing the set of tweets mapped to each specific county for a given seism.

Table 1. Features used for our tracking system

Feature	Description
NUMBER OF TWEETS	Number of tweets in the data batch
TWEETS NORM	Fraction of tweets over county population
AVERAGE WORDS	Average length of tweets in number of words
AVERAGE LENGTH	Average length of tweets in number of chars
QUESTION MARKS	Fraction of tweets with question marks
EXCLAMATION MARKS	Fraction of tweets with exclamation marks
UPPER WORDS	Fraction of tweets with uppercase words
HASHTAG SYMBOLS	Fraction of tweets containing the # (hashtag) symbol
MENTION SYMBOLS	Fraction of tweets containing the @ (mention) symbol
RT SYMBOLS	Fraction of tweets containing the "RT" symbol
CONTAINS EARTHQUAKE	Fraction of tweets containing the word "earthquake"
POPULATION	Number of inhabitants in the county

3.2   Estimation of a Region of Interest

Our method starts detecting the region of interest from where county data batches will be used to infer Mercalli intensities. This step of the method separate counties into two classes. We do this using a 0/1 classifier trained over county-seismic data batches pairs. These data batches were labeled according to the actual Mercalli intensity reported into two disjoint classes. The 0 class represents an earthquake that was not perceived (not reported in the Mercalli scale) and the 1 class represents an earthquake that was effectively perceived by people with an intensity value in the Mercalli scale. Each data batch is represented by a vector of features, using the features defined in Table 1. Once the 0/1 classifier was trained, our method is ready to detect the region of interest on new earthquakes at county level.

The set of counties labeled in the 1 class is used as an input to characterize the event in the third and last stage of our method: the inference of the maximum intensity in the Mercalli scale. Data batches labeled in the 1 class are provided to our method for a regression procedure, where the Mercalli intensity at each county will be estimated.

3.3   Maximum Mercalli Intensity Estimation

Reinforced Mercalli Estimation. The maximum Mercalli intensity estimation starts by inferring a reinforced Mercalli variable at county level. Let i be the index of a county in the set of counties distributed in the region of interest of the seism. The social support $s(i) \in [0, 1]$ at the i-th county is defined as the ratio between earthquake observers (people who have posted at least one message related to the seismic movement during the period of observation) and Twitter users at the county. We combine $m(i)$ with its support $s(i)$ to provide a reinforced Mercalli support estimation. The Reinforced Mercalli support estimation consist only of supported Mercalli point estimates. Unsupported Mercalli point estimates will be discarded combining both factors in a soft minimum bivariate function defined as follows:

$$\text{REINFORCED MERCALLI SUPPORT}(i) = \frac{2 \cdot \overline{m}(i) \cdot s(i)}{\overline{m}(i) + s(i)}, \tag{1}$$

where $\overline{m}(i)$ is the Mercalli point estimate at the i-th county, constrained to the $[0,1]$ interval. To achieve a Mercalli point estimate in the $[0,1]$ interval, we normalize the estimate from the Mercalli intensity scale in $\{1,12\}$ to $[0,1]$ as $\overline{m}(i) = \frac{m(i)-1}{11}$. Then, as $\overline{m}(i)$ and $s(i)$ range in $[0,1]$, the Reinforced Mercalli support function also ranges in $[0,1]$.

The Reinforced Mercalli support only rates high supported events with high intensity in the Mercalli scale. The rationale of the Reinforced Mercalli support is to limit the effect of false positives. Then, only relevant events will be included in the tracking system at the cost of diminishing the effect of supported low intensity earthquakes. This cost is marginal for the tracking system as Mercalli intensity reports are only critical for high intensity earthquakes. Low intensity earthquakes can be characterized using only the Richter magnitude scale. As for low intensity events the impact in terms of damages is very limited, it is not necessary to characterize this kind of earthquakes on the damage scale.

Adjusted Mercalli at County Level and Maximum Intensity Detection. Our method considers each county as a sensor. Data aggregation at county level unleashes people reactions in front of an earthquake aggregating different reaction signals. We model the activation of a sensor at county level using an activation function. We do this using a sigmoid function $\frac{1}{1+e^{-x}}$ to model the level of activation of a county in front of a given earthquake. The activation of the function is fixed to Mercalli intensity 3, as is defined as the first level of the Mercalli scale where the event is felt. Then, a county is considered as active starting from level 3, as from this level to up, the earthquake will be reported. We do this by applying the sigmoid function to the Reinforced Mercalli support, scaling the function to $\{1,12\}$ and shifting it to 3 $(11 \cdot \text{RE.M.S.} + 1)_{\{1,12\}} - 2$. As the sigmoid function ranges in $[0,1]$, by combining it with the Mercalli point estimate of the county $m(i)$ we obtain an Adjusted Mercalli intensity at the county, denoted by $M_{adj}(i)$ and obtained from the following expression:

$$M_{adj}(i) = m(i) \cdot \text{SIGMOID}\left(11 \cdot \left[\frac{2 \cdot \overline{m}(i) \cdot s(i)}{\overline{m}(i) + s(i)}\right] - 1\right). \tag{2}$$

The value of the Adjusted Mercalli intensity is retrieved from a surface that comprises a collection of sigmoid functions in the Mercalli scale, stretching the sigmoid according to the Mercalli point estimate.

Adjusted Mercalli intensities are at some extent noisy signals of the event, as the quality and quantity of information per county varies. However, by looking for the maximum intensity, our method is able to detect the county with the highest support and consequently, the high confidence data to provide a fast estimation of the maximum intensity in the Mercalli scale. Then, our method for maximum intensity detection ends as follows, looking for the maximum adjusted Mercalli in the region of interest (ROI) for a given earthquake:

$$\text{MAX. INTENSITY} = \text{MAX}\{M_{adj}(i) \mid i \in ROI\}.$$

4   Experiments

4.1   Dataset

A collection of 825310 tweets was retrieved from Twitter. These tweets were collected using keywords as "quake", "earthquake" and "seismic movement" (in Spanish). The collection comprises a year and a half of Twitter data, matching the keywords during 2016 and the first semester of 2017. From these tweets, only 2200 include the geolocation field, representing only the 0.26% of the data. The collection was posted by 309749 users where 207015 records a location field in their profiles, representing the 66.8% of the users recorded in the data.

As only a very small fraction of the tweets is geo-located, we inferred the tweet location using the user location. From the set of 207015 users with user location in our dataset, 57546 matched Chile in the country field. Then we used approximate matching to associate this field with a Chilean county. To do this we used fuzzy string matching, implemented in *Fuzzy wuzzy*[1]. Using an 80% of fuzzy confidence level, a total of 41885 Chilean users were mapped to Chilean counties. These users record in the dataset a total of 190249 tweets mapped to the 345 different counties in Chile.

A second database was used to conduct a cross match between Twitter and earthquake records. We used data collected by the National Seismological Center of Chile, comprising 331 records of earthquakes in Chile during the observation period, ranging magnitudes in Richter from 2.2 Mw to 7.6 Mw. In addition, the National Seismological Center of Chile provided Mercalli reports for these events along the Chilean territory.

The cross match between our tweet collection and the Mercalli earthquake records was conducted over the county field. Only county batches that record tweets until 30 min after an earthquake were studied, accounting for a total of 6790 county-Mercalli pairs with Twitter activity. A total amount of 6548 county batches unmatched a Mercalli report, indicating the presence of tweets that mention earthquake keywords in counties where it was unperceived. In summary, our Twitter-Mercalli dataset comprises 331 earthquakes with 187317 tweets distributed over 345 Chilean counties during 18 months of Twitter activity, with county-earthquake pairs separated into 6790/6548 perceived/not-perceived earthquake data batches.

From the total amount of 331 earthquakes, 264 were selected for training and exploratory issues, reserving the remaining 68 earthquakes for testing and

---

[1] Fuzzy wuzzy is a Python string matching library that uses the Levenshtein Distance to calculate differences between string sequences. It is available in: https://github.com/seatgeek/fuzzywuzzy.

validation tasks, representing a training/testing split of 80/20 percent. The training/testing splitting process was conducted using random sampling over earthquakes according to each Mercalli level, keeping the same proportions between intensities in training and testings folds, avoiding over/under representations of low/high intensity earthquakes in training and/or testing folds. Training/testing proportions of instances according to the maximum Mercalli intensity report of each earthquake are shown in Table 2.

Table 2. Training/testing instance partitions according to the maximum Mercalli intensity of each seism

Partition	II	III	IV	V	VI	VII
Training	11	105	103	39	4	2
Testing	3	26	26	10	2	1
Overall	14	131	129	49	6	3

4.2 Estimating the Region of Interest

Training/testing county data batches accounts for 10491/2847 instances at county level. To study the problem of perceived/not-perceived earthquakes at county level, we train a 0/1 classifier. In the training fold 5021 instances accounts of the 0 class (unreported Mercalli) and 5470 for the 1 class (reported Mercalli). Training was conducted using 5 folds cross validation, using an SVM of C-SVC type for classification with a radial basis function as a kernel implemented in Weka 3.7. As the focus of the problem is the detection of the 1 class, we used cost sensitive learning, penalizing false negatives in the 1 class to maximize the recall, at the cost of a high FP rate. More learning algorithms were tested among them naive Bayes or a Multilayer Perceptron but SVM was the one with the best results, with 7325 correctly classified instances, representing in overall a 69.82% of accuracy. The detailed accuracy by class is shown in Table 3

Once model selection is evaluated, applying the model built on training instances on the testing instances, the performance of the classifier reaches 1867 correctly classified instances over a total amount of 2847 instances, achieving a 65.57% of accuracy. These results show that the classifier generalizes well,

Table 3. Training accuracy by class using 5-folds cross validation

Class	FP rate	Precision	Recall	F-measure	ROC area
0 (unreported)	0.189	0.736	0.575	0.646	0.693
1 (reported)	0.425	0.675	0.811	0.737	0.693
Weighted avg.	0.312	0.705	0.698	0.693	0.693

as overall accuracies between training and testing partitions are similar. However, what is really important is that the recall in the testing partition remains high, showing good properties in terms of predictability for the 1 class. The results disaggregated by class are shown in Table 4.

Table 4. Testing accuracy by class

Class	FP rate	Precision	Recall	F-measure	ROC area
0 (unreported)	0.184	0.765	0.517	0.617	0.667
1 (reported)	0.483	0.594	0.816	0.687	0.667
Weighted avg.	0.323	0.685	0.656	0.650	0.667

Table 4 shows that the 0/1 classifier is able to recover the region of interest for each earthquake. The results show that each region of interest is over-estimated as the low precision for class 1 shows but achieving a good coverage of the actual region as its high recall shows. To better understand how the 0/1 classifier behaves, we disaggregate matching/mismatching testing instances according to the actual level of Mercalli intensity.

Table 5. Matching/mismatching instances according to the actual Mercalli intensity

Actual	Predicted	-	I	II	III	IV	V	VI	VII
0	0	790	-	-	-	-	-	-	-
0	1	737	-	-	-	-	-	-	-
1	0	-	66	85	62	25	5	-	-
1	1	-	130	234	351	198	65	95	4
Instances		1527	196	319	413	223	70	95	4
Error rate		0.48	0.33	0.26	0.15	0.11	0.07	-	-

As Table 5 shows, the false negative rate is very low, and as the intensity of the earthquake increases, the error rate decreases. High intensity earthquakes (V to up) show an almost perfect performance. The thick part of this error occurs in low intensity earthquakes (III to down), which is natural for this kind of phenomena as in this part of the Mercalli scale many people do not recognize the event as an earthquake, being felt only under very favorable conditions (for instance, on upper floors of buildings). On the other hand, for the 0/1 classifier it is hard to distinguish counties where the earthquake is reported but it is unperceived. The over estimation of the region of interest will not affect the performance in the task of maximum intensity detection in the Mercalli scale of damages, as by looking for the maximum intensity in this area, the method will discard noisy data recorded in many counties.

Detecting the maximum intensity of an earthquake in he Mercalli scale

Now we compare our method with the state of the art in the specific task of maximum intensity prediction in the Mercalli scale of damages. Results based on mean absolute error measures for each earthquake in the testing set are shown in Table 6

Table 6. Averaged MAE of the maximum Mercalli intensity for each earthquake.

	Proposal	Cresci *et al.* [3]	Baseline
II	$\langle 1.00,+1.0,-1.0 \rangle$	$\langle 2.00,+0.0,-0.0 \rangle$	$\langle 4.12,+0.6,-0.4 \rangle$
III	$\langle 0.69,+1.3,-0.6 \rangle$	$\langle 1.05,+0.9,-1.0 \rangle$	$\langle 3.04,+1.0,-0.8 \rangle$
IV	$\langle 1.25,+2.7,-1.2 \rangle$	$\langle 0.20,+0.7,-0.2 \rangle$	$\langle 1.27,+1.4,-0.8 \rangle$
V	$\langle 0.55,+0.4,-0.5 \rangle$	$\langle 0.73,+1.2,-0.7 \rangle$	$\langle 0.94,+0.8,-0.6 \rangle$
VI	$\langle 1.00,+1.0,-1.0 \rangle$	$\langle 0.5,+0.5,-0.0 \rangle$	$\langle 1.00,+1.0,-1.0 \rangle$
VII	$\langle 1.00,+0.0,-0.5 \rangle$	$\langle 11.00,+0.0,-0.0 \rangle$	$\langle 1.00,+0.0,-0.0 \rangle$

As Table 6 shows, our method performs well in the specific task of maximum intensity prediction, being very competitive with the state of the art. The method of Cresci *et al.* [3] performs better than our method in IV level intensity earthquakes but at the cost of poor results on low and high energy earthquakes. The improvement of our method over the baseline is important. Note that the baseline correspond to a regression over the lexical features at county level, picking the maximum value detected in each earthquake. Our proposal applies the adjusted Mercalli at county level to improve the baseline, picking the maximum over the set of adjusted Mercalli estimates defined in Eq. 2. The results show that adjusted Mercalli variable is useful for maximum intensity detection.

5   Conclusion

In this paper we have proposed the method that predicts the maximum Mercalli intensity of an earthquake using social media features. The state of the art shows efforts in earthquake detection, namely where an earthquake was felt and which was its maximum intensity. Our proposal performs well in this specific task, being very competitive with the state of the art [3] in earthquake detection and maximum intensity prediction tasks using less features. However, the specific contribution of our proposal is to provide a new Mercalli estimate, named adjusted Mercalli, combining supported estimates at county level for the regression method. Our proposal discards the use of geological models of the ground or the inclusion of signals captured from spatially distributed seismographs as was done in previous work [2] The simplicity of our method favors its application in many countries, avoiding the need to build huge networks of seismographs to track the effects of earthquakes. Our method shows that social media provides valuable information helpful for the task of Mercalli damages reports, providing accurate and fast reports of maximum Mercalli intensity.

Currently, we are extending our method to work with more features. The inclusion of time-based features helps to characterize the tweet stream (e.g. tweet interval rate), a valuable source of information for earthquake detection task. We think that these features will also be helpful in the elaboration of spatial intensity reports. In addition, we are working with network-based features (e.g. RT depth). Preliminary experiments using these features show promissory results.

At last but not least, the design of a system for early tracking of earthquake damages is the next step of this project. How to efficiently use our method to provide spatial real-time damage reports is one of our most challenging tasks in the near future. The pursuit of this goal involves efforts in data integration and visualization, among other challenging tasks four our group.

Acknowledgements. M. Mendoza was funded by Conicyt PIA/Basal FB0821. This work was also supported by the Millennium Nucleus Center for Semantic Web Research under Grant NC120004.

References

1. Avvenuti, M., Cresci, S., Marchetti, A., Meletti, C., Tesconi, M.: Ears (earthquake alert and report system): a real time decision support system for earthquake crisis management. In: Proceedings of the 20th ACM SIGKDD International Conference on Knowledge Discovery and Data Mining, pp. 1749–1758. ACM (2014)
2. Burks, L., Miller, M., Zadeh, R.: Rapid estimate of ground shaking intensity by combining simple earthquake characteristics with tweets. In: 10th US National Conference on Earthquake Engineering, Frontiers of Earthquake Engineering, Anchorage, AK, USA, 21–25 July 2014
3. Cresci, S., La Polla, M., Marchetti, A., Meletti, C., Tesconi, M.: Towards a timely prediction of earthquake intensity with social media. IIT TR-12/2014. Technical report. IIT: Istituto di Informatica e Telematica, CNR (2014)
4. D'Auria, L., Convertito, V.: Real-time mapping of earthquake perception areas in the Italian region from Twitter streams analysis. In: D'Amico, S. (ed.) Earthquakes and Their Impact on Society, pp. 619–630. Springer, Switzerland (2016). https://doi.org/10.1007/978-3-319-21753-6
5. Doan, S., Vo, B.-K.H., Collier, N.: An analysis of Twitter messages in the 2011 Tohoku earthquake. In: Kostkova, P., Szomszor, M., Fowler, D. (eds.) eHealth 2011. LNICST, vol. 91, pp. 58–66. Springer, Heidelberg (2012). https://doi.org/10.1007/978-3-642-29262-0_8
6. Earle, P., Guy, M., Buckmaster, R., Ostrum, C., Horvath, S., Vaughan, A.: Omg earthquake! Can Twitter improve earthquake response? Seismol. Res. Lett. **81**(2), 246–251 (2010)
7. Robinson, B., Power, R., Cameron, M.: A sensitive Twitter earthquake detector. In: Proceedings of the 22nd International Conference on World Wide Web, pp. 999–1002. ACM (2013)
8. Sakaki, T., Okazaki, M., Matsuo, Y.: Earthquake shakes Twitter users: real-time event detection by social sensors. In: Proceedings of the 19th International Conference on World Wide Web, WWW 2010, pp. 851–860. ACM (2010)
9. Zhang, X., Shasha, D.: Better burst detection. In: Proceedings of the 22nd International Conference on Data Engineering, ICDE 2006, pp. 146–146. IEEE (2006)

Use of Personal Color and Purchasing Patterns for Distinguishing Fashion Sensitivity

Takanobu Nakahara(✉)

School of Commerce, Senshu University,
2-1-1 Higashimita, Tama-ku, Kawasaki-shi, Kanagawa, Japan
nakapara@isc.senshu-u.ac.jp

Abstract. In this research, we focus on customers with high or low fashion sensitivity to determine how they differ regarding their color choices and the garments that they purchase. Customers with high fashion sensitivity tend to purchase items that are more expensive than those purchased by customers with low fashion sensitivity, and therefore the high-sensitivity customers contribute more to sales. Furthermore, the purchasing characteristics of customers with high fashion sensitivity represent important information for product development. We ascertain these features by comparing two customer groups based on purchasing data from e-commerce apparel sites, including customer ID and garment color information, and also questionnaire data. Specifically, we enumerate emerging patterns to make a classification model of high or low fashion sensitivity by adding information about personal color preferences based on color psychology using the concept of four seasons.

Keywords: Emerging patterns · Hierarchical structure
Sensory marketing · Classification model

1   Introduction

Our objective herein is to ascertain the predilection for purchasing garments on e-commerce apparel sites by using color information of the garments from the perspective of sensory marketing. Sensory marketing is an emerging paradigm for both businesses and consumers. It is intended to aid or influence a person's thinking both consciously and unconsciously through the five senses of sight, sound, smell, touch, and taste. Of these, we focus on sight, which is considered to be one of the strongest human senses [1] with regard to sensory stimuli and perception, to determine the features of fashion sensitivity.

Sensory perception is known to change according to experience. For example, bitterness is a signal of things that are inherently poisonous, so preferences for coffee, beer, and other bitter food and beverages arise via a transition from discomfort to pleasure through repeated consumption. Also, by listening repeatedly to music that is initially of no interest, it can become favorable. As in such

© Springer International Publishing AG, part of Springer Nature 2018
G. Meiselwitz (Ed.): SCSM 2018, LNCS 10914, pp. 258–267, 2018.
https://doi.org/10.1007/978-3-319-91485-5_20

phenomena, it is known that sensitivity in general changes with the number of contacts, which is known as the "mere-exposure effect" [2].

In this research, we focus on customers with either high or low fashion sensitivity to assess how they differ in color preferences and the garments that they purchase. Customers with high fashion sensitivity tend to purchase items that are more expensive than those purchased by customers with low fashion sensitivity, and therefore the former customers contribute more to sales. Furthermore, the purchasing characteristics of customers with high fashion sensitivity represent important information for product development. We ascertain these features by comparing two customer groups based on purchasing data from e-commerce apparel sites, including customer ID and garment color information, and also questionnaire data. The purchasing data covers approximately 100,000 customers whose purchasing behavior was collected each day for 12 months. The questionnaire data were collected from 3,000 customers and comprise their answers to roughly 100 questions. These data are courtesy of a data analysis competition held in fiscal 2016 and sponsored by the Joint Association Study Group of Management Science (JASMAC).

In computational experiments, we construct a classification model that distinguishes between customers with high or low fashion sensitivity by using emerging patterns (EP) [3]. These distinguish between high and low fashion sensitivity by adding information about personal color preferences based on color psychology using the concept of four seasons [4]. Ultimately, the goal is to make customers with low fashion sensitivity more interested in fashion and thereby increase the sales of e-commerce sites by repeatedly applying external stimuli based on their preferences.

In Sect. 2, we present the method for enumerating EPs and we construct a classification model using them. In Sect. 3, we present and analyze the results obtained by applying our proposed method to actual data. Finally, in Sect. 4, we draw conclusions and suggest future work.

2 Methodology

In this section, we describe the personal color used as color information for the first time and describe the pattern mining method and model construction using the color information.

2.1 Personal Color

The color information registered in the data does not completely represent the color of the clothes. For example, even if clothes has multiple colors, only the most representative color among them is registered by text. However, it is difficult to express the color of clothes with only monochrome text information, and even if analyzed using only that information, it is impossible to obtain correct results. Therefore, to handle uncertain color information, we use the personal color to classify colors into groups, and analysis is performed based on those groups.

Figure 1 shows the personal color based on the four seasons method. In personal color, we handle two basic colors: a warm base such as yellow that represents a warm feeling, and a cool base such as blue that expresses a cold feeling. All colors belong to those two bases, except for white, gray, and black. Furthermore, when a certain color belongs to the warm base, it is divided into seasons of spring or autumn, and if it belongs to the cool base it is divided into summer or winter. Therefore, a certain color can be represented by the hierarchical structure of the bases of warm and cool and the four seasons. Note that one color may belong to multiple seasons; for example, burgundy belongs to both spring and autumn.

In fact, each color belongs to the Practical Color Coordinate System (PCCS), which is a discrete color space indexed by hue and tone as developed by the Japan Color Research Institute [5]. Figure 2 shows an image of the PCCS, which categorizes colors by tone. It divides individual hues into 12 tones (vivid, soft, pale, etc.) based on the impressions that they impart in terms of vividness. A vivid tone is a grouping that is close to a pure color. Raising the lightness produces a pale, light tone and lowering the lightness produces a deep, dark tone described as dark grayish [6].

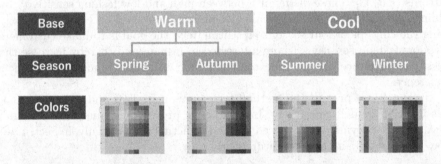

Fig. 1. Personal color. One color is represented by the hierarchical structure of the bases of warm and cool and the four seasons. (Color figure online)

For the color information of the data to be used, the color names of the clothes are entered as text such as "Burgundy," and it is necessary to match those names with the PCCS color information handled by the four seasons method. In this research, matching was done using a "color search dictionary"[1], which is a color search site. We inputted the color name and acquired the PCCS information from the website.

Table 1 lists the personal characteristics of the colors belonging to each season from the study of color psychology.

Spring. People of spring type have cute, fun, and lovely clothes with youthful bright colors.

---

[1] http://www.colordic.org/search/.

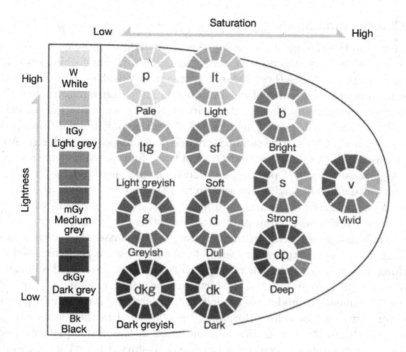

Fig. 2. Image of the Practical Color Coordinate System (PCCS), which categorizes colors by tone (adapted from [6]).

Table 1. Characteristics of each season based on color psychology

	Personality impressions
Spring	bright, youthful, active, lovable, pop
Autumn	calm, trendy, mature, gorgeous, sophisticated
Summer	quiet, refreshing, gentle, trendy, precise
Winter	sharp, vivid, clear, luxury, modern

Autumn. People of autumn type have an adult atmosphere with a natural feeling and wear natural clothes that look nice and calm.

Summer. People of summer type wear clothes that give a soft, elegant atmosphere, cool and sweet impression.

Winter. People of winter type generally have a distinctive and unique atmosphere and their clothes colors look sharp and full of contrast.

Using these characteristics of the seasons, we can estimate a person's personality from the colors that they choose.

2.2 Classification Model

In constructing the classification model, this study defines those customers with high fashion sensitivity as a positive class and those with low fashion sensitivity as a negative class and uses these classes as dependent variables. Fashion sensitivity is determined by the scores of the four questionnaire items relating to fashion consciousness. Specifically, the four items are "Clothing is one way of showing my personality," "Fashion is part of my lifestyle," "I often look at what other people are wearing," and "One's value can be increased or decreased by clothing." A person who answered "yes" to more than three of these four questions was defined as having high sensitivity (678 people), and someone who answered "no" to all four questions was defined as having low sensitivity (471 people).

The explanatory variables are created by EPs, namely itemsets whose supports change considerably from one class to another, capturing discriminating features that sharply contrast instances between the classes. Specifically, when extracting an EP as an explanatory variable, the item used is the purchased item of clothing and its color pair. For example, if a customer purchased a burgundy T-shirt, the item is expressed as "T-shirts_burgundy." To enhance the reliability of the classification model, we exploit the hierarchical structure in personal colors. For instance, the color "Burgundy" is contained in the season category "winter," and these two categories are contained in the base category "Cool." Therefore, if a customer purchases a burgundy T-shirt, then "T-shirts_winter" and "T-shirts_cool" are also created as items in addition to "T-shirts_burgundy."

To find characteristic purchasing patterns for high and low fashion sensitivity, we use the support and growth rate to enumerate EPs whose values are greater than or equal to the minimum support and minimum growth rate, respectively, which are the thresholds for the user. With the positive transaction set expressed as D_p and the negative set as D_n, the support of the positive class regarding two items a and b is defined as follows:

$$support_{D_p}(a, b) = \frac{|Occ_p(a, b)|}{|D_p|}, \tag{1}$$

In the above formula, $Occ_p(a, b)$ represents the transaction set D_p in which items a and b co-occur.

The growth rate of the positive class relative to the negative class for two items a and b is defined as follows:

$$GR_{D_n \to D_p}(a, b) = \frac{support_{D_p}(a, b)}{support_{D_n}(a, b)}. \tag{2}$$

This formula represents the ratio of the co-occurrence probability of the positive class to that of the negative class. If the ratio is greater than 1.0, we can say that items a and b have a co-occurrence pattern that is distinctive of the positive class (an EP). The same method is applied to the negative class, and EPs that are characteristic of the positive and negative classes are enumerated.

Explanatory variables other than the pattern for constructing the model utilize the fashion questionnaire and demographic attributes such as gender and age. Table 2 lists the explanatory variables used.

Table 2. List of explanatory variables

Explanatory variables	Description
Questionnaire (97 questions)	About participation in events, fashion issues, fashion view, fashion change, a feeling of happiness, life values, fashion consciousness, purchase timing, etc.
Percentage purchased in each season	Total purchase percentage for each customer will be 100%
Emerging patterns	Minimum growth rate = 1.1, minimum support = 0.1
Gender	Male, Female
Age	Under 20, early 20s, late 20s, early 30s, late 30s, 40 and over

The classification model is a logistic regression model. With the dependent variable in the classification model as $y \in \{0,1\}$ (0: negative, 1: positive) and with the p explanatory variable vectors as $\mathbf{x} = (x_1, x_2, \cdots, x_p)$, the logistic regression model is expressed as Eq. (3):

$$\Pr(y = 1|\mathbf{x}) = f\left(\boldsymbol{\beta}^\top \mathbf{x} + \beta_0\right), \tag{3}$$

where $f(\cdot)$ is a logistic function defined as $f(a) = 1/(1 + \exp(-a))$; $\boldsymbol{\beta} \in \mathbb{R}^p$ is a regression coefficient vector and $\beta_0 \in \mathbb{R}$ is a constant term, both of which are estimated from the training samples. Moreover, we use variable selection that involves selecting a set of relevant explanatory variables from many candidates and using them to construct a statistical model. This procedure facilitates interpretation of the subsequent analysis of the statistical model and enhances the model's predictive performance by preventing overfitting [7]. We apply to a stepwise method that begins with no explanatory variables, whereupon the variable that leads to the largest decrease in Akaike's information criterion is added or eliminated iteratively.

3   Calculation Experiment

A model is constructed by the method described in the previous section for 678 customers with high fashion sensitivity and 471 with low fashion sensitivity. In the model construction, the training and test data were randomly sampled from the overall data set at a ratio of 9:1. The prediction accuracy is 78.1%[2], indicating a relatively high correct rate.

Table 3 gives the results for the explanatory variables selected by the stepwise method. Positive coefficients are the choice factors of high-sensitivity customers,

---

[2] The correct-answer rate when randomly selecting two customer sets was 59%.

and those with negative coefficients are the choice factors of low-sensitivity customers. As a remarkable explanatory variable for customers with high sensitivity, fashion perspective is an important factor. For example, high awareness of fashion appeared from statements such as "fashion expresses my identity," "I care how my fashion is viewed by other people," "I try new fashion as much as possible," and "Buying clothes makes me feel refreshed and relieves my stress." These customers incorporate information about new fashion, they enjoy fashion, and they find that it relieves stress. In short, fashion is part of their lives. Meanwhile, customers with low sensitivity are not conscious about fashion because they cannot judge what to wear by themselves and instead choose clothes by listening to the opinions of friends and shop assistants. Also, the life values of those customers are different. Customers with high sensitivity value wealth (Q3_17), evaluation (Q3_16), and personality (Q3_6), whereas customers with low sensitivity value history (Q3_14), latest technology (Q3_7), and stability (Q3_1). These factors are important features representing the differences between the two customer groups, but it is difficult to apply them directly to sale promotion because it is difficult to change personality and fashion consciousness. Therefore, by interpreting the EP that expresses characteristic purchasing behavior, we obtain a hypothesis that drives customers with low sensitivity to purchase.

Figure 3 shows the EPs resulting from the selected explanatory variables of Table 3. Each row represents a pattern and each column represents an item that could be included in the pattern, such as cloth and its color. Patterns 1–4 are characteristics of customers with high sensitivity and patterns 5 and 6 are characteristics of customers with low sensitivity.

	T-shirt/Cut and sewn		Knit/Sweater	Shirt/Blouse	Pants			Sneaker	Parker
	Winter	White	Gray	Summer	Gray	Black	Beige	Warm	Cool
Pattern 4				●		●			
Pattern 2								●	●
Pattern 3	●			●					
Pattern 1							●		
Pattern 5		●							
Pattern 6			●						

Fig. 3. Emerging patterns (EPs) in the model. Each row is a pattern and each column is an item that could be included in the pattern, such as cloth and its color.

"Pattern 5" involves a cut-and-sewn white T-shirt, which is an EP for low-sensitivity customers. "Pattern 6" involves a gray knitted sweater as an EP for low-sensitivity customers. Both involve the purchase of only a single item of clothing, and its chosen color is white or gray, which are common colors. It is understood that customers with low fashion sensitivity select safe colors that are easy to match with any clothes.

Table 3. Explanatory variables used in the model

	Explanatory variable	Coefficient	
Q8_8	Fashion view: Fashion is an expression of my identity.	2.6851	*
Q8_14	Fashion view: I care how my fashion is viewed by other people.	2.1953	*
Q8_5	Fashion view: I am sensitive to fashion trends	1.7848	*
Q8_11	Fashion view: I try new fashion as much as possible.	1.5204	*
Q8_16	Fashion view: Buying clothes makes me feel refreshed and relieves my stress.	1.3391	*
Q8_6	Fashion view: I have a different style from the people around me	1.2825	*
Q8_12	Fashion view: I am careful to choose clothes that are suitable for the time, place, and occasion	1.1888	*
Pattern 1	High sensitivity: Pants_beige	1.0774	*
Q8_20	Fashion view: If I like the quality and function of the clothes, their brands do not matter	0.9669	*
Q8_29	Fashion view: I have specific favorite brands	0.8722	*
Pattern 2	High sensitivity: Sneakers_Warm & Parka_Cool	0.7384	
Pattern 3	High sensitivity: T-shirt/Cut-and-sewn_Winter & Pants_Gray	0.7084	
Q8_19	Fashion view: I use clothes carefully even when it becomes old	0.7074	*
Q8_3	Fashion view: When deciding what to wear, I am conscious of the opposite sex	0.6686	
Q8_27	Fashion view: For some items I like, I have the same item in different colors	0.6594	*
Pattern 4	High sensitivity: Shirt/blouse_Summer & Pants_Summer & Pants_Blue	0.6342	
Q2_11	Conscious: Conscious of brands when shopping	0.6038	*
Q3_6	Life values: Personality (uniqueness, heterogeneity, special, etc.)	0.591	*
Q8_1	Fashion view: I have imitated celebrity fashion	0.59	*
Q8_17	Fashion view: I do not care about dress casually at home	0.5352	*
Q1_3	Event: I have participated in barbecues and outdoor events	0.5318	*
Q3_16	Life values: Emphasize evaluation from people close to me	0.4553	*
Q3_17	Life values: Wealth (money, assets, savings, money making, etc.)	0.3132	
Q6_7	Purchase time of clothes: before travel/event	0.2607	
Q9_1	Fashion change: Interest in fashion	0.2126	
Q3_14	Life values: History (old, established, traditional, ruins, culture, etc.)	−0.218	
Q3_7	Life values: State of the art (science, technology, IT, pioneers, etc.)	−0.2982	
Q2_12	Conscious: I try to find cheap things even if it takes time	−0.3433	
Pattern 5	Low sensitivity: T-shirt/Cut-and-sewn_White	−0.4705	
Gender	Male	−0.5912	
Q3_1	Life values: Stable (secure, conservative, etc.)	−0.6723	*
Q1_4	Event: I have participated in a music festival	−0.6783	*
Q8_2	Fashion view: When choosing clothes, I decide after listening to the opinions of friends and shop assistants.	−0.708	*
Q7_5	Fashion problem: I do not know what to wear depending on the situation.	−0.7081	*
Q1_8	Event: I have participated in a Halloween party	−0.7413	
Pattern 6	Low sensitivity: Knit/Sweater_Gray	−0.7883	*
Q7_2	Fashion problem: I do not know the differences among brands	−1.1195	
Ratio	Purchase ratio in Summer	−6.4492	*

*: p-value < 0.05

By contrast, the common feature of the outstanding patterns for fashion-sensitive customers is selecting multiple clothing categories and various colors, and different levels of item appear in these patterns. This is due to the effect of hierarchical classification. For example, "Pattern 4" shows that high-sensitivity customers purchases shirts/blouses of summer colors and blue pants. Also, "Pattern 2" shows that high-sensitivity customers purchase sneakers of the warm category and parkers of the cool category. In this way, customers with high fashion sensitivity purchase clothes that show consciousness of total coordination considering personal color and do not purchase white or gray clothing alone.

It is important to change the consciousness of customers with low fashion sensitivity by referring to the purchasing behavior of customers with high fashion sensitivity and repeatedly presenting coordination to which customers with low fashion sensitivity can refer.

4   Conclusion

In this research, we have used purchasing-history data from e-commerce apparel sites and proposed a method of using EPs by personal color as a method of handling color information. Handling of colors is a problem with apparel-based data, and it is difficult to obtain appropriate results from analyzing monochrome text information as-is. We showed that we can construct a highly accurate classification model that can interpret meaning by using hierarchical classification using personal color as proposed in this research. Customers with high fashion sensitivity are highly conscious about trends and coordination, and it became clear that they are selecting colors while being conscious of multiple fashion items and personal colors. Meanwhile, customers with low fashion sensitivity find that they cannot choose clothes according to the situation and instead buy clothes without referring to the opinions of friends or shop assistants. In addition, the item to be purchased belongs to a single category and has common colors such as white and gray are selected. To improve sales, it is necessary to provide fashion information considering total coordination and personal color and to develop sensibility about fashion by making contact repeatedly.

Acknowledgments. This study was supported in part by CREST of the Japan Science and Technology Agency and by JSPS KAKENHI Grant Numbers JP15K17146 and 16H02034.

References

1. Hultén, B.: Sensory Marketing: Theoretical and Empirical Grounds (Routledge Interpretive Marketing Research), p. 175. Taylor and Francis (2015)
2. Zajonc, R.B.: Attitudinal effects of mere exposure. J. Pers. Soc. Psychol. **9**(2, Pt. 2), 1–27 (1968). https://doi.org/10.1037/h0025848
3. Dong, G., Li, J.: Efficient mining of emerging patterns: discovering trends and differences. In: Proceedings of the Fifth ACM SIGKDD International Conference on Knowledge Discovery and Data Mining, pp. 43–52 (1999)

4. Kamiyama, Y., Nagasaka, N.: The Textbook of Personal Color. Shinkigensha (2004). (in Japanese)
5. https://en.wikipedia.org/wiki/Practical_Color_Coordinate_System
6. http://art-design-glossary.musabi.ac.jp/tone/
7. Guyon, I., Elisseeff, A.: An introduction to variable and feature selection. J. Mach. Learn. Res. **3**, 1157–1182 (2003)

Multimodal Negative-Attitude Recognition Toward Automatic Conflict-Scene Detection in Negotiation Dialog

Shogo Okada[1]([✉]), Akihiro Matsuda[2], and Katsumi Nitta[2]

[1] Japan Advanced Institute of Science and Technology, Nomi, Japan
okada-s@jaist.ac.jp
[2] Tokyo Institute of Technology, Tokyo, Japan

Abstract. This study aims to present a method for detecting conflict scenes during negotiations through negative-attitude recognition using multimodal features of face-to-face negotiation interactions. This research is conducted based on a new multimodal data corpus that includes annotation of the participants' negative attitude during the conflicting scenes. Semantic orientation (positive/negative) of the words used by the participants in the negotiation dialog, as well as the speaking turn and prosodic features, are extracted as multimodal features. Linear SVM was used to fuse the multimodal features, and the latest fusion technique was used for estimating the negative attitudes. When the proposed method is applied, classification accuracy of negative/non-negative attitude detection reaches 62.8%; recall is 0.641, precision is 0.645, and F value is 0.638.

Keywords: Impression · Multimodal interaction · Negative attitude
Negotiation

1 Introduction

Negotiation is a discussion between participants supporting conflicting views on a specific issue that is aimed at reaching an agreement. It is the process through which the participants attempt to resolve the points of difference or to suggest an idea that satisfies various interests. A conflict scene is often observed during such a negotiation, and the negotiators must try to solve the conflict to reach an agreement. Trainees in the negotiation tactics (such as Law School students) need to experience various kinds of conflict scenes to learn how to negotiate with other people under different practical scenarios. Automatic detection of the conflict scene during negotiation can be a useful technique for developing negotiation educational material and implementing an interactive negotiation system. In this

G. Meiselwitz (Ed.): SCSM 2018, LNCS 10914, pp. 268–278, 2018.
https://doi.org/10.1007/978-3-319-91485-5_21

study, we analyze the negative attitude of participants in conflict scenes during face-to-face negotiation by constructing a classification model of the scene, where participants assume opposing attitudes. In other words, this study addresses the modeling of participants' negative attitude as a social signal.

A new multimodal data corpus, including the annotation of negative or non-negative attitudes in negotiation, is collected through dyadic negotiation interaction, which is a role-play task. Two participants were assigned the roles of a local resident and city government officer. In the task scenario, the resident collected a large amount of garbage in the garden. The neighbors complained that the garbage collected by the resident was stinking. The city government officer wanted the residence to be clean of the garbage. Conversely, the local resident would like to keep the garbage. Information regarding each stance in this negotiation was secretly disclosed to each participant, i.e., the local resident and city government officer. This led to a discussion and conflict between the participants on specific topics. Throughout the negotiation, we observed the conflict scenes and analyzed the negative attitudes of the participants during these conflict scenes.

We extracted multimodal features (verbal and nonverbal) from a manual transcription of the spoken dialog and from various nonverbal patterns, including the timing of turn taking, utterance overlap, and prosodic features. We developed the prediction model of the conflict scenes from these multimodal features using machine learning techniques and evaluated it.

2 Related Work

The social signal processing is a hot topic on the HCI research area. The target social-signal variables are public speaking skills [1,2], speaking assessment [3,4] for non-native speakers, persuasiveness in social media [5], sentiment analysis in social media [6], communication skills in job interviews [7–9], leadership skills [10], and communication skills [11] in group discussion. Assessment modeling techniques have been applied in speaking and communication skill training systems [12–14]. These variables are mainly related with positive attitudes. In this research, we focus on modeling negative attitudes in negotiation interactions.

Previously, many studies have applied communication analysis on negotiation interactions [15,16]. Some researches focused on the social signal processing in negotiations. Park et al. presented a prediction model of the reaction (accept or deny) in dyadic negotiation under various incentive scenarios [17]. In [17], the model was trained to predict whether an offer of a participant would be accepted or rejected by the interaction partner (counterpart participant). Park et al. focused on extracting not only acoustic features, including speaking turn and prosody, and visual features, such as gaze behavior, posture, and smiling, but also mutual behaviors as defined from the nonverbal behavior of both participants. The main difference between [17] and the proposed study is summarized in the

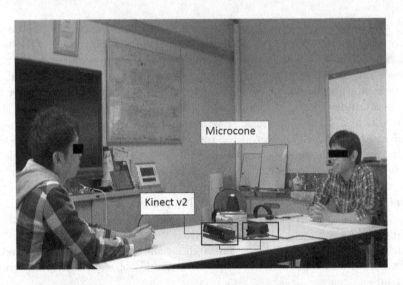

Fig. 1. Environment of dyadic negotiation interaction

following two points. First, the proposed study focuses on predicting the conflict scene and the negative attitude of the participants. Second, we extracted not only nonverbal features but also linguistic features, including semantic orientation of spoken words, to predict the negative attitude in conflict scene. This study shows that linguistic semantic features are effective for predicting the conflict features. In this paper, we address a previously unaddressed challenge that is negative-attitude recognition toward conflict-scene detection in negotiation using verbal and nonverbal multimodal information.

The negative or positive attitude of the user is a target variable in sentiment analysis. Sentiment analysis is an important topic in natural language processing [18,19]. The objective of sentiment analysis is to identify whether the content of a target document including online reviews is negative or positive based on the frequency (or probability distribution) of negative or positive words in the document. In recent year, sentiment analysis in social media (such as online review) from multimodal cues was conducted by [6], but there have not been any studies focusing on face-to-face negotiation.

3   Negative Attitude Recognition

Conflict scenes are defined as the scenes where a participant is showing non cooperative and negative attitude toward the counterpart participant. For example, a conflict scene in a dyadic negotiation interaction is shown in Fig. 1. In this study, the attitude is recognized using multimodal information including linguistic features and prosodic features. The procedure followed to estimate the attitude is shown in Fig. 2.

We set a microphone array (Microcone [20]) at the center between the two participants and record the speaking voices. Next, the dialog transcript is prepared manually. Two third-party coders watch the video of the interaction and annotates the degree of conflict after each utterance turn. Since the degree of conflict is annotated for each utterance, we estimate the conflict scene for each utterance. Multimodal features are extracted from the speech signal and manual transcription observed within the utterance. Finally, the model that estimates the degree of conflict is developed from these multimodal features by machine learning techniques. In this experiment, we evaluate the utterance-recognition accuracy in each conflict scene. Visual information is captured by Microsoft Kinect v2. In this research, the visual signal is not processed, because we focus on the language and audio features of the negotiation.

4   Multimodal Negotiation Dialog Corpus

The new multimodal data corpus, including the annotation of negative attitudes, is collected through the dyadic negotiation role-play task.

4.1   Setting of Negotiation Task

In the task scenario, the local resident collects a large amount of garbage in the garden. The neighbors express their concern that the garbage collected by the local resident smells badly. The city government officer would like the residence to be clean of garbage. However, the local resident would like to keep the garbage. Secret information is conveyed to both the parties independently regarding their

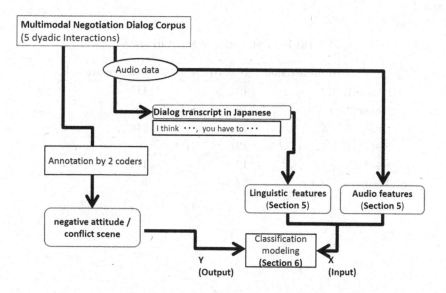

Fig. 2. Overview of storytelling assessment modeling

role in the negotiation. The setting leads to an exciting discussion and there is conflict on some topics. Through the negotiation interaction, we could often observe the conflict scenes and collect social signal multimodal data on such conflict scenes. Ten participants, eight women and two men, were recruited for this experiment. Two of them were in their twenties and eight in their fifties. On the task setting of Sect. 4.2, five sessions (two participants were coupled in each session) are recorded. In five sessions, 942 utterances are recorded using the microphone array. The interaction is also captured using a video camera for annotating the negative attitudes.

4.2   Negative Attitude Annotation and Task Setting

The negative attitudes of the participants are annotated manually. At first, the coders annotate utterances by observing the conflict scenes in the videos and in manual transcription of the dialogs. The degree of the negative attitudes is annotated on a five-point scale: 5: very cooperative attitude, 4: a little cooperative attitude, 3: neutral, 2: a little non-cooperative attitude, 1: very non-cooperative attitude. After the annotation, we classified the five scales into binary classes (cooperative or non-cooperative) by merging 1–3 into the non-cooperative (negative) class and 4–5 into the cooperative (positive) class. Two coders annotate the cooperation degree into utterances in one session. Cohen's kappa value was 0.826 in between binary class annotations by two coders after converting each five-scale value into a binary value. The agreement is sufficient. The utterances in the other four sessions are annotated by a coder. At a result of the annotation process, the total negative utterances are 466 and total positive utterances are 476. The frequency of positive/negative utterances per session is shown in Table 1.

Table 1. Statistics of annotations

	Total utterance	Negative attitude	Positive attitude
Session1	274	164	110
Session2	76	18	58
Session3	137	21	116
Session4	197	127	70
Session5	258	136	122
Mean	188.4	93.2	95.2

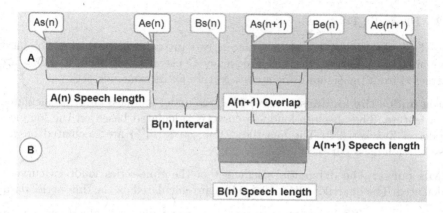

Fig. 3. An example of speaking turn feature extraction

5   Multimodal Features

The multimodal features that are input to the model are extracted from the manual transcription and audio signal of the utterances of both participants.

5.1   Speaking Turn Features

The turn-taking between two participants sometime captures the degree of conflict on the conversation. Analyzing turn-taking in conversation is important. The overlap between utterances observed from two participants shows that discussion becomes exciting or a participant talks with a non-familiar attitude. Since the overlap and interval between consecutive utterances are correlated to the frequency of negative attitudes and occurrence of a conflict scene, the features can predict such an attitude or scene effectively. As speaking-turn features, we extract the overlap, interval, and speaking length from the speaking activity.

Figure 3 shows the method of extracting these turn-taking features from the speaking activity. The vertical axis denotes the time and label segments A (upper) and B (bottom) are the utterance segments observed from the two participants. Thoroughly observing the data corpus, we obtain the following findings that show the importance of turn-taking features for modeling negative attitudes. The overlap of utterances is likely observed in conflict scenes. A long pause is likely observed when a participant assumes a negative attitude, while the other participant maintains a non-negative attitude. The statistical processing of the features is as follows.

Overlap: Overlap features are calculated as the overlapped length of utterance. Let $Be(n)$ is the end point (timestamp) of utterance t by participant B in consecutive utterances, and $As(n+1)$ is the start point of the following utterance by participant A. Then, the overlap length is as follows.

Interval: The pause length between consecutive utterances is defined as an interval.

Speech length: The total length of an utterance is defined as the speech length

5.2 Prosodic Features

Based on the speech length, we obtain two prosodic features, i.e., loudness and pitch. In addition, the Mel Frequency Cepstral Coefficients (MFCC) are extracted from the signal using speech feature extraction code [21].

Area under the loudness curve: The area under loudness curve is calculated as a feature. The absolute loudness curve is calculated based on the loudness model of Zwicker [22]. The quartiles (25%, 50%, 75%) are calculated on the absolute loudness curve.

RMS curve: The differential coefficient of the time-series loudness curve is calculated. The quartiles (25%, 50%, 75%) are calculated on the time-series data.

Mean of specific loudness range: The mean value of loudness on the frequency range of 510 Hz–3.7 kHz is calculated.

Pitch: The maximum, mean, and variance are calculated from F0 samples obtained within an utterance.

Mel Frequency Cepstral Coefficients (MFCC): The MFCC features (13 dimensions) are calculated as audio features.

5.3 Linguistic Features

We used a dictionary constructed by the spin model [23] to extract the features used here to assign the words spoken by the participants to a positive (desirable) or negative (undesirable) orientation. The dictionary consists of transactions. Each transaction includes a Japanese word and an indication of the degree of desirability ranging from -1 to $+1$.

The linguistic features are extracted from the transcriptions using Japanese morphological analysis tool MeCab [24], which segments sentences into word sequences and automatically gives the part of speech (PoS) for each word in the word set. If word w_i, which is contained in the utterance, is found in the dictionary word set through matching, the value $v(w_i)$ of the word orientation is stored. The feature F_j of semantic orientation j in the utterance is calculated using $v(w_i)$ in the following equation. In this equation, N_j denotes the total number of w_i in utterance j. We removed words with multiple pronunciations from the candidate words list, because the morphological analysis sometime fails to estimate the pronunciation of such words accurately.

6 Experiments

To confirm the effectiveness of the proposed method for estimating whether the participant has a negative or a positive attitude, we evaluated the trained model using machine learning techniques. Soft margin linear SVM was used as

Table 2. Experimental results

		Audio (Nonverbal)	Language (Verbal)	Latest fusion
Accuracy [%]		60.7	56.7	62.8
Negative (Uncooperative)	Recall	0.575	0.807	0.725
	Precision	0.598	0.571	0.616
	F-measure	0.586	0.668	0.666
Positive (Cooperative)	Recall	0.622	0.405	0.557
	Precision	0.599	0.682	0.674
	F-measure	0.610	0.509	0.610
Average	Recall	0.598	0.606	0.641
	Precision	0.599	0.626	0.645
	F-measure	0.598	0.589	0.638

a machine learning model for the negative/positive attitude classification, as a binary classification task. Five interaction sessions were included in the negotiation dialog corpus. In the experiments, the dataset obtained from the four sessions were used as the training data, and dataset obtained from the remaining one session was used as the test data. Therefore, five-fold cross validation testing was performed. Table 2 shows the classification accuracy of the test data. The accuracy is compared among models that are trained from audio features, linguistic features, and multimodal features (audio and linguistic).

For the audio model, the mean classification accuracy of all sessions was found to be 60.7%. In addition, recall was 0.598, precision was 0.599, and F value was 0.598. For the linguistic model, the mean accuracy of all sessions was found to be 56.7%. Recall was 0.606, precision was 0.626, and F value was 0.589. The results show that classification is more effective using nonverbal features than linguistic features. The Table also shows that fusing multimodal features improves the accuracy of the audio model by 2% and that of the linguistic model by 6%. The accuracy for all sessions was found to be 62.8%. Latest fusing (score fusion of the output of the two models) is performed in this experiment. Recall was 0.641, precision was 0.645, and F value was 0.638.

7 Discussion and Future Works

The classification accuracy differs between the sessions. Session 5 was less accurate (45.6%) in comparison with the other sessions, when only the audio features are used. In session 5, the participant's voice level was very low when the participant has a negative attitude. Conversely, the voice level increased in the other sessions under the same situation. Consequently, the audio model could not predict the negative attitude of participants in session 5. Estimation of negative attitude in session 5 was less accurate (35.8%), when only the linguistic features were used. In session 3, negative words (e.g., "fire") were often observed in the transcription. For example, the participant said "I am concerned that your house will catch fire..". Although in this case, the participant did not take a negative attitude, but the semantic orientation becomes negative on the word level. The fusing of audio and linguistic features is an effective method for improving the classification accuracy in the case of session 3 and 5. The accuracy is improved to 63.9% in session 5, and to 52.6% in session 3 using the fusing technique. Sentence structure analysis is promising for analyzing the semantic orientation of utterances on the sentence level. In the other sessions, semantic orientation calculated using the method proposed by [23] was found to be effective in improving the accuracy.

Here, we have summarized the limitations of this work. The number of participants is limited to 10, and more samples are required to improve the accuracy. To apply this framework to an automatic system, linguistic features must be extracted in an automated manner using speech processing techniques such as the one proposed by [25]. More types of linguistic features, such as PoS and BoW, should be explored to improve the classification accuracy. In this research, we focused on an effective analysis using audio and linguistic features. It is well known that visual nonverbal features are also important in capturing the social signal. In future, we aim to combine the visual features obtained from a depth sensor and RGB data into audio features. In this research, audio and linguistic feature were combined using the latest fusion method. The co-occurrence of the linguistic and audio feature could not be captured using this method. In future, we aim to extract the co-occurrence relationship between mutual activity or multimodal activity by co-occurrence mining of multimodal information [26].

8   Conclusions

In this study, we have proposed a method for detecting conflict scenes in negotiations through negative-attitude recognition using multimodal features of face-to-face negotiation interactions and have evaluated it. For this analysis, a new multimodal data corpus, which included the annotation of the participants' negative attitude during conflict scenes, was collected through dyadic negotiation interaction. Semantic orientation (positive/negative) of the words spoken in a negotiation dialog was used as verbal features. Speaking turn and prosodic features were extracted as nonverbal features. Linear SVM was used to combine the multimodal features, and the latest fusion technique was used to estimate the negative attitude. Classification accuracy of negative/non-negative attitude was 62.8%, recall was 0.641, precision was 0.645, and F value was 0.638.

Acknowledgment. The authors would also like to thank the anonymous referees for their valuable comments and helpful suggestions. The work is supported by Japan Society for the Promotion of Science (JSPS) KAK-ENHI (15K00300, 15H02746).

References

1. Wörtwein, T., Chollet, M., Schauerte, B., Morency, L.P., Stiefelhagen, R., Scherer, S.: Multimodal public speaking performance assessment. In: Proceedings of ACM International Conference on Multimodal Interaction (ICMI), pp. 43–50 (2015)
2. Ramanarayanan, V., Leong, C.W., Chen, L., Feng, G., Suendermann-Oeft, D.: Evaluating speech, face, emotion and body movement time-series features for automated multimodal presentation scoring. In: Proceedings of ACM International Conference on Multimodal Interaction (ICMI), pp. 23–30 (2015)
3. Chollet, M., Prendinger, H., Scherer, S.: Native vs. non-native language fluency implications on multimodal interaction for interpersonal skills training. In: Proceedings of ACM International Conference on Multimodal Interaction (ICMI), pp. 386–393 (2016)
4. Zhang, Y., Weninger, F., Batliner, A., Hönig, F., Schuller, B.: Language proficiency assessment of english l2 speakers based on joint analysis of prosody and native language. In: Proceedings of ACM International Conference on Multimodal Interaction (ICMI), pp. 274–278 (2016)
5. Park, S., Shim, H.S., Chatterjee, M., Sagae, K., Morency, L.P.: Computational analysis of persuasiveness in social multimedia: A novel dataset and multimodal prediction approach. In: Proceedings of ACM International Conference on Multimodal Interaction (ICMI), pp. 50–57 (2014)
6. Context-dependent sentiment analysis in user-generated videos. In: Proceedings of Annual Meeting of the Association for Computational Linguistics, ACL 2017 (2017)
7. Nguyen, L.S., Frauendorfer, D., Mast, M.S., Gatica-Perez, D.: Hire me: Computational inference of hirability in employment interviews based on nonverbal behavior. IEEE Trans. Multimed. 16(4), 1018–1031 (2014)
8. Rasipuram, S., Pooja Rao S.B., Jayagopi, D.B.: Asynchronous video interviews vs. face-to-face interviews for communication skill measurement: a systematic study. In: Proceedings of ACM International Conference on Multimodal Interaction (ICMI), pp. 370–377 (2016)
9. Naim, I., Tanveer, M.I., Gildea, D., Hoque, M.E.: Automated analysis and prediction of job interview performance. IEEE Trans. Affect. Comput. PP(99), 1 (2016)
10. Sanchez-Cortes, D., Aran, O., Mast, M.S., Gatica-Perez, D.: A nonverbal behavior approach to identify emergent leaders in small groups. IEEE Trans. Multimed. 14(3), 816–832 (2012)
11. Okada, S., Ohtake, Y., Nakano, Y.I., Hayashi, Y., Huang, H.H., Takase, Y., Nitta, K.: Estimating communication skills using dialogue acts and nonverbal features in multiple discussion datasets. In: Proceedings of ACM International Conference on Multimodal Interaction (ICMI), pp. 169–176 (2016)
12. Tanaka, H., Sakti, S., Neubig, G., Toda, T., Negoro, H., Iwasaka, H., Nakamura, S.: Automated social skills trainer. In: Proceedings of International Conference on Intelligent User Interfaces (IUI), pp. 17–27 (2015)
13. Damian, I., Baur, T., Lugrin, B., Gebhard, P., Mehlmann, Gregorand Andre, E. In: Games are Better than Books: In-Situ Comparison of an Interactive Job Interview Game with Conventional Training, pp. 84–94 (2015)
14. Hoque, M.E., Courgeon, M., Martin, J.C., Mutlu, B., Picard, R.W.: MACH: my automated conversation coach. In: Proceedings of ACM International Joint Conference on Pervasive and Ubiquitous Computing (Ubicomp), pp. 697–706 (2013)

15. Barry, B., Oliver, R.L.: Affect in dyadic negotiation: a model and propositions. Organ. Behav. Hum. Decis. Process. **67**(2), 127–143 (1996)
16. Barry, B., Friedman, R.A.: Bargainer characteristics in distributive and integrative negotiation. J. Pers. Soc. Psychol. **74**(2), 345 (1998)
17. Park, S., Scherer, S., Gratch, J., Carnevale, P., Morency, L.P.: Mutual behaviors during dyadic negotiation: automatic prediction of respondent reactions. In: Proceedings of Affective Computing and Intelligent Interaction (ACII), pp. 423–428. IEEE (2013)
18. Hu, M., Liu, B.: Mining and summarizing customer reviews. In: Proceedings of ACM SIGKDD International Conference on Knowledge Discovery and Data Mining, pp. 168–177 (2004)
19. Dave, K., Lawrence, S., Pennock, D.M.: Mining the peanut gallery: opinion extraction and semantic classification of product reviews. In: Proceedings of International Conference on World Wide Web, pp. 519–528 (2003)
20. [Computer program], V.: Microcone: Intelligent microphone array for groups (2013). http://www.dev-audio.com
21. Fernandez, R., Picard, R.W.: Classical and novel discriminant features for affect recognition from speech. In: Ninth European Conference on Speech Communication and Technology (2005)
22. Zwicker, E., Fastl, H.: Psychoacoustics: Facts and Models, vol. 22. Springer, Heidelberg (2013). https://doi.org/10.1007/978-3-540-68888-4
23. Takamura, H., Inui, T., Okumura, M.: Extracting semantic orientations of words using spin model. In: Proceedings of Annual Meeting on Association for Computational Linguistics, pp. 133–140 (2005)
24. Kudo, T., Yamamoto, K., Matsumoto, Y.: Applying conditional random fields to japanese morphological analysis. Proc. Empirical Methods Nat. Lang. Process. (EMNLP) **4**, 230–237 (2004)
25. Okada, S., Komatani, K.: Investigating effectiveness of linguistic features based on speech recognition for storytelling skill assessment. In: Recent Trends in Applied Artificial Intelligence, 26th International Conference on Industrial, Engineering and Other Applications of Applied Intelligent Systems, IEA/AIE 2018 (2018)
26. Okada, S., Aran, O., Gatica-Perez, D.: Personality trait classification via co-occurrent multiparty multimodal event discovery. In: Proceedings of ACM International Conference on Multimodal Interaction (ICMI), pp. 15–22 (2015)

Speed Dating and Self-image

Revisiting Old Data with New Eyes

Eleonora Peruffo[1,2], Sofia Bobko[1,2], Brian Looney[1,2],
Bernadette Murphy[1,2], Magie Hall[1,2(✉)], Quinn Nelson[1,2],
and Simon Caton[1,2]

[1] National College of Ireland-NCI, Dublin, Ireland
[2] University of Nebraska-Omaha, Omaha, USA
mahall@unomaha.edu

Abstract. In this paper we perform a variety of analytical techniques on a speed dating dataset collected from 2002–2004. There have previously been papers published analyzing this dataset however we have focused on a previously unexplored area of the data; that of self-image and self-perception. We have evaluated whether the decision to meet again or not following a date can be predicted to any degree of certainty when focusing only on the self-ratings and partner ratings from the event. Further to this we have examined how the decisions received after the session can affect the stability of one's self-image over time. We also performed some general exploratory analysis of this dataset in the area of self-image and self-perception; evaluating the importance of these attributes in the grand scheme of attaining a positive result from a 4 min date.

Keywords: Speed dating · Logistic regression · Attractiveness attribute
Regression tree · Random forest · ANOVA

1 Introduction

The very relatable prospect of having only 4 min in which to make a favourable impression on a prospective dating partner is a format which has proved particularly rich for research in a variety of psychological fields in the last ten years since the widely cited original publication in [1] and original experimental study.

The objective in revisiting this original dataset by this team was to bring to this older though still widely popular dataset, a fresh set of data mining students willing to apply their highly analytical and predictive approaches, and consider whether a new angle for study could be developed.

The dataset and related work already conducted brought us to assess firstly whether predictive techniques more recently developed in data mining would have predictive insight when applied to this older dataset.

Secondly this paper focuses on the area of self-image and self-representation taking into account a number of different methodological approaches to analyzing this aspect of the data and its support or otherwise about self-image and self-perception of individuals remaining stable over time and after receiving positive or negative feedback.

© Springer International Publishing AG, part of Springer Nature 2018
G. Meiselwitz (Ed.): SCSM 2018, LNCS 10914, pp. 279–297, 2018.
https://doi.org/10.1007/978-3-319-91485-5_22

The structure of this paper is:

Section 2 considers an overview of literature and related work on the developing area of speed dating research from the seminal paper published by [1] 's work and the insights research has provided on stability of self-representation in particular.

Section 3 discusses in detail the methodological approaches taken to this revisit of the dataset, including initial inspection of the data, cleaning and processing transformations, and the implementation of each approach to generate an analytical or predictive result.

Section 4 evaluates the process and results, considering applied success criteria.

Section 5 completes the paper with an overall conclusion on the research performed in this paper and an evaluation of the success in revisiting this dataset.

2 Related Work

The data collected by [1] has been used to explore gender mating preferences in heterosexual dyads and to learn insight on race preferences in dating [2]. The dataset has also been the object of an exploratory analysis by [3]. However, even if the information that can be extracted from this dataset may seem to have been exhausted, the self-rating score that participants gave to themselves across time hasn't been explored and findings about self-rating and therefore self-image would contribute to the stream of studies on self-image and self-representation in contemporary society. Since the participants of the speed-dating event organized by [1] were asked to rate their own attributes over time, before, during, and after the event after finding out if they got any dates, this allows for an exploration of the impact on self-perception after an event of acceptance or rejection as noted in [4, 5]: "an individual's identity does not remain completely stable over time. Instead, individuals begin to reevaluate their ideas about themselves following a triggering event, such as a change in life status (e.g., a divorce or a move), or when a desire for personal growth arises (e.g., desire to learn more about oneself or to find a potential mate)"; [4] found that positive responses gave a boost to the participants' self-perception thus determining the importance of others' feedback for human self-perception. There was no relevant difference between men and women.

During the first decade of the 21st century, the exploration of speed dating behavior was used to gain a better understanding of human mating practices while at the same time self-image and self-representation gathered new interest from the academic world due to the fact that various ways of internet dating were blossoming. There is a vast literature on self-image and self-representation which goes back to 1930 and the birth of psychology. The papers we are going to consider to inform our study, however, will be limited to those which explore self-image and self-perception. As described in [4] "whether their sense of themselves changes while they are participating in the activity of Internet dating, and what effect online and offline feedback from other users has upon the identity creation and re-creation process within this context".

The main difference between speed dating and internet dating is the fact that one is an offline activity while the other takes place online before any actual face-to-face meeting and self-projection can be contrived better [6]. As pointed out by [7], speed dating reduces "lies" on physical aspect because these traits are immediately verifiable during the event while it might be that people are not so explicit about personal circumstances; their study on a UK dataset, contributed to elicit the elimination of self-image lies on physical aspect during speed-dating and focus more on self-built truth about other aspects of an individual such as previous dating history or current situation, financial means (however it can be argued that clues from dressing style and way of behaving and speaking can be good indicators) and character; moreover it's easy to project a certain type of personality during a short 4 min encounter. [8] State that men give more weight on physical attractiveness than women, and less on characteristics linked to the possession of status and resources. [9] Found out that own preferences in speed dating are mostly subsided by the fact that participants pick their potential matches based on attractiveness but they observe that own preferences are still a determinant in the type of events people, in this case speed daters, choose to attend. The same result was obtained by [10, 11] Picked up the invite of [1, 7, 9] to study a sample over time and followed the romantic life of their study participants for one year. Using a different sample from that of [9], where the interest to forge a long term relationship was stated by participants, [11] found that attractiveness was still a major factor in speed daters choices for both men and women but women also valued other attributes such as level of education, income, openness to experience and shyness. In relation to the age attribute, according to [11]'s experiment men become choosier with age while the opposite is true for women.

A different analysis was produced and described by [12] where also face-to-face dates were studied. During this study they realized that not only self-image and physical aspect matter on mate selection, but also words matter and how these are delivered. Women proved "clicking" with men who used appreciative language and sympathy and who interrupted women during their conversations. Successful dates were related with women being the focal point and engagement in the conversation, and men showing alignment with and understanding of the woman. Also during this study it was discovered that women are more selective that men and they reported lower rates of "clicking" than men.

Another study that was produced and analyzed by [8] has also agreed with [12] about women being choosier than men. They have also found that judgments of attractiveness and vitality perceptions were the most accurate and were predominant in influencing romantic interest and decisions about further contact. But women were less likely to want further contact because they perceived their partners as possessing less attractiveness/vitality and as falling shorter of their minimum standards of attractiveness/vitality. They also found that women underestimated their partner's romantic interest, whereas men exaggerated it.

The effect of "similarity-attraction" on dating was also analyzed. Findings by [13] reveal that perceived, but not actual, similarity significantly predicted romantic liking in this speed-dating context.

[14] states that while the role of self-assessments for human mating decisions has been proposed repeatedly by various authors and supportive correlations have sometimes been demonstrated, detailed descriptions of how such cognitive representations might come about and influence mating decisions are almost absent from the literature. But the most interesting study from the point of view of our analysis is the one by [15] who found out that self-perception plays a role in mate selection in the sense that speed

daters choice is guided by their own self-perceived value in a "like-attract" relationship or as it's commonly said "in their league".

These papers informed our analysis and gave us the indication that there was a space for this type of analysis on the dataset prepared by [1].

3 Methodology

We used the KDD method for our analysis.

3.1 Data Processing

This is a dataset obtained from a designed experiment among university students, data were collected about their background and also after the event as follow up questions on the number of dates following a successful match. There were sessions from 2002 to 2004 and data from 14 has been included in the dataset. Participants met for 4 min and had to record the accept/decline option for each partner.

We examined the dataset using the dataQualityR and Amelia packages, as suggested by [3] the categorical and numeric variables were examined. The dataset has quite a few missing values: "half of the dataset has 20% missing values. Even worse 30% of the dataset has more than 50% missing values! This is due to a main reason: the participants had to fill a bunch of paper forms concerning their preferences, personal details, etc. The forms were very long, and there probably was some resilience to fill them up entirely—especially when they concerned personal details."

- Career has already been coded in the original dataset into 17 categories- use "career_cd"
- Field of study has already been coded in the original dataset into 18 categories- use "field_cd"

As suggested by [3] the following actions have been taken:

- Input the only missing id using the "iid" of the person,
- ten missing "pids" 1, input missing values using the partner and wave features, it's always the same person 128. One missing value remaining, otherwise the person met with himself.
- replaced 355 "0" with "NA",
- replaced 1064 "blanks" with "NA"

The following transformations have also been made on the data to ensure availability of complete data for required analysis:

- Attribute "shared interests" was excluded from our analysis as contained a high number of missing data (more than a 1000).
- For machine learning analyses only complete instances were included for selected variables ("attractive", "sincere", "intelligent", "fun", "ambitious" and "expected happiness"). Rows with "NA" were excluded, from 8378 instances we ended up with 6853. 1525 rows were deleted which represent 18.20% of all data. More than 6 k rows left for the analysis are still considered very representative.

- For descriptive analyses only complete instances were included for selected variables ("attractive", "sincere", "intelligent", "fun", and "ambitious"). These variables are different to the once mentioned above as these were collected at different times. So rows with "NA" were excluded, from 8378 instances we ended up with 3724. More than 3 k rows left for the analysis are still considered very representative.
- Some attributes were measured in a scale from 1–10 and some other in percentage up to 100. Those represented by percentages were normalised to a 1–10 scale.
- Some people erroneously gave ratings above 10, those ratings were reduced to a maximum of 10.

3.2   Further Dataset Processing

- The 3724 subset of the original dataset was further processed to add unique record count numbers and links for individual and partner ids, to aid easier identifiers for those who self-rated their attractiveness higher or lower than their date partners and to identify offers received, which is not explicit in the original dataset and can be derived by matching offers made to all pairs to identify offers received for each individual. This identified that the full data set does not have an individual record for participant number 118, which is presumed to be an error in the compilation of the original dataset. This absence has no material impact on results, though for processing the offers made by others to this missing record were retained.
- For the second ANOVA analysis the 3724 lines items representing the individual dates (one row per date) in the dataset have been summarised into 245 lines which is the number of individual participants in this dataset.
  - The information about number of matches, offers made and received, and total number of dates has been retained but summarised by participant
  - The information regarding the inputs on self-image in each time period are then only visible once per participant, instead of being repeated on each line for each new date which that participant had. This permits the split into subset groups by participants who have had a match or not had a match, those who have received a higher rating from their speed date partners than they assessed for themselves etc.
  - If this processing is not done then for example, the split by match will not correctly interpret the line items for match and no match. This will result in participants' data who have received a match being incorrectly included in the grouping for no match.
    - The opportunity was also taken to add a number of additional fields to the dataset for to give scope for further analysis.
      - Num_offers_received: a binary 1,0 field. This is not present in the original dataset but is calculable by processing the dataset to match all of the offers made with the date pairs, summarising the results per participant into new variables for 'offers made' and 'offers received'.
      - Flag_MatchNoMatch: Binary flag - 1 if this person has received any dates, Otherwise zero
      - Count of match 2: a field to record the number of dates which a participant has had.

- Flag_Others_Higher: Binary flag - 1 if this person has a rating from others which is higher than his own self rating for attractiveness, otherwise zero
- Avge_attr_others: Average rating for the recipient of the scores for attractiveness received from Others i.e. their dates

All our implementation was done with R.

3.3 Analysis and Results

Logistic Regression Analysis

Some of the reviewed papers have described how they performed logistic regression analysis in their dating analysis. [7] Constructed a model where they tried to identify what socio-demographic characteristics of potential partners determined the decisions about further contact. So, they tried to predict decisions about further contact (yes/no) based on these socio-demographic characteristics of potential partners.

Another study [12] has also adopted the logistic regression approach but their variables were different. They also tried to predict decisions about further contact as a function of perceptions of attractiveness/vitality and perception standard matching.

In our analyzed dating event all participants rated attributes of each person they met and indicated if they would like to meet him/her again. We applied logistic regression model for predicting decisions made by participants of a dating event under study based on the ratings they gave to their partner's attributes. These attributes were "attractive", "sincere", "intelligent", "fun" and "ambitious".

In this dating event women were sitting while men were rotating. We choose to do our analysis from two points of view. In our first model we tried to predict decisions of women about future contact based on attributes ratings they gave to their male partners; and the other way around we tried to predict decision of men about future contact with women based on attributes ratings they gave to the women. In this first model we have also included one available variable that was "expected happiness" that represented how happy a participant expects to be with the people will meet. This variable was not available for male partners. In these models we tried to identify what variables (attributes) were the most influencing on participant's dating decisions.

Subsets of the dataset (that was left after transformations described in methodology section above) were created selecting relevant variables for both logistic regression analyses. This subset was divided into training and test data sets, each one without altering the original distribution of the dataset.

In our first logistic regression model (women model) women's decision about future contact was a dependent variable, where "attractive", "sincere", "intelligent", "fun", "ambitious" and "expected happiness" were independent variables represented by ratings given to their male partners. Alternative models were run (including/excluding variables based on their significance level and keeping track on performance and other measurements) and the best model was selected. It was very surprising to discover in our final model that the attribute "intelligence" resulted as non-significant. All other variables were left as included in our final model. The model was tested by building a prediction model on test data, achieved accuracy of predictions done on a test data is 0.75. The area under the ROC curve is 0.82. Final (women's) logistic model that was built is shown below in Eq. (1):

$$Ln\left((P\,meet\,again)/(P\,dont't\,tmeet\,again)\right) = -5.37 + 0.55Attr - 0.05Sinc + 0.37Fun$$
$$- 0.12Amb + 0.06\,Exphappy$$

$$(1)$$

We can assess βetas coefficients to determine whether a change in a predictor variable makes the event (accept meeting again) more likely or less likely. Positive coefficients indicate that the event becomes more likely as the predictor increases. Negative coefficients indicate that the event becomes less likely as the predictor increases. So, an increase in attractiveness, or fun, or expected happiness would increase chances of women saying yes to future contact. The opposite is true for increase in sincerity, and ambition would decrease chances of further contact.

A similar model was created from a point of view of men's, where decision of men about future contact depended on ratings men gave to women's attributes. Alternative models were run (including/excluding variables based on their significance level and keeping track on performance and other measurements) and the best model was selected. As in the previous model, the attribute "intelligence" resulted as non-significant. The model was tested by building a prediction model on test data, achieved accuracy of predictions done on a test data is 0.74. The area under the ROC curve is 0.81. Final (men's) logistic model that was build is shown below in Eq. (2):

$$Ln\left((P\,meet\,again)/(P\,dont't\,tmeet\,again)\right) = -5.19 + 0.59Attr - 0.06Sinc + 0.33Fun$$
$$- 0.19Amb$$

$$(2)$$

Results (coefficients) of this model (men's) are similar to the first model (women's). Some authors described in their studies that they discovered gender differences, like women are more selective and that women value more intelligence than man [12]. However, we did not discover this and can conclude that in this analysis we didn't find any gender difference between women and men in their decision of future contact based on the attribute ratings given to their partners.

Decision Trees

We decided to test the same models with a C 5.0 tree and with a regression tree to check if we could find a better accuracy. The dataset was again split in two for decision taken by the speed-dater based on attributes rating and the decision of the partner based on attributes rating. Each of the datasets was split into train and test datasets with a percentage of 75% and 25% respectively. To make the analysis replicable the seed was set at '1234'.

The C5.0 tree resulted in a 74% accuracy prediction for both the men and the women.

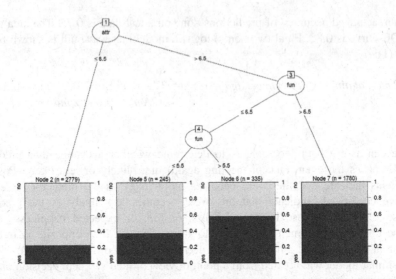

Fig. 1. C5.0 tree - own rating model

As in the regression, in both models the attractiveness and fun attributes were the ones which were considered most important in order to predict the likelihood of deciding to meet with the person. The two attributes also are sufficient to explain the C5.0 tree and consequently the choice of participants. People's decision to meet was based first and foremost on their attraction to the speed-dater and secondly on the perception of them being a fun-loving person (Fig. 1).

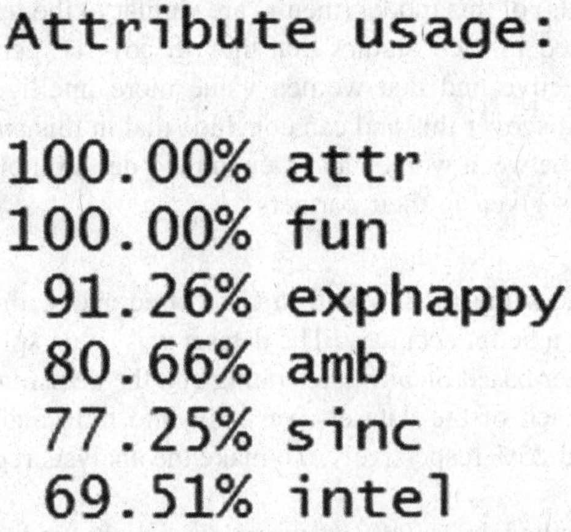

Attribute usage:

```
100.00% attr
100.00% fun
 91.26% exphappy
 80.66% amb
 77.25% sinc
 69.51% intel
```

Fig. 2. Attribute usage C5.0 own model 1

For the regression tree model we tried first a simple rpart model which gave a low accuracy, 74% in the own model and 73% in the partner model, so we decided to use a more complex version of the tree by setting complexity at 0 and minimum split at 2. The accuracy for this model was much higher: 85% in the own rating model and 86% in the partner's rating model. By increasing the split, the accuracy started to go down. The regression tree performed better than the logistic regression in predicting the likelihood of deciding to meet in terms of accuracy and confirmed that the main predictors are attractiveness and fun. The interesting thing to notice was that the two models we run, one for own decision and one for partner's decision didn't give different results, the most valued attributes in both cases were attractiveness and fun. It could well be that in the setting of a speed-dating event these are two of the most easily identifiable features since first impression plays a big part in a meeting programmed to last only 4 min (Fig. 2).

Random Forest

Another approach which we tested was to run the model through 2 different types of random forest. The first type was a Random Forest a la Breiman's approach [16] using the RandomForest function in the RandomForest R package. The second type is an extension of the conditional inference tree run in the previous section; a conditional inference forest using the function cForest from the package party in R.

Random forests were a logical progression from our use of decision trees on the model as due to the Law of Large numbers they do not have the same overfitting issue as decision trees and they are an effective tool in prediction [16].

For the ownModel; the accuracy of the randomforest was 73% and of the conditional inference forest was 74%.

For the partnerModel; the accuracy of the randomforest was 72% and of the conditional inference forest was 74%.

Model	Seed level	# Trees	Accuracy
Random Forest	500	1000	72.99
		2000	72.82
	1000	1000	72.82
		2000	72.99
	2000	1000	72.59
		2000	72.76
Conditional Inference Forest	500	1000	74.04
		2000	74.28
	1000	1000	74.34
		2000	74.22
	2000	1000	74.1
		2000	74.22

Fig. 3. Seeds analysis

Due consideration was given to the seed level and number of trees to be used in the Random Forest. An analysis was performed to ascertain the significance that different seed levels and different numbers of trees within the forest would have on the accuracy of the model. The results of this analysis can be seen in Fig. 3.

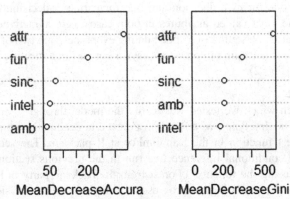

Fig. 4. Variable importance - own model

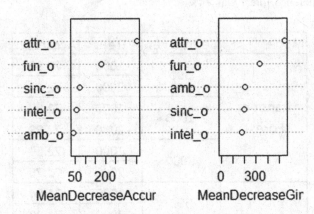

Fig. 5. Variable importance - partner model

Figures 4 and 5 show the importance of each variable in the model as measured by the randomforest and displayed on a dot chart. As we can see from the plots, both in the own and partner models the forests confirm that the most important attributes are attractiveness and fun as our other model approaches have also inferred.

Descriptive Analysis

As we have mentioned already, during this dating event all participants were asked to rate their own attributes a few times, once at signup time (Time 0), then the day after participating in the event (Time 1) and finally 3–4 weeks after participants had been sent their matches (Time 2). The attributes used were the same as in our other analysis, "attractive", "sincere", "intelligent", "fun" and "ambitious".

We wanted to analyse how self-image (self-perception) of the participants is changing (if changing) with time.

We believed that the change in self-image of participants would differ between those people who got matches and for those who did not get a match. We would expect to see that self-image of those people who got matches would increase with time, and for those who did not get matches would decrease.

We will show below an example of the graphs produced for the "attractiveness" attribute. Similar graphs were produced for all other considered attributes. These graphs intend to show a change in ratings given over time. People were divided in two groups, people who got matches (represented by the yellow graph below) and people who did not get matches (represented by the blue graph below).

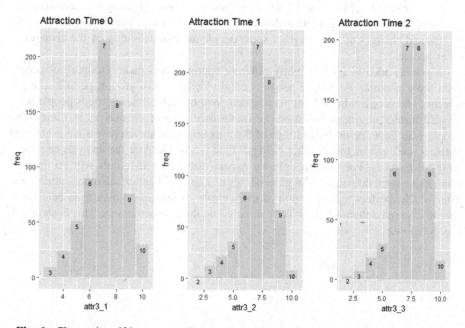

Fig. 6. Change in self-image over time for people who got matches (Color figure online)

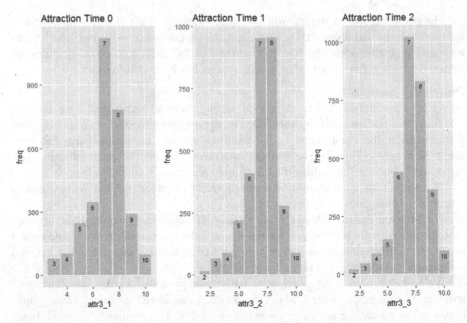

Fig. 7. Change in self-image over time for people who didn not get matches (Color figure online)

As we can see in Figs. 6 and 7 we do not observe expected changes in self-image of the participants. We believe that this is due to the fact that people are analysed in groups and changes of self-image of some people are compensating by the changes of other people.

It was decided to apply slightly different approach and make grouping of people even more granular. As before people were divided into two groups, people who got matches and people who did not get a match. Within each of these groups people were divided further in groups based on the rating they gave of themselves at Time 0 (so people who gave themselves ratings of 5 for example on attractiveness were grouped together, and similar for all other ratings). First we will look at the analysis of people who got matches. For these people we expect that their self-ratings will go up with time. There were no people who gave ratings of 1 or 2 to their attractiveness attribute at Time 0. The perfect representation of the increase in self-image over time belongs to people who rated themselves at Time 0 with 3 on their attractiveness. This is shown in Fig. 8 below.

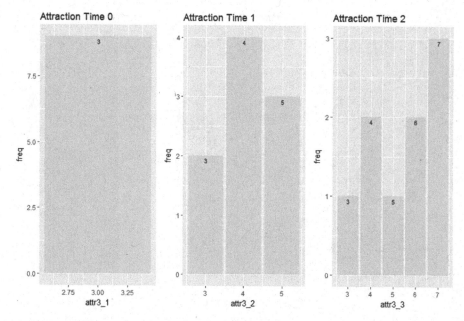

Fig. 8. Change in self-image over time for people who got matches and original rate 3

In this graph we can clearly observe an increase in rating at Time 1 and further more at Time 2. This is held practically true for starting ratings of 4–8, but some very insignificant cases of reduction in ratings given were observed; overall a tendency of increase in ratings given over time is observed. This does not hold true for ratings starting at 9 and 10. A reduction of those ratings over time was noted, this might be due to the fact that people originally gave themselves exaggerated ratings.

Next we analysed people who did not get matches. For these people we expect that their self-ratings would go down with time. There were no people who gave ratings of 1 or 2 to their attractiveness attribute at Time 0. The perfect representation of the decrease in self-image overtime belongs to people who rated themselves at Time 0 with 10 on their attractiveness. This is shown in Fig. 9 below.

In this graph we can clearly observe a decrease in rating at Time 1 and further more at Time 2. This is held practically true for starting ratings equal to 9, 8 and 4, but some very insignificant cases of increase in rating were observed; the overall tendency is a decrease in ratings given over time. This holds partially true for original ratings of 7, 6 and 5 where some decreases and increases in ratings over time were observed. This does not hold true for starting ratings equal to 3. An increase from that rating was noted over time, this might be due to the fact that people originally given themselves very low rating.

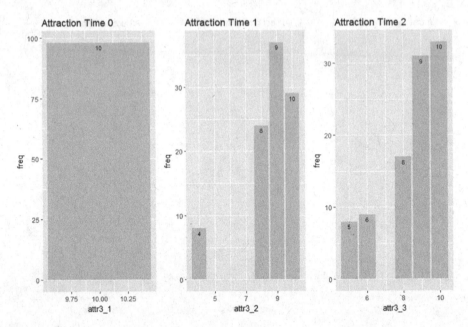

Fig. 9. Change in self-image over time for people who did not get matches and original self-rating = 10

We can conclude that we have found evidence of a tendency of people increasing their self-image perception over time if they got matches; in the opposite tendency to decrease level of self-image if they did not get a match.

It was worth exploring this further and finding out if there was a statistically significant difference among the self-rating measurements at time1, time2 and time3. We run a one-way ANOVA for each of the 5 attributes: attractiveness, sincerity, fun, intelligence, ambition. For participants who received a match, sincerity and intelligence ratings differences were statistically significant even when checking with the Tukey test. Attractiveness was significant in the ANOVA but not significant when checking for the adjusted p-value.

For participants who didn't receive a match, all the attributes rating differences were statistically significant except for the 'fun' attribute. The post-Tukey test confirmed that the differences among the rating time for intelligence and sincerity was between time1 and time2; and time1 and time3, i.e. between the event sign up time and after the event but before finding out about the match; and between the sign-up time and after finding out about the results. For the attractiveness attribute, there was a difference in rating after finding out about the results of the speed dating. For ambition, there was a shift between time1 and time2. We can conclude that for this dataset the observations of [4] hold true: a positive response influenced the self-rating for intelligence and sincerity rating while a negative response had an impact on a wider range of attributes: attractiveness, sincerity, intelligence and ambition.

Descriptive Analysis 2 and ANOVA 2

As previously described, within the process of conducting speed dating events, all participants were requested to rate their own attributes a number of times, once at signup time (Time 1), then the day after participating in the event (Time 2) and finally 3–4 weeks after participants had been sent their matches (Time 3). The attributes used were the same considered in the other analyses, "attractive", "sincere", "intelligent", "fun" and "ambitious" and are the same as those used by the speed daters to rate their date partners.

We wished to analyse how self-image (self-perception) of the participants is changing (if changing) over time, with respect to whether their self-rating for attractiveness (the strongest explanatory variable) was higher or lower than the average attractiveness rating they received from all of their date partners (reflecting on their self-perception and self-confidence), and whether the participants ultimately received any match/multiple matches.

We speculated in advance that there would be movement in the ratings over time, especially in those groups who had received quite different ratings from others than their own assessment, and that there would be movement in self-ratings over time in particular for those who had received no matches, which may decrease their self-rating levels.

This was explored by running a number of ANOVA analyses to identify if there was any statistically significant difference among the self-rating measurements at time 0, time 1 and time 2 (in the dataset tables and certain extracts these may be referred to as time 1 (T1), time 2(T2) and time 3(T3).

The first ANOVA was run in a similar manner to the ANOVA in the previous section i.e. a one-way ANOVA for each of the 5 attributes: attractiveness, sincerity, fun, intelligence, ambition, on the processed dataset described in the methodology section.

It was run on a further processed version of the dataset which reconfigures the 3724 lines (one line for each date) and summarises the data into 245 lines (one record for each participant) which ensures the R code will use the field 'match' to split the groups between participants with a match and no match (and not for each speed-date record which resulted in a match vs no match), fields for validation checks, and for other variables to be added intended for use in further analyses.

The resulting ANOVA analysis was performed using R found that there were no statistically significant variances in movement in the ratings of any of the five attributes for any subgroup over time at the $p < 0.05$ confidence level chosen.

Given the lack of statistical significance to these results these have not been reproduced in this paper, but the result supports the stability of self-image of the participants over time, in particular for the variable 'attractive', considered by [11] the most important of explanatory and predictive variables for speed dating analysis.

It should be noted that the original descriptive analysis, appeared to indicate that there was a more substantial movement in the self-rating over time, as it was run using the field 'match' to split the data set without recognising that this will not correctly split the dataset.

A further ANOVA was also run using a new variable Flag_Other_Higher calculated in the data transformation process. This permits the ANOVA to break the dataset into groups which represent the grouping of people who have received a higher rating from their date partners than their own self-rating, and the grouping who have received a lower self-rating from their date partners than they have given themselves.

The resulting ANOVA analysis was performed using R found that there were no statistically significant variances in movement in the ratings of any of the five attributes for any subgroup over time at the $p < 0.05$ confidence level chosen for these groups with the potential to have difference self-image and resulting self-confidence levels.

Given the lack of statistical significance to these results these have not been reproduced in this paper, but again this statistical confirmation supports the stability of the self-image based over time, in particular those based on the variable 'attractive'.

On further investigation of the results and the dataset we identified a number of relevant explanations

- Central tendency of the self-ratings of participants, especially for attractiveness.
  - The corollary of this is that the 'outlier' groups one may have expected are quite small or non-existent. No participant self-rated their attractiveness as one or two, and in the entire processed dataset only 5 people rated as a three, and nine people self-rated as a ten.
- Central tendency of the ratings from others, which tends to group towards the central tendency of the self-ratings, partly attributable to averaging but may perhaps be a 'wisdom of crowds' effect.

We also re-evaluated our variable 'match' to consider whether it was fully capturing all relevant information in the decision processes. The variable match is a binary variable where 1 means that both parties have made an offer of a match. This variable therefore does not capture whether either party has made an offer, but not received a corresponding offer.

The full dataset and the extracted dataset were reconsidered, reconsidering the 'match' field versus the existing fields of 'dec' i.e. decision = 1,0 with 1 meaning that the participant has made an offer for a match, and also a new variable introduced from this evaluation process called 'Offer received', again a binary 1,0 field. This is not present in the original dataset but is calculable by processing the dataset to match all of the offers made with the speed date pairs, summarising the results per participant into new variables for 'offers made' and 'offers received' as already explored by

There was insufficient time to complete statistical analysis on the dataset using these variables but they are described here to support the following observations which arose from reconsidering the full dataset.

Out of the 551 participants group, 99 participants recorded no matches (gender split 53% women, 47% men), of which 31 of these participants had made no offers at all for a match (gender split 70% women, 30% men), and by expressing no preference could therefore never be in a position to obtain a full 'match', despite offers were received. Only 22 out of the full dataset received no offers, (reducing to 7 in the extract used for self-rating.

At the other extreme of expressing no preference, 20 participants made an offer to every date partner on the night, with all except two receiving at least one match with this strategy (gender split 30% women, 70% men).

These all suggest that these may leave scope for further work to be performed on these fields and their insight into dating behaviour.

4 Evaluation

We evaluated all our analysis. For the predictive methods:

1. Logistic regression: the evaluation was performed through the calculation of the ROC curve which gave results of 0.82 for the women's model and 0.81 for the men's model. Models were tested by building a prediction model on test data, achieved accuracy of predictions done on a test data was 0.75 for the women's model and 0.74 for men's model.
2. Regression tree, C 5.0 tree, the randomforest and conditional inference forest. These two steps were taken before running the analysis: (a) Randomisation of training and test data, 75% train and 25% test data (b) Models built for the own decision model and for the partner's model. Most models accuracies were in the range 72%–74% while the regression tree with added complexity gave an accuracy of 85% and 86%.

For the analysis of variance: we used the Fligner-Killeen test [17, 18] to check for homogeneity of variances, performed the one-way ANOVA and used Tukey post-hoc to check if differences occurred between groups (Table 1).

Table 1. Regression Tree Analysis

Method	Metrics		
	Accuracy own model	Accuracy partner's model	Evaluation
Logistic regression	0.75	0.74	Test data
Regression tree	0.73	0.74	Test data
Regression tree complex	0.85	0.86	Test data
C 5.0 tree	0.74	0.74	Test data
Random forest	0.73	0.72	Test data

For participants who received a match: (a) the attribute sincerity gave a result of a statistically significant result for $F_{(2,1965)} = 0.000293$, $p <= 0.05$. Post hoc Tukey tests revealed that there was a significant difference found between time 3 and time 1 and between time 2 and time 1. No difference was found between time 2 and time 3. (b) the attribute intelligence gave a statistically significant result for $F_{(2,1965)} = 0.0264$, $p <= 0.05$. Post hoc Tukey tests revealed that there was a significant difference found between time 2 and time 1.

For participants who didn't receive a match: (a)Attractiveness: this analysis produced a statistically significant result for $F_{(2,9201)} = 0.00445$, $p <= 0.001$. Post hoc Tukey tests revealed that there was a significant difference found between time 3 and time 1; (b) Sincerity: This analysis produced a statistically significant result for F $(2,9201) = 1.77e-07$, p $<= 0.001$. Post-hoc Tukey tests revealed that there was a significant difference found between time 2 and time 1 and time 3 and time 1; (c) Intelligence: this analysis produced a statistically significant result for $F(2,9201) = 0.000113$, p $<= 0.001$. Post hoc Tukey tests revealed that there was a significant difference found between Time 2 and time 1 and time 3 and time 1; (d) Ambition: this analysis produced a statistically significant result for $F(2,9201) = 0.0628$, p $<= 0.1$. Post hoc Tukey tests revealed that there was a significant difference found between time 2 and time 1.

For the analysis of variance run under analysis 2: this was run in a similar manner as the first ANOVA, i.e. used the Fligner-Killeen test [17, 18] to check for homogeneity of variances, and performed a one-way ANOVA, but this these were run on a a a re-processed data set which was adapted for the changed processing and additional fields. Tukey post-hoc are not required as there were no statistically significant differences identified by the ANOVA in the initial results.

5   Conclusion

This revisit to this long existing speed-dating dataset with the fresh eyes of this group of data mining students and a combination of analytical and predictive approaches has given us evidence that it is indeed possible to obtain new insights from old data, in this particular case around the concept of self-image, and the perception of that image by others.

The predictive methods for regression gave evidence to consider that there is considerable predictive value in self-ratings information, especially attractiveness, as to whether or not a successful decision to meet again will arise, based only on the examination of self-ratings and partner's ratings recorded from the speed dating event.

The high level of accuracy obtained in the results, especially in the regression trees is a satisfactory result obtained for the application of these methodologies.

The descriptive analyses and analysis of variance gave evidence for the stability of those self-images and self-perceptions over time and their consistency throughout the receipt of both positive and negative feedback following dating events. This insight suggests that further development of most successful predictive techniques focused on newer datasets may be a suitable, and possible profitable (in a number of senses) area for further exploration.

References

1. Fisman, R., Iyengar, S.S., Kamenica, E., Simonson, I.: Gender differences in mate selection: evidence from a speed dating experiment*. Q. J. Econ. **121**, 673–695 (2006)
2. Fisman, R., Iyengar, S.S., Kamenica, E., Simonson, I.: Racial preferences in dating. Rev. Econ. Stud. (2008)
3. Leverger, C.: Analysis of a Speed Dating Experiment dataset with R, Gephi and Neo4j (2016)

4. Yurchisin, J., Watchravesringkan, K., Mccabe, D.B.: An exploration of identity re-creation in the context of internet dating. Soc. Behav. Personal. Int. J. **33**(8), 735–750 (2005)
5. Schau, H.J., Gilly, M.C.: We are what we post? self- presentation in personal web space. J. Consum. Res. **30**(3), 385–404 (2003)
6. Lawsonm, H.M., Leck, K.: Dynamics of Internet Dating A Brief History of Dating Practices
7. Belot, M., Francesconi, M.: Can Anyone Be 'The' One? Evidence on Mate Selection from Speed Dating (2006)
8. Fletcher, G.J., Kerr, P.S., Li, N.P., Valentine, K.A.: Predicting romantic interest and decisions in the very early stages of mate selection standards, accuracy, and sex differences. Pers. Soc. Psychol. Bull. **40**(4), 540–550 (2014)
9. Kurzban, R., Weeden, J.: Do advertised preferences predict the behavior of speed daters? Pers. Relatsh (2007)
10. Todd, P.M., Penke, L., Fasolo, B., Lenton, A.P.: Different cognitive processes underlie human mate choices and mate preferences. Proc. Natl. Acad. Sci. USA **104**(38), 15011–15016 (2007)
11. Asendorpf, J.B., Penke, L., Back, M.D.: From dating to mating and relating: predictors of initial and long-term outcomes of speed-dating in a community sample. Eur. J. Pers. **25**(1), 16–30 (2011)
12. Brooke, D.: New Stanford Research on Speed Dating Examines What Makes Couples 'Click' in Four Minutes (2013)
13. Tidwell, N.D., Eastwick, P.W., Finkel, E.J.: Perceived, not actual, similarity predicts initial attraction in a live romantic context: evidence from the speed-dating paradigm. Pers. Relat. **20**(2), 199–215 (2013)
14. Penke, L., et al.: How self-assessments can guide human mating decisions. In: Mating Intelligence: Sex, Relationships, and the Mind's Reproductive System, pp. 37–75 (2007)
15. Buston, P.M., Emlen, S.T.: Cognitive processes underlying human mate choice: the relationship between self-perception and mate preference in Western society. Proc. Natl. Acad. Sci. USA **100**(15), 8805–8810 (2003)
16. Breiman, L.: Random forests. Mach. Learn. **45**(1), 5–32 (2001)
17. Conover, W.J., Johnson, M.E., Johnson, M.M.: A comparative study of tests for homogeneity of variances, with applications to the outer continental shelf bidding data. Technometrics **23**(4), 351–361 (1981)
18. R Documentation: Fligner-Killeen Test of Homogeneity of Variances. https://stat.ethz.ch/R-manual/R-devel/library/stats/html/fligner.test.html. Accessed 28 Apr 2016

Forecasting the Chilean Electoral Year: Using Twitter to Predict the Presidential Elections of 2017

Sebastián Rodríguez[1]([✉]), Héctor Allende-Cid[1], Wenceslao Palma[1],
Rodrigo Alfaro[1], Cristian Gonzalez[2], Claudio Elortegui[3], and Pedro Santander[3]

[1] Escuela de Ingeniería Informática,
Pontificia Universidad Católica de Valparaíso, Valparaíso, Chile
{sebastian.rodriguez.o,hector.allende,wenceslao.palma,
rodrigo.alfaro}@pucv.cl
[2] Instituto de Ciencias de Lenguaje y Literatura,
Pontificia Universidad Católica de Valparaíso, Valparaíso, Chile
cristian.gonzalez@pucv.cl
[3] Escuela de Periodismo,
Pontificia Universidad Católica de Valparaíso, Valparaíso, Chile
{claudio.elortegui,pedro.santander}@pucv.cl

Abstract. Failures of traditional survey methods for measuring political climate and forecasting high impact events such as elections, offers opportunities to seek alternative methods. The analysis of social networks with computational linguistic methods have been proved to be useful as an alternative, but several studies related to these areas were conducted after the event (post hoc). Since 2017 was the election year for the 2018–2022 period for Chile and, moreover, there were three instances of elections in this year. This condition makes a good environment to conduct a case study for forecasting these elections with the use of social media as the main source of Data. This paper describes the implementation of multiple algorithms of supervised machine learning to do political sentiment analysis to predict the outcome of each election with Twitter data. These algorithms are Decision Trees, AdaBoost, Random Forest, Linear Support Vector Machines and ensemble voting classifiers. Manual annotations of a training set are conducted by experts to label pragmatic sentiment over the tweets mentioning an account or the name of a candidate to train the algorithms. Then a predictive set is collected days before the election and an automatic classification is performed. Finally the distribution of votes for each candidate is obtained from this classified set on the positive sentiment of the tweets. Ultimately, an accurate prediction was achieved using an ensemble voting classifier with a Mean Absolute Error of 0.51% for the second round.

Keywords: Election forecasting · Sentiment analysis
Machine learning · Ex Ante forecasting

© Springer International Publishing AG, part of Springer Nature 2018
G. Meiselwitz (Ed.): SCSM 2018, LNCS 10914, pp. 298–314, 2018.
https://doi.org/10.1007/978-3-319-91485-5_23

1   Introduction

2016 was a year in which two classic institutions undoubtedly failed: the media and surveys of public opinion. They failed in their capacity to probe important socio-political dynamics and in their predictive capacity, regarding high impact events. These events include the 2016 US presidential election, the Brexit poll and the Colombian 2016 peace agreement referendum. For this reason, new alternatives to measure the political climate have arisen to meet these needs.

Nowadays, the massive use of social networks has allowed for multiple interactions between users, who express their opinions on different topics, people, events and brands. Moreover, the use of social networks in election years, people tend to comment about the candidates, either by their proposals or by their performance in media related events. To extract the relevant information from these political opinions, different Computational Linguistic methods can be applied, such as Sentiment Analysis (SA) on these interactions to get the overall political sentiment.

Twitter is one of the most influential social networks for sharing political messages. This platform is a micro-blogging site, which allows users to broadcast short messages with a maximum of 140 characters (recently increased to 280 characters) called "tweets". With over 328 million monthly active users and 500 million tweets generated per day [23], Twitter has the potential of becoming a valuable source when analyzing sentiment, and, even more so, political related sentiment in an election year.

During 2017, Chile had three instances of elections (Primaries, First and Second round) providing a rich environment to measure the political climate. Furthermore, given that several studies related to forecasting of elections have been conducted, [6,8,12,20,27,31], a similar exercise may be undertaken into the Chilean reality so as to examine the outcome of similar methods. Not only that, but several of these studies have been conducted post hoc, so they cannot be taken as true forecasting.

Given all this, the main goal of this study is to make three predictions Ex Ante of each instance of the 2017 Chilean presidential elections. The approach taken to make these predictions is using Supervised machine learning algorithms with Sentiment Analysis techniques. First a number of experts do a manual pragmatic sentiment labeling over tweets collected over a period of time before the elections, which serves as the input for the different classification models. The tweets then are collected ten days before the prediction day, classified, and the distribution of the overall preferences of those tweets is analyzed to make the prediction. Finally, these predictions are contrasted with the true results of the elections after the event has occurred.

The paper is organized as follows: in Sect. 2, the state of the art is presented. Section 3 describes both the problem and the data used in this study. In Sect. 4, the methodology is shown, detailing the process carried out through the 3 elections, data processing, the metric to be used and the models applied. Finally, in Sect. 5 the results are discussed and in Sect. 6 the conclusions of the study are presented.

2   State of the Art

The development and use of social networks lets millions of users to generate knowledge and, in turn, share it in an easy way that has allowed widespread growth. Given this phenomenon, there is an interest in finding methods to monitor public opinion and behavior, regarding a wide variety of topics. [18, 21, 28]. Such topics include the areas of health, economy, and politics, the latter being the one pertaining to this research.

One way of carrying out this monitoring is by means of Sentiment Analysis, which consists of the use of natural language processing tools, in addition to computational linguistics, in order to assign a polarity value to a document [25]. In the social network context, it has been observed that Twitter may be used as a corpus to which these techniques could be applied, therefore extracting useful information [1, 17, 24].

Regarding the exercise of making political predictions employing social networks, one of the first seminal studies was the one proposed by Tumasjan et al. [31], in which the German federal elections of 2009 were analyzed. In that research it was found that the number of messages analyzed reflected the distribution of the votes in the election. It is worth noting that, albeit this was an initial approach, there are certain studies that thereupon detected particular problems with this method.

Therefore, Gayo-Avello et al. [14,15] identified certain problems regarding inconsistencies in various studies dealing with predictions carried out using the social network Twitter. These problems ranged from methodological flaws in which the studies were not predictive (post hoc prediction), statistical flaws in which the samples were not representative, and issues related to the training of the models, among other concerns. Furthermore, there are several authors that likewise detected conflictive results and shortcomings in the prediction process using Twitter [10, 16, 20, 22].

With the aforementioned further research, new studies arose, which took into consideration the deficiencies detected by [14]. An example of this is Bermingham and Smeaton [3], where a study applied to the general elections of Ireland as a case study was carried out, integrating sentiment analysis to the prediction process. The authors conclude that Twitter possesses in fact some predictive power, and that it becomes marginally improved when sentiment analysis is incorporated.

Other studies that have used sentiment analysis to make predictions of electoral results have been: the U.K. general elections of 2010 [12], the Dutch senate elections of 2011 [27], the French elections of 2012 [8], the U.K. general elections of 2015 [6] and the U.S. presidential elections of 2016 [29]. All these studies agree that Sentiment Analysis boosts the predictive power of their methods. However, the issue of these studies remains, in their inability to tackle all the problems identified by Gayo-Avello [14]. Regarding this, Beauchamp [2] made a study concerning the extrapolation and interpolation of vote intention in the US presidential elections of 2012 dealing with most of these problems. Nonetheless, it still presents the problematic that is a post hoc prediction, instead of real forecasting.

3   Problem and Dataset

In this section the definition of the problem of this study is introduced: 2017 as an election year for Chile. Also, the dataset and the candidates who participated in each of the elections are described.

3.1   Problem Definition

During 2017 in Chile, there were several presidential referendums for the upcoming presidential period 2018–2022, which were divided up throughout the year in primary elections, first round and second round.

Two political coalitions participated in the primary elections: "Chile Vamos" (Right-wing coalition) and "Frente Amplio" (Left-wing coalition). In the right-wing coalition, there were three participants: Sebastián Piñera, Felipe Kast and Manuel José Ossandón. In the left-wing coalition, there were only two: Beatriz Sánchez and Alberto Mayol. Since there were only two coalitions, this election was taken as two independent elections on the same day, which was July 2nd. The winners of these elections were Sebastian Piñera for "Chile Vamos", and Beatriz Sánchez for "Frente Amplio".

In the first round, the two winners of the primary elections participated in an election with six other candidates. These were: Alejandro Navarro, Eduardo Artés, José Antonio Kast, Carolina Goic, Marco Enríquez Ominami and Alejandro Guillier. This election was held on November 19th and the winners of that election were Sebastián Piñera and Alejandro Guillier.

Finally, the second round was carried out on December 17 with the winners mentioned above. The results of this election were that Sebastian Piñera won over Alejandro Guillier with 54.57% of the voting preferences.

Given the sustained growth of social networks in Chile and Latin America [30], these instances presented an interesting test case to do automated SA over the social networks. An immediate application is to analyze the behavior and opinions of the users and their messages on social networks given the participation of the candidates in media events. Although Facebook is the social network with the largest number of interactions, the Twitter API turned out to be more permissive at the time to track the interactions of users. This allows to check derived interactions of media events related to the candidates.

For this particular reason, it is very interesting to track the opinions of the people in presidential election years and find the opinions and preferences of the people regarding the participating candidates. This in order to find indicators/variables that can help in the prediction process. In recent years, traditional instruments (surveys) have failed worldwide to make predictions in different political events [7,32], Chile in year 2017 being another example of this.

Given all this, the question arises whether the methods based on machine learning using data from social networks serve as a reliable predictor. Conveniently, the nature of this year, allowed to perform three prediction exercises related to this area. Therefore, this study is expected to be valuable for the body of knowledge related to predictions using social networks, providing an insight into the merits and challenges of the applied approach.

3.2 Dataset

The dataset used for this study corresponds to a compilation of tweets generated during the presidential campaigns of the year 2017 by all the users that made mention of either the presidential candidate's account, or the name and surname of each candidate in the messages. In total, there has been tracking to 11 candidates, from May 14th to December 19th of 2017, being this last date on the day of the Second round of the presidential elections in Chile.

The first thing to mention is that we have gathered two kinds of tweets: the original message and the Retweets (RT). As its name implies, the first consists of a message in which the user wants to express something related to a certain candidate. The number of original messages obtained during each period of time is presented in Table 1.

Table 1. Original tweets collected through the elections periods of the year.

Corpus	Primary elections	First round	Second round
Alberto Mayol	53957	-	-
Felipe Kast	66914	-	-
Manuel Jose Ossandon	94564	-	-
Alejandro Navarro	-	29524	-
Eduardo Artes	-	10070	-
Jose Antonio Kast	-	259248	-
Carolina Goic	-	130196	-
Marco Enriquez Ominami	-	147941	-
Beatriz Sánchez	115605	197415	-
Alejandro Guillier	-	34316	117258
Sebastian Piñera	195566	572222	304911
Total	526606	1380932	422169

On the other hand, the RT consists of an action by means of which the users are able to replicate an original message as it was written, without adding any content to it. Although this message is the same as the original, it is delivered by another user, providing information for the tweets Sentiment Analysis. The total number of RTs for each candidate during the different election periods are presented in Table 2.

Table 2. Retweets collected through the elections periods of the year.

Corpus	Primary elections	First round	Second round
Alberto Mayol	110514	-	-
Felipe Kast	138577	-	-
Manuel Jose Ossandon	140451	-	-
Alejandro Navarro	-	74143	-
Eduardo Artes	-	14645	-
Jose Antonio Kast	-	760073	-
Carolina Goic	-	355002	-
Marco Enriquez Ominami	-	368539	-
Beatriz Sánchez	263900	421894	-
Alejandro Guillier	-	99164	344868
Sebastian Piñera	414321	1225873	799633
Total	1067763	3319333	1144501

Regarding both tables, the tracking was limited for each candidate to the round they participated in. One of the features observed is that there is a higher amount of RTs than original messages. This could be relevant, since doing tracking of the RTs, possible influencers of this social network could be detected. Another thing to take into account is that along as the different instances of elections were being conducted, the participation in this social network increased. Finally, the candidate with the highest number of messages was Sebastián Piñera and the one with the lowest number of messages was Eduardo Artes.

As for the manual sentiment analysis of tweets, it was carried out by six experts trained to detect the polarity of the messages. They conducted this labeling process mainly in particular time schedules related to certain media events (interviews, debates, etc.) Concerning the possible sentiments, they correspond to three labels: Positive, Negative, and Neutral. It should be noted that the sentiment analysis approach is based on a pragmatic labeling, rather than a semantic one. This means that the polarity is labeled over the context of the tweet, instead of the semantic polarity of the words composing the tweet. Given this, if a tweet was labeled as positive for a candidate, the feeling is transferred only to it because of the context. Table 3 shows the total volume of tweets tagged for each of the candidates tracked through the three elections.

In the case of labeled tweets, there is no balance between positive and negative classes for each candidate. Regarding neutral tweets, these correspond to most of the tweets labeled for all candidates, with the exception of Manuel Jose Ossandon. Finally, the candidate that generated the most labeled activity was Sebastian Piñera, while the one with least labeled activity was Eduardo Artes.

Table 3. Labeled tweets obtained from the experts from all the elections.

Corpus	Positive	Negative	Neutral
Alberto Mayol	5994	3711	14932
Felipe Kast	17635	5382	7383
Manuel Jose Ossandon	4786	13827	11908
Alejandro Navarro	772	3581	9432
Eduardo Artes	161	2356	5570
Jose Antonio Kast	11327	4070	55787
Carolina Goic	7980	2901	25217
Marco Enriquez Ominami	15495	4421	35170
Beatriz Sánchez	17579	25008	56648
Alejandro Guillier	6876	9262	63925
Sebastian Piñera	37082	51587	102459
Total	125687	126106	388431

4   Methodolody

Since throughout the year, three election processes were held, the methodology that was proposed for each of these share certain foundations. The idea behind it is that in order to make the prediction, first a set of tweets is taken before the election and is separated into two parts: Training set and Prediction set.

The first set consists of all the tweets that have a label within a certain date range, which will serve to train a supervised learning algorithm. The prediction set, on the other hand, consists of all the tweets regardless of the label, then all the tweets are selected from the final date of the Training set to the date when the elections will be held. This range of dates is called the prediction window.

Regarding the labels that were used, as the manual classification was carried out with a pragmatic approach in which the sentiment is directed to the candidate, positive labels were used for the training of the algorithm. This is mainly done by transcribing a positive label to "positive-candidate". With the new labels, the classifier will be trained for all the candidates that participated in that election, with the tweets labeled for the training set. Once the classifiers have been trained, they are applied to the total volume of messages within the prediction window. This allows to make the prediction for this gross amount, obtaining a number of preferences of candidates, which are then converted into the percentages of the prediction.

For the primary elections, a prediction window of 10 days was adopted, and because it was the first predictive exercise, a delta of 3 days before the election was taken. This was mainly done by what is described in [9], where the authors indicate that while daily monitoring of social networks is indeed convenient, there is some evidence that the prediction can be made days before the event.

With the results obtained from the primary elections, a post-election prediction process was carried out. This process had as a goal to be able to adjust the models and obtain some information on how the different prediction windows and training sets behaved. This was done with the aim of applying this knowledge in the following elections.

For the elections of the first round a prediction window of 10 days was adopted, as well as for the primary elections. Using the results obtained with the primary exercise, a date for the prediction 6 days before the election was chosen.

Finally, with the results obtained, for the second round it was decided to monitor daily the political preferences. This was done due to the unsatisfactory results obtained with the election methods of the first round, as it will be detailed in the results section. In this sense, the 10-day prediction window was also kept, but the prediction date was changed to the day before the elections. Table 4 provides a summary of what was described above for all elections.

Table 4. Date range for the training/prediction sets and how many days before the election the prediction was made.

Election	Training set dates	Prediction set days	Days before election
Primaries	June 1st–June 19th	June 19th–June 29th	3 days
First round	June 1st–November 6th	November 6th–November 16th	5 days
Second round	June 1st–December 6th	December 6th–December 16th	1 days

In the case of the tweets themselves, a preprocessing of the texts was performed. This includes the elimination of stopwords in Spanish, normalization of text, eliminating accent marks, removing scores and symbols. Moreover, the entire messages were taken in lower case. In addition to this, web links and all mentions to Twitter user accounts (username) were removed as well. The latter was done because they can give noise to the automatic training process of the classifiers. Finally, it was also decided to leave the emojis in the tweets (a digital image used to express an emotion or idea.) Hashtags (#word) were also kept, because they can provide useful distinctive information to carry out the classification among candidates.

In order to confirm that the predictions are correct, the results of the elections will be used as ground truth. The metric used to measure this comparison is the Mean Absolute Error (MAE) of the prediction. The MAE is computed as:

$$MAE = \frac{1}{n} \sum_{i_1}^{n} |y_i - x_i| \tag{1}$$

Where y_i corresponds to the prediction made for the i element, and x_i corresponds to the real i element value (in this case, the percentage of the electorate for a candidate).

In order to carry out the training and prediction, the texts were transformed using a word bag approach and a unigram representation. In the case of the algorithms used, they will be detailed in the following subsection.

4.1   Algorithms

In this subsection the six different baseline models are presented. These are: AdaBoost, Support Vector Machines (SVM), Decision Trees, Random Forest and two voting classifiers. The selection of these models is due to the following: SVM's have been used since the very beginning of SA, and have proved to deliver good performances with different data sets [19]. On the other hand, there are not many studies detailing the performance for political sentiment analysis with AdaBoost and Random Forest, therefore the decision to use them corresponds to the desire to contribute to the general knowledge on how these algorithms perform, in this context.

AdaBoost. AdaBoost (Adaptive Boosting) [13] is an ensemble method that combines weak classifiers which have relatively good classification accuracy, in order to make one strong classifier. This process begins with training N classifiers with modified versions of the data. Subsequently, the individual predictions are merged through a weighted majority vote to make the final prediction. These weights are updated on each iteration of the boosting algorithm, where erroneous instances are weighted up and vice versa. A Decision tree was used as the weak classifiers and the hyperparameter tuned in this study corresponds to the amount of estimators used for the creation of the ensemble.

Support Vector Machines. Support Vector Machines [11] algorithms are a supervised learning method for both classification and regression. SVMs represent the training data as points in a space and use hyperplanes to make separations between classes. Afterwards, by using that space new examples are projected, and the prediction is made according to which side of the separation the projection falls on. To make the hyperplanes, the SVMs implements kernels which allows them to make both linear (linear kernel) and nonlinear separations (polynomial and radial basis function kernels). For this study, a SVM with a linear kernel was implemented, which provides one hyperparameter to tune: cost of misclassification of the data on the training process (C).

Decision Trees. Decision Trees (DTs) are a non-parametric supervised learning method used for classification and regression. Their aim is to create a model that predicts the value of a target variable by learning simple decision rules inferred from the data. The algorithm used in this study was Classification and Regression Tree (CART) [4], which is based on the C4.5 tree algorithm. The hyperparameters that were tuned were the separation criteria and the depth of the tree.

Random Forest. Random forest (RF) [5] is an ensemble method, which uses bootstrap aggregation (bagging) over the original features to build a large number of decision trees. The main objective for RF is to deal with over-fitting and also reduce the variance between trees. The classification then is computed with the mode of the outputs of each tree within the forest. For RF, the hyperparameter is the number of DTs used, depth and separation criteria when building the DTs.

Voting Classifier. Voting Classifier is an ensemble method, which groups several classifiers to do the classification. These are trained in parallel with the same data, and then vote to make the prediction of a sample. In this study, two voting classifiers have been used, these being Hard and Soft voting classifiers. The Hard Voting (V1) classifier makes a classification by majority vote, while the latter (v2) makes the prediction selecting the largest sum of the predicted probabilities for all classifiers. These voting classifiers are built with each method described previously (AdaBoost, DT, SVM and RF).

4.2 Implementation

In this study, the library scikit-learn [26] was used for the implementation of the different algorithms described in the previous subsection. For the primary elections, the machine used had an i5 3.2 processor and 16 GB RAM. For the first and second round elections, the machine used had an i9-7900X processor with 128 GB RAM.

5 Results

As detailed in Sect. 3.1, two coalitions participated in the primary elections, so the prediction exercise was carried out as if they were two separate elections. For this reason, the MAE is reported separately, one for each coalition. The MAE obtained with the different algorithms for all elections is presented in Table 5.

Table 5. Mean Absolute Error for each algorithm in each election.

Algorithm	P. Chile Vamos	P. Frente Amplio	First round	Second round
AdaBoost	**0.0346**	0.1684	0.0656	0.0117
LSVM	0.1533	**0.1587**	0.0909	0.1923
DT	0.1036	0.2363	0.0737	0.1706
RF	0.0860	0.2560	0.0648	0.1750
V1	0.0563	0.2878	**0.0635**	0.0150
V2	0.1042	0.2057	0.0700	**0.0051**

As stated before, the primary elections were divided into two blocks (P. Chile Vamos and P. Frente Amplio). The best results were obtained by AdaBoost with an MAE of 3.46% for Chile Vamos' primary elections. For the Frente Amplio the lowest MAE was obtained LSVM with a 15.87%. Next, in Fig. 1 the percentages predicted with AdaBoost vs the actual results of the election are presented.

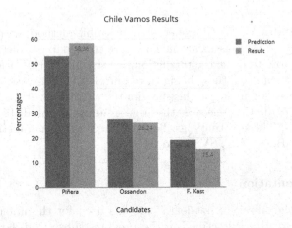

Fig. 1. Percentages of the prediction versus the real values for Chile Vamos election using AdaBoost

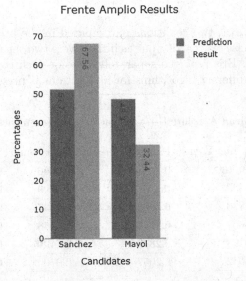

Fig. 2. Percentages of the prediction versus the real values for Frente Amplio election using Linear Support Vector Machine

On the other hand, Fig. 2 presents the results obtained for the Frente Amplio's primary elections, where there is a clear deviation of the prediction versus the real electoral votes.

Regarding the elections of the first round, the Hard voting classifier (V1) obtained the best results, achieving an MAE of 6.35%. Although all the classifiers obtained an MAE under 10%, it was found that when comparing the results of the prediction versus the real values, all the classifiers had a tendency to give a greater favoritism to J. Kast than to Guillier (runner up of the first round).

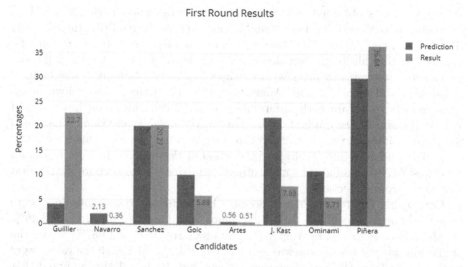

Fig. 3. Percentages of the prediction versus the real values for first round election using Hard Voting Classifier

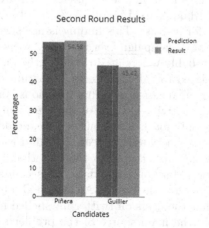

Fig. 4. Percentages of the prediction versus the real values for second round election using Soft Voting Classifier

The Fig. 3 shows what was indicated previously in the prediction obtained by the best classifier for this election.

Finally, for the second round the soft voting classifier (V2) obtained an MAE of 0.51%, being the lowest MAE throughout the study. Figure 4 shows the prediction results for V2, detailing the narrow margin between the prediction and the real election result.

6 Discussion

With the results obtained, AdaBoost had an overall good performance. This is because it obtained the best result in one ofjavascript:void(0); the elections (P. Chile Vamos), and results closer to the lowest MAE consistently. It must be pointed out that although each election was made throughout the year 2017, the use of social networks in political campaigns was used in a more active manner in the first and second round. Apart from this, the immediate deployment of these models to obtain early information in the primaries could have worked against the aim of the study. For this reason, the primary elections served as a good prediction exercise, both to get an idea of what the elections were going to be like, and to refine the hyperparameters of the models in order to obtain the lowest MAE possible. The main objective of this was to prepare for the first and second round elections.

Concerning the results of the primary elections, the results obtained were close to the real percentages for Chile Vamos' election, with an MAE of 3.46%. On the other hand, Frente Amplio's results had an MAE of 15.87%, where the prediction was correct on who was going to win (Fig. 2), although the values were further apart from the real electorate percentages. It should also be noted that the best classifier for Frente Amplio's prediction (LSVM), although it obtained the lowest MAE for that prediction, in the other elections its performance was the worst in all cases. This is an important issue for future research.

For the first round although an MAE of 6.35% was obtained, the results of the prediction show a bias for J. Kast. This finding is not surprising, given that he had a strong social network campaign over the last 2 weeks before the elections. Due to this, the bias probably was related to one of the main concerns proposed by Gayo-Avello [14] related to "all the tweets are assumed to be trustworthy". Regarding this, Post hoc analysis showed that malicious behavior related to false accounts generating fake activity in favor of the Candidate (Astroturfing), and that it was present not only for J. Kast, but other candidates as well.

Regarding these same elections, the reason behind the bad performance of candidate Guillier (runner up), is due to the fact that there was an error in capturing the interactions for said candidate. This candidate had two official accounts and apart from this, the name "Alejandro Guillier" was tracked. The problem with this is that the users on twitter misspelled his name as "Guiller"; hence this was possibly the main source of the problems of the prediction for this candidate. This tracking error was detected and fixed in the previous days of the election day. This can be seen in the total of tweets for the candidate in the first round versus the second round (Tables 1 and 2). Finally, this increase may also be due to the fact that the different left-wing candidates urged their voters to give their support to Guillier, potentially increasing the discussion on twitter.

On the other hand, the results obtained in the second round were close to the electoral results, with an MAE of 0.51% for the Soft Voting Classifier (V2). These results are attributed to the amendment for tracking the candidate Guillier.

Regarding the raw volume of tweets prediction discussed in [31], low MAE results were obtained for the primaries. However, when this method was applied to the other two elections, the prediction distances from the true values as detailed in Table 6.

Table 6. Mean Absolute Error for raw volumes of tweets analysis.

P. Chile Vamos	P. Frente Amplio	First round	Second round
0.0278	0.0221	0.0673	0.1765

Finally, the lack of a description of the demography of the users, and the approach that each tweet of the prediction set is a vote towards a candidate after the classification process are issues also detailed by Gayo-Avello. This could explain easily the high values of MAE obtained in the Frente Amplio's primary elections, or the predicted value for J. Kast for the first round. Taking the latter as an example, it could be seen that J. Kast had a very strong political campaign at the end of the last two weeks before the election as mentioned before. This increases the number of tweets that express a positive sentiment towards the candidate. For that reason, it is urgent to look for a better method to model the vote intention for the Twitter users.

7 Conclusions

In this study three electoral predictions were made through several months of the year 2017 for the Chilean presidential elections. Supervised learning algorithms were trained with pragmatic sentiment labeled tweets and predicted the distribution over a prediction set. Hyperparameters tuning for both algorithms and training/prediction set were conducted from the primaries election and first round, resulting in a final accurate prediction with an MAE of 0.51% with the Soft voting classifier.

One of the main motivations of this study was to make Ex Ante predictions of the elections, which resulted in a challenging problem. Example of this was to discover after the first round the error of tracking towards the account of the candidate Guillier, or the failed estimate towards the candidate J. Kast giving him the second majority. Mostly, this could be one of the flaws detailed by Gayo-Avello cite Gayo1, which mentions that "all the tweets are trustworthy".

This leads to the fact that in the event of making a prediction of this style, factors such as astroturfing and the use of social bots must be taken into account; as well as the need to make a review of the demographics of users.

As future work, the use of other labels is proposed to improve the performance of predictive models. As such, this presents a source of information not exploited in the present study. Other relevant information that can be obtained from the total tagged corpus, is the use of topic models both to be able to see the political discourse and the opinions of the Twitter users changing through the electoral period. In addition to this, it would be interesting to use the topic model words obtained and assigning them a greater weight when making predictions.

Acknowledgments. This work was supported by the "Proyectos Interdisciplinarios" Grant of VREIA - Pontificia Universidad Católica de Valparaíso. Héctor Allende-Cid's work was supported by the "Fondecyt Initiation into Research 11150248" of Conicyt, Chile.

References

1. Agarwal, A., Xie, B., Vovsha, I., Rambow, O., Passonneau, R.: Sentiment analysis of Twitter data. In: Proceedings of the Workshop on Languages in Social Media, pp. 30–38. Association for Computational Linguistics (2011)
2. Beauchamp, N.: Predicting and interpolating state-level polls using Twitter textual data. Am. J. Polit. Sci. **61**(2), 490–503 (2017)
3. Bermingham, A., Smeaton, A.: On using twitter to monitor political sentiment and predict election results. In: Proceedings of the Workshop on Sentiment Analysis Where AI Meets Psychology (SAAIP 2011), pp. 2–10 (2011)
4. Breiman, L., Friedman, J.H., Olshen, R.A., Stone, C.J.: Classification and Regression Trees. Chapman & Hall, New York (1984)
5. Breiman, L.: Random forests. Mach. Learn. **45**(1), 5–32 (2001). https://doi.org/10.1023/A:1010933404324
6. Burnap, P., Gibson, R., Sloan, L., Southern, R., Williams, M.: 140 characters to victory?: Using Twitter to predict the UK 2015 general election. Elect. Stud. **41**, 230–233 (2016)
7. By, R.T., Ford, J., Randall, J.: Changing times: what organizations can learn from brexit and the 2016 us presidential election. J. Change Manag. **17**(1), 1–8 (2017). https://doi.org/10.1080/14697017.2017.1279824
8. Ceron, A., Curini, L., Iacus, S.M., Porro, G.: Every tweet counts? How sentiment analysis of social media can improve our knowledge of citizens' political preferences with an application to Italy and France. New Media Soc. **16**(2), 340–358 (2014). https://doi.org/10.1177/1461444813480466
9. Ceron, A., Curini, L., Iacus, S.M.: Politics and Big Data: Nowcasting and Forecasting Elections with Social Media. Taylor & Francis, New York (2016)
10. Chung, J., Mustafaraj, E.: Can collective sentiment expressed on twitter predict political elections? In: Proceedings of the Twenty-Fifth AAAI Conference on Artificial Intelligence, pp. 1770–1771, AAAI 2011. AAAI Press (2011). http://dl.acm.org/citation.cfm?id=2900423.2900687
11. Cortes, C., Vapnik, V.: Support-vector networks. Mach. Learn. **20**(3), 273–297 (1995)

12. Franch, F.: (Wisdom of the crowds)2: 2010 UK election prediction with social media. J. Inf. Technol. Polit. **10**(1), 57–71 (2013). https://doi.org/10.1080/19331681.2012.705080
13. Freund, Y., Schapire, R.E.: A short introduction to boosting. In: Proceedings of the Sixteenth International Joint Conference on Artificial Intelligence, pp. 1401–1406. Morgan Kaufmann, San Francisco (1999)
14. Gayo-Avello, D.: No, you cannot predict elections with Twitter. IEEE Internet Comput. **16**(6), 91–94 (2012)
15. Gayo Avello, D., Metaxas, P.T., Mustafaraj, E.: Limits of electoral predictions using twitter. In: Proceedings of the Fifth International AAAI Conference on Weblogs and Social Media. Association for the Advancement of Artificial Intelligence (2011)
16. Jungherr, A., Jürgens, P., Schoen, H.: Why the pirate party won the German election of 2009 or the trouble with predictions: a response to Tumasjan, A., Sprenger, T.O., Sander, P.G., & Welpe, I.M. "predicting elections with Twitter: what 140 characters reveal about political sentiment". Soc. Sci. Comput. Rev. **30**(2), 229–234 (2012). https://doi.org/10.1177/0894439311404119
17. Kouloumpis, E., Wilson, T., Moore, J.D.: Twitter sentiment analysis: the good the bad and the OMG!. ICWSM **11**(538–541), 164 (2011)
18. Lariscy, R.W., Avery, E.J., Sweetser, K.D., Howes, P.: Monitoring public opinion in cyberspace: how corporate public relations is facing the challenge. Public Relat. J. **3**(4), 1–17 (2009)
19. Liu, B., Zhang, L.: A survey of opinion mining and sentiment analysis. In: Aggarwal, C., Zhai, C. (eds.) Mining Text Data, pp. 415–463. Springer, Boston (2012). https://doi.org/10.1007/978-1-4614-3223-4_13
20. Lui, C., Metaxas, P.T., Mustafaraj, E.: On the predictability of the us elections through search volume activity (2011)
21. Madge, C., Meek, J., Wellens, J., Hooley, T.: Facebook, social integration and informal learning at university: 'it is more for socialising and talking to friends about work than for actually doing work'. Learn. Media Technol. **34**(2), 141–155 (2009)
22. Metaxas, P.T., Mustafaraj, E., Gayo-Avello, D.: How (not) to predict elections. In: 2011 IEEE Third International Conference on Privacy, Security, Risk and Trust (PASSAT) and 2011 IEEE Third Inernational Conference on Social Computing (SocialCom), pp. 165–171. IEEE (2011)
23. Omnicore Agency: Twitter by the numbers: stats, demographics & fun facts. https://www.omnicoreagency.com/twitter-statistics/. Accessed 28 Dec 2017
24. Pak, A., Paroubek, P.: Twitter as a corpus for sentiment analysis and opinion mining. In: LREc, vol. 10 (2010)
25. Pang, B., Lee, L., et al.: Opinion mining and sentiment analysis. Foundations and trends®. Inf. Retriev. **2**(1–2), 1–135 (2008)
26. Pedregosa, F., Varoquaux, G., Gramfort, A., Michel, V., Thirion, B., Grisel, O., Blondel, M., Prettenhofer, P., Weiss, R., Dubourg, V., Vanderplas, J., Passos, A., Cournapeau, D., Brucher, M., Perrot, M., Duchesnay, E.: Scikit-learn: machine learning in Python. J. Mach. Learn. Res. **12**, 2825–2830 (2011)
27. Sang, E.T.K., Bos, J.: Predicting the 2011 Dutch senate election results with Twitter. In: Proceedings of the Workshop on Semantic Analysis in Social Media, pp. 53–60. Association for Computational Linguistics, Stroudsburg, PA, USA (2012). http://dl.acm.org/citation.cfm?id=2389969.2389976
28. Shirky, C.: The political power of social media: technology, the public sphere, and political change. Foreign Aff. **90**(1), 28–41 (2011)

29. Singh, P., Sawhney, R.S., Kahlon, K.S.: Forecasting the 2016 US presidential elections using sentiment analysis. In: Kar, A.K., Ilavarasan, P.V., Gupta, M.P., Dwivedi, Y.K., Mäntymäki, M., Janssen, M., Simintiras, A., Al-Sharhan, S. (eds.) I3E 2017. LNCS, vol. 10595, pp. 412–423. Springer, Cham (2017). https://doi.org/10.1007/978-3-319-68557-1_36

30. Fung Global Retail & Technology: Deep dive social media in Latin America. Technical report, May 2016. https://www.fbicgroup.com/sites/default/files/Social

31. Tumasjan, A., Sprenger, T.O., Sandner, P.G., Welpe, I.M.: Predicting elections with Twitter: what 140 characters reveal about political sentiment. ICWSM **10**(1), 178–185 (2010)

32. Valentino, N.A., King, J.L., Hill, W.W.: Polling and prediction in the 2016 presidential election. Computer **50**(5), 110–115 (2017)

Evaluation of Network Structure Using Similarity of Posts on Twitter

Yusuke Sato[1(✉)], Kohei Otake[2], and Takashi Namatame[3]

[1] Graduate School of Science and Engineering, Chuo University,
1-13-27 Kasuga, Bunkyo-Ku, Tokyo 112-8551, Japan
al3.bthf@chuo-u.ac.jp
[2] School of Information and Telecommunication Engineering, Tokai University,
2-3-23, Takanawa, Minato-Ku, Tokyo 108-8619, Japan
otake@indsys.chuo-u.ac.jp
[3] Faculty of Science and Engineering, Chuo University,
1-13-27 Kasuga, Bunkyo-Ku, Tokyo 112-8551, Japan
nama@indsys.chuo-u.ac.jp

Abstract. Social networking service (SNS) is very popular in our lives, with expanding internet environments and mobile device. Through the SNS, user can submit their opinion or reputation freely, anytime and anywhere. These activities are getting great attention on a various business scenes in recently. Twitter is one of the most popular SNS, and used by numerous people in the world. In addition, since various information is posted on Twitter, it is expected to be utilized as a business strategy, and there have been many studies on the marketing using Twitter data. Moreover, we can get some information about user's network in Twitter. In this research, we attempt to evaluate the network structure using similarity of post on Twitter. We created the user network using similarity of posts mentioned about four titles of Japanese TV drama, and we grasped the post categories that is easy to get user's interest. From the result, we discussed the difference between TV drama and suggestions for promotion strategies of TV drama production company.

Keywords: Social Networking Service · Twitter · Network analysis
Graph representation · Natural Language Processing

1 Introduction

Social Networking Service (SNS) is very popular in our lives, with the development of information technology and mobile device such as smart phone. By using SNS, it is possible that user can share various information through their friends freely anywhere and anytime. From this reason, the information transmission between consumer on SNS is actively performed, and it sometimes affects the real world. Therefore, SNS have gotten a lot of attention in the business scene as a promotion and marketing tool in recent years [1, 2]. Furthermore, SNS is regarded as an important tool that make it possible to transmit information to many people efficiently in various industries such as retailers, EC sites, political activities and so on.

© Springer International Publishing AG, part of Springer Nature 2018
G. Meiselwitz (Ed.): SCSM 2018, LNCS 10914, pp. 315–329, 2018.
https://doi.org/10.1007/978-3-319-91485-5_24

Twitter is one of the most popular SNS in the world. By using Twitter, users can perform various actions such as "Tweet" and "Retweet". Moreover, there are a variety of information such as user's opinion and reputation on Twitter. Using Twitter or other SNS data including those information, we can elucidate various phenomenon occurring on Twitter (e.g. information diffusion and network of friendship between users). Therefore, there have been numerous research related to marketing activities using SNS data. To understand the user behavior on SNS, various researchers have applied studies [3–5]. In these studies, they targeted posts data about specific products or the structure of SNS itself and analyzed the SNS data. On the other hand, regarding the TV drama targeted in this research, it is inferred that there are various phenomenon caused by audience (such as post activity in real time of broadcasting time or the period from the episode to the next episode). Therefore, it can be said that elucidation of post activities on Twitter by audience is important analysis for the promotion strategies of TV drama.

2 Related Studies and Our Purpose

In this section, first, we introduce some related studies about SNS analysis. Next, we show the objective of this study.

Yang et al. [3] analyzed the information diffusion phenomenon on Twitter. Especially, they proposed model that able to capture the three main specifics of information diffusion (speed, scale and range) using survival analysis. As the result, they found that some specific of the tweets can predict the diffusion phenomenon. Matsumura et al. [4] proposed an influence diffusion model that express how articles and words were diffused. As the result, using the above model, they identified influencer who post information that gets interest of others and words reflecting consumer insights. Matsuo et al. [5] investigated the network structure of the user networks created on the largest SNS site in Japan. Moreover, they confirmed the structure of the community formed by the relationship of users on the network.

In this study, we focus on Twitter data and attempt to evaluate the network structure among users using similarity of posts. For the analysis, we used tweet data posted about four Japanese TV drama. The information about TV drama are frequently posted on Twitter by audience, and its contents are various (e.g. contents about story, actor or actress, etc.). Focusing on its situation, we also try to evaluate the users' interest to post categories by dividing the user network into several communities and comparing network indexes. From these results, it is possible to identify the post category that is easy to get interest among users, and it is expected to obtain a useful suggestion for the promotion strategy performed by the TV drama production company.

3 Data Summary

We targeted four titles of Japanese TV drama and collected tweets data posted about these titles. In this study, we selected these four titles based on broadcasting period and evaluation by ranking site. We used hashtags and keywords (drama titles) and collected the data by using the application programing interface (API) of Twitter. Consequently, we collected about 577,000 tweets in total. These tweet data were posted during broadcasting period of each TV drama and include User ID, tweet date and time, tweet text, the number of favorite and Retweets and so on. Summary (e.g. broadcast period (time zone: JST) and category of each title) of targeted TV drama and collected tweets data are shown Tables 1, 2 and 3.

Table 1. Broadcast period, frequency and time of targeted Japanese TV drama

Title	Period	Broadcast frequency	Broadcast time
A	2017/04/18 ~ 2017/06/20	Every Tuesday	PM10:00 ~ PM11:00
B	2017/04/17 ~ 2017/06/26	Every Monday	PM09:00 ~ PM10:00
C	2017/04/14 ~ 2017/06/16	Every Friday	PM10:00 ~ PM11:00
D	2017/04/16 ~ 2017/06/18	Every Sunday	PM09:00 ~ PM10:00

Table 2. Summary of targeted Japanese TV drama

Title	Content	Category
A	TV drama based on Japanese comic	Love romance
B	TV drama based on Japanese novel	Mystery
C	TV drama based on Japanese novel	Mystery
D	TV drama created by Japanese TV station originally	Drama

Table 3. Summary of collected tweet data

Title	The number of tweets	The number of unique posted users
A	65,153	13,417
B	288,004	27,425
C	104,557	25,007
D	92,147	17,430

4 Evaluation of Network Structure Using Posts Similarity

In this study, we performed analysis in 3 steps. In the 1st step, we extracted representative 50 keywords by each drama and classified these keywords into 13 post categories by Natural Language Processing. In the 2nd step, we visualized the network that express the posting relationship between users and post categories. Especially, we created incidence matrix and bipartite graph by using the weight which means user's posting importance for each category. In the final step, we divide the above network into several communities. Targeting these communities, we grasped post categories which is mainly posted by users of each community and compared the network indicators such as network density between communities. From above results, we discuss the user's interest for each post category.

4.1   Identify and Classify Keywords

In the 1st step, we identify the keywords of each title and classify these keywords into post categories. Firstly, we performed morphemes analysis to divide all tweet texts of each title into columns of morphemes (minimum elements constituting sentences). Morphological analysis is a commonly used method for dividing the natural language (text data) into morphemes and discriminating parts of speech and the like of each morpheme. It is need became all of letters are connected Japanese sentence. In the morphological analysis, information such as parts of speech words defined in grammar and the dictionary is used for dividing process. In this study, we used the R language to perform morphological analysis. Moreover, the dictionary used for analysis was Mecab [6], a Japanese morpheme dictionary.

Targeting terms extracted by morphological analysis, we selected three parses (nouns, verbs, adjectives), and identified the keywords of each title using the *tfidf* method [7]. The *tfidf* method is a type of index of word weighting and is calculated by the product of *tf* (term frequency) and *idf* (inverse document frequency). The *tfidf* values of word i in the document j is calculated by the following equations.

$$tfidf_{i,j} = tf_{i,j} \times idf_i \tag{1}$$

$$tf_{i,j} = \frac{n_{i,j}}{\sum_s n_{s,j}} \tag{2}$$

$$idf_i = \log \frac{|D|}{|\{d : d \in t_i\}|} \tag{3}$$

where $n_{i,j}$ is the occurrence frequency of word i in document j, $\sum_s n_{s,j}$ is the summation of count of all the words in document j, $|D|$ is the total number of documents, $|\{d : d \in t_i\}|$ is the number of documents that contain word i.

We defined the top 50 words which have high *tfidf* values as the keywords for each title and classified these keywords into 13 post categories based on the its meaning. The names of those post categories and its description are shown in Table 4. The number of keywords of each post category by each title are shown in Table 5.

From above result, it is found that whether post categories were posted or not differ depending on title of TV drama. For example, "Location" and "Other_TVshow" are posted on only title D, "Broadcasting_station" were posted on title A and B.

4.2   Bipartite Graph of Users and Categories

In the 2nd step, we created the network among users by the bipartite graph to grasp posting relationship between users and post categories. In this research, we targeted the top 500 users who posted frequently during each title of TV drama. The network graph was constructed by using the relationship which is whether user posted the post category or not and their weight for post category based on post ratio. In particular, we created the incidence matrix and the bipartite graph as the following procedure.

Table 4. The names of categories and its description

Category name	Description
L_character	Leading character
S_character	Supporting character
L_actor/actress	Leading actor or actress
S_actor/actress	Supporting actor or actress
Location	Shooting place of each title
Emotion (+)	Positive emotion about content of each episode
Emotion (−)	Negative emotion about content of each episode
Main_topic	The main topic of each title
Contents_topic	The subtopic of each title
Broadcasting_type	The word meaning the type of episode (e.g. the end of the series of each title)
Broadcasting_station	The station which broadcasts each title
Other_TVshow	TV programs other than each TV drama
Title	Title of each TV drama

Table 5. The number of keywords in each title[a]

Category	Title			
	A	B	C	D
L_character	4	7	6	5
S_character	2	1	1	0
L_actor/actress	7	5	10	11
S_actor/actress	4	1	3	3
Location	0	0	0	1
Title	4	2	2	1
Emotion (+)	10	14	9	10
Emotion (−)	4	5	4	2
Main_topic	2	3	1	4
Contents_topic	9	12	3	11
Broadcasting_type	6	8	16	7
Broadcasting_station	1	1	0	0
Other_TVshow	0	0	0	2

[a] Each keyword can belong to multiple categories.

1. We calculated the frequency of keywords for all tweet texts of targeted users. Here, we counted presence or absence of keywords, without consideration for that same keywords appears more than once in a tweet text.
2. Based on the post category of each keyword, we calculated the post ratio of each post category by each user.
3. We calculated a weight $W_{i,j}$ for post category j of user i in accordance with the following conditions. In addition, we defined the matrix constituted by these

weights as the incidence matrix X which means the user's posting importance for each post category.

$$W_{i,j} = \frac{r_{i,j} \times 100}{T_j} \tag{4}$$

where $r_{i,j}$ is post ratio of post category j of user i, T_j is the number of terms which belongs to post category j.

4. Based on the incidence matrix X, we created bipartite graphs in which nodes are users and post categories, weights of edges are $W_{i,j}$.

For the visualizing the bipartite graph, we used Fruchterman-Reingold algorism [8]. Fruchterman-Reingold algorism is a method based on dynamic model for visualizing network. This algorithm has a feature to arrange the connected nodes close to and to locate unconnected nodes far from each other. Figures 1, 2, 3 and 4 shows the bipartite graph of users and post categories of each title of TV drama. In the bipartite graph of each title, dark gray edges express high weighted edges which have weights in the top quartile points of all weights. In addition, Fig. 5 shows the result of calculating the number of edges connected to the category node that is the number of users who posted the category by weight.

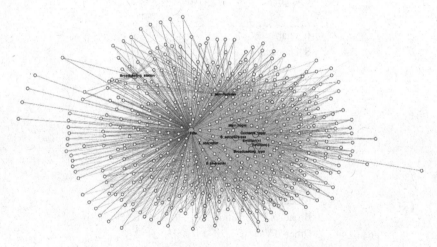

Fig. 1. Bipartite graph of users and post categories of title A

From above result, it turned out that the category most posted from users on title A and B is the "Title", users of title C and D frequently post about "L_character". In addition, about those categories, we can see that the number of edges with high weight (dark gray edge on the bipartite graph) is more than the number of edges with low weight (light gray edge on the bipartite graph). On the other hand, we can see that other posting categories show reverse trends, in other words, most edges are composed of edges with low weight.

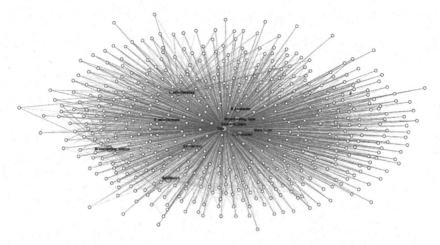

Fig. 2. Bipartite graph of users and post categories of title B

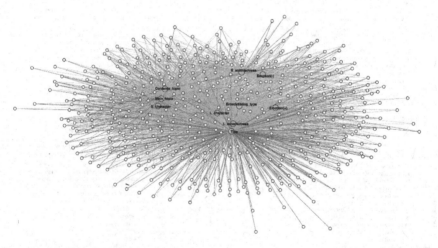

Fig. 3. Bipartite graph of users and post categories of title C

4.3   User Network Using Similarity of Post Categories

In the final step, we created user network based on their similarity of the post categories. Moreover, dividing the network into several communities, we evaluated the relationship between network indicator and post categories of each community. In order to detect some communities from the network, we need adjacency matrix rather than incidence matrix. Firstly, by using incidence matrix X of previous step, we created new incidence matrix X'. Especially, we created the new matrix based on following condition to more strictly define whether the user posted each posting category or not.

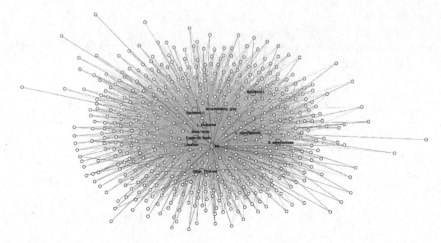

Fig. 4. Bipartite graph of users and post categories of title D

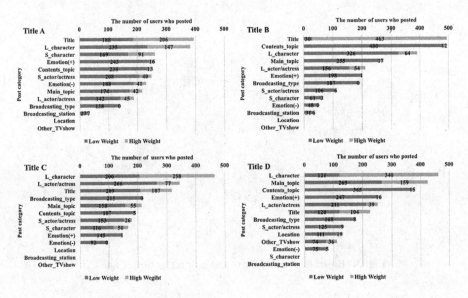

Fig. 5. The number of users who posted about each category

$$W'_{i,j} = \begin{cases} 1 & W_{i,j} \geq V_{3rd,j} \\ 0 & W_{i,j} < V_{3rd,j} \end{cases} \tag{5}$$

where $W_{i,j}$ is the weight for post category j of user i which is calculated on the previous step, $V_{3rd,j}$ is the weight in the top quartile points of all weights of post category j.

So, incidence matrix X' means that only users with high weights for each category are redefined as "Users who posted the category".

Next, we created adjacency matrix Y which describes the similarity of post categories between users. Especially, we converted the incidence matrix X' into adjacency matrix Y as follows.

$$Y = X'X'' \tag{6}$$

where X'' is the transposed matrix of incidence matrix X', the all diagonal elements of the adjacency matrix Y are 0.

Table 6. The network indexes using user similarity of post category of each title

Title	The number of nodes	The number of edges	Density
A	499	57,745	0.462
B	487	51,223	0.410
C	491	56,260	0.450
D	498	60,743	0.486

From the above transformation, the adjacency matrix Y means the similarity of the post category between users. Using adjacency matrix Y, we created and visualized the user network. For the visualization, we deleted isolated nodes not connected to any other node out of 500 user nodes of each title. The network indexes of user network of each title are shown Table 6.

From Table 6, regarding the user network of all titles, we can see that the nodes are connected with moderate density compared to general social networks of friendship between users.

Finally, we divide these user network into several communities by spin glass method [9]. Spin glass method is one of the most popular method to detect communities from the network. This method assigns each node to the community so as to minimize the Hamiltonian function expressed by the following equation.

$$\mathcal{H}(\{\sigma\}) = -\sum_{i \neq j} a_{ij} Y_{ij} \delta(\sigma_i, \sigma_j) + \sum_{i \neq j} b_{ij}(1 - Y_{ij})\delta(\sigma_i, \sigma_j)$$
$$+ \sum_{i \neq j} c_{ij} Y_{ij}[1 - \delta(\sigma_i, \sigma_j)] - \sum_{i \neq j} d_{ij}(1 - Y_{ij})[1 - \delta(\sigma_i, \sigma_j)] \tag{7}$$

where Y_{ij} denotes the adjacency matrix of the graph, σ_i denotes the group index of node i in the graph, and $a_{ij}, b_{ij}, c_{ij}, d_{ij}$ denote the weights of the individual contributions.

In addition, we determined the number of communities by using the modularity Q [10]. This value is an index for evaluating the accuracy of community detection and calculated by following equation. We can define the dividing the community has high Q value as appropriate detection.

$$Q = \frac{1}{2M} \sum_{i \neq j} \left[A_{ij} - \frac{k_i k_j}{2M} \right] \delta(\sigma_i, \sigma_j) \qquad (8)$$

where k_i is the number of edges of node i, M is summation of the number of edges which exist in the network.

Table 7. The result of community detection of each title

Title	The number of communities	Modularity
A	3	0.22
B	3	0.28
C	3	0.23
D	3	0.22

We can conduct appropriate community detection by adopting the division result of high Q value. As the result of community detection based on modularity Q, the user network was divided three community by each title. The modularity Q of community detection of each title are shown Table 7.

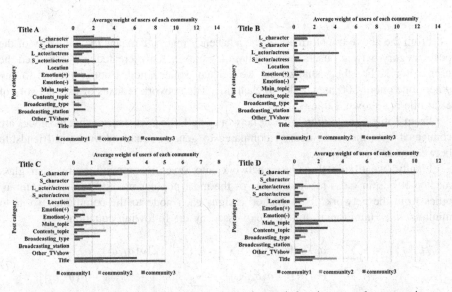

Fig. 6. The average weight for each category of users belonging to each community

Targeting these communities, we attempt to grasp the post categories which is mainly posted by users who belong to each community. Figure 6 shows the user's average weight to each post category by each community of each title. By using result of Fig. 6, we defined the topic of each community shown in Table 8.

Table 8. The topic of each community of each title

Title	Community	Topic
A	1	TV Drama Headline
	2	Content
	3	Character and Emotion ($\pm$)
B	1	Content and Emotion ($+$)
	2	Content
	3	TV Drama Headline
C	1	Content and Emotion ($+$)
	2	Content
	3	TV Drama Headline
D	1	Content and Emotion ($+$)
	2	Content
	3	TV Drama Headline

From Table 8, it turned out that the topics posted for each community differed. Considering all the communities, there are six types of topics. "TV Drama Headline" is user group posted tweet including words that have potential of becoming headline of TV drama such as title names of TV drama and actor or actress names. "Content" is the topic which is mainly posted both content of episode and character of TV drama. "Character and Emotion" or "Content and Emotion" is the topic that have been posted the characters of TV drama or content of the episode with user's emotion. However, there are not only positive emotions but also negative emotions. Regarding community 2 of title D, this community posted about actor or actress names, title names and other TV show names. Therefore, we defined this community as the user group which mainly posted about promotion of TV drama, named "Promotion". Moreover, about community 3 of title D, we named "Actor or Actress" which is user group posted only leading actor or actress. As the total tendency, it is found that user's emotion, regardless positive or negative, are posted with content of episode or character of TV drama.

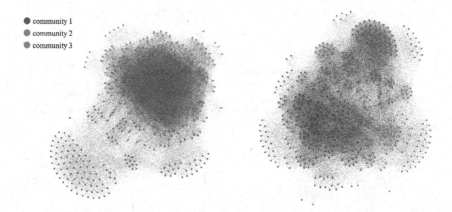

● community 1
● community 2
● community 3

Fig. 7. User network using similarity of post category of title A (left) and B (right)

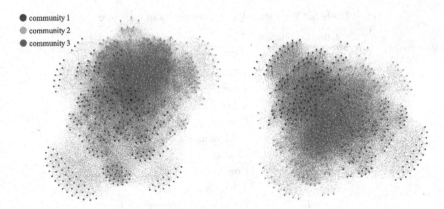

Fig. 8. User network using similarity of post category of title C (left) and D (right)

Table 9. Network indexes of each community of title A

	Community 1	Community 2	Community 3
Topic	TV Drama Headline	Content	Character and Emotion (±)
The number of nodes	165	148	186
Density	0.662	0.912	0.777
Average degree	108.727	134.202	143.817

Table 10. Network indexes of each community of title B

	Community 1	Community 2	Community 3
Topic	Content and Emotion (+)	Content	TV Drama Headline
The number of nodes	206	153	128
Density	0.676	0.894	0.907
Average degree	138.660	135.921	115.296

Table 11. Network indexes of each community of title C

	Community 1	Community 2	Community 3
Topic	Character and Emotion (±)	Content	TV Drama Headline
The number of nodes	142	198	151
Density	0.771	0.777	0.804
Average degree	108.845	153.111	120.649

Table 12. Network indexes of each community of title D

	Community 1	Community 2	Community 3
Topic	Content and Emotion (−)	Promotion	Actor or Actress
The number of nodes	153	223	122
Density	0.916	0.645	1.000
Average degree	139.268	143.210	121.000

Furthermore, by using network indexes of each community, we evaluate the user's interest to the topic of each community. We used the density and average degree of network as network indexes. Figures 7 and 8 show the user network using similarity of post category and Tables 9, 10, 11 and 12 show network indexes calculated by each community by each title.

From the above result, even if it is a similar topic, it turned out that there is a difference in the density of the user for each title, that is, the degree of user's interest for the topic. In terms of the result of each title, in the title A, there are not many users who pay attention to "TV Drama Headline" and "Character and Emotion (±)". On the other hand, topic of "Content" gets a lot of interest of users. About title B, it turned out that "Content" and "TV Drama Headline" topic is posted by relatively large number of users.

Regarding title C, there aren't so many users who pay attention to "Character and Emotion (±)" and "Content" topic. In the title D, many users are interested in "Content and Emotion (−)" and "Actor or Actress". Moreover, a few users posted about topic "Promotion".

5 Discussion

First of all, we discuss posting categories that attract users' interest in each title. In the title A, since the users of community 2 are most closely connected, the topic on the contents of the TV drama attracts audience's interest. It is inferred that the reason for this result is that title A is a love romance TV drama including elements such as affair, and audience actively posts the contents of the episode because its story is unpredictable.

From the results of title B, the community 3 has the highest density among users. Therefore, it can be said that topics (such as title names and actor names) that have potential of becoming headlines of TV drama are actively posted. It is assumed that users are pay attention to the actor or actress and its title name rather than the contents of the TV drama because the leading actor of title B is a member of a popular idol group in Japan.

About title C, the densities of all the communities are moderate, there is not much different between them. In other words, it can be said that users are post the same importance degree for any topic and it is a TV drama that has been posted about various kinds of topics in well balance.

Regarding title D, community 1 and 3 has the high user's density. Since this title was broadcasted in the TV slot that is broadcast station and broadcast time zone in which many masterpiece TV dramas were broadcasted in the past, it is inferred that the audience pay much attention to the topic related to contents of episodes and actor or actress. In particular, community 3 is the user group that emphasizes on only actor or actress, it is assumed that there are a certain number of fans of actor or actress in the community. Furthermore, in title D, there is only one user community that mainly post the topic related to the promotion of TV drama. However, it turns out that the density of that community is not so high. This is presumably because not only whether the user is a fan of the actor (actress) but also whether user watch other TV programs also influences the importance of "Promotion" topic.

In addition, as the overall knowledge, it was found that the emotion (regardless of positive or negative) of the users was posted with the contents of episodes and characters of the TV drama. It can be said that this is a natural result as a topic on which audience express opinions.

6   Conclusion

In this research, targeting four titles of Japanese TV drama, we evaluated the network structure using similarity of posts on Twitter. Even if users posted about same title of TV drama, it turned out that there are differences of user's importance to post categories among communities by dividing the user network into several communities. Moreover, it also found that there are differences among four titles of TV drama as well. It is expected to utilize these results for the strategies for promotion or marketing on Twitter of companies related to each TV drama.

As the future work of our research, we need to evaluate user's interest for topic from various point of view such as "who are the main users in the same community?" and "what topics are easy to post simultaneously with other topics?". In addition, it is possible to obtain more useful suggestions as a promotion strategy by using follow or follower relations data among users on Twitter in combination.

Acknowledgment. We thank Rooter Inc. for providing valuable datasets and for their useful comments.

References

1. Elisabeta, I., Ivona, S.: Social media and its impact on consumers behavior. Int. J. Econ. Pract. Theor. 4(2), 295–303 (2013)
2. Sitaram, A., Bernardo, A.H.: Predicting the future with social media. Computing 25(1), 492–499 (2010)
3. Jiang, Y., Scott, C.: Predicting the speed, scale, and range of information diffusion in Twitter. In: Proceedings of the 4th International AAAI Conference on Weblogs and Social Media, ICWSM, vol. 10 (2010)

4. Matsumura, N., Yamamoto, H., Tomozawa, D.: Finding influencers and consumer insights in the blogosphere. In: International Conference on Weblogs and Social Media, Seattle, Washington (2008). (in Japanese)
5. Matso, Y., Yasuda, Y.: How relations are built within a SNS World: Social network analysis on Mixi. Trans. Jpn. Soc. Artif. Intell. **22**(5), 531–541 (2007). (in Japanese)
6. MeCab. http://taku910.github.io/mecab/. 23 Feb 2018
7. Ricardo, A.B., Berthier, A.R.: Modern Information Retrieval: The Concepts and Technology Behind Search, 2nd edn. Addison-Wesley Professional, Harlow (2011)
8. Thomas, M.J.F., Edward, M.R.: Graph drawing by force-directed placement. Softw. Pract. Experience **21**(11), 1129–1164 (1991)
9. Joerg, R., Stefan, B.: Statistical mechanics of community detection. Phys. Rev. E **74**(1), 016110 (2006)
10. Mark, E.J.N., Michelle, G.: Finding and Evaluating Community Structure in Networks. Phys. Rev. **69**(2), 026113 (2004)

Estimating Speaker's Engagement from Non-verbal Features Based on an Active Listening Corpus

Lei Zhang[1], Hung-Hsuan Huang[1,2(✉)], and Kazuhiro Kuwabara[1]

[1] College of Information Science and Engineering,
Ritsumeikan University, Kusatsu, Japan
hhhuang@acm.org
[2] Center for Advanced Intelligence Project, RIKEN, Kyoto, Japan

Abstract. The elderly who live alone are increasing rapidly in these years. For their mental health, maintaining their social life with others is reported useful. Our project aims to develop a listener agent who can engage active listening dialog with the elderly users. Active listening is a communication technique that the listener listens to the speaker carefully and attentively. The listener also ask questions for confirming or showing his/her concern about what the speaker said. For this task, it is essential for the agent to evaluate the user's engagement level (or the attitude) in the conversation. In this paper, we explored an automatic estimation method based on empirical results. An active listening conversation experiment with human-human participants was conducted for corpus collection. The speakers' engagement attitude in the corpus was subjectively evaluated by human evaluators. Support vector regression models dedicated to the periods when the speaker is speaking, the listener is speaking, and no one is speaking are built with non-verbal features extracted from facial expressions, head movements, prosody and speech turns. The resulted accuracy was not high but showed the potential of the proposed method.

Keywords: Active listening · Elderly support
Multimodal interaction

1 Introduction

The population of elderly people is growing rapidly in developed countries. If they do not maintain social life with others, they may feel loneliness and anxiety. For their mental health, it is reported effective to keep their social relationship with others, for example, the conversation with their caregivers or other elderly people. There are already some non-profit organizations recruiting volunteers for engaging "active listening" with the elderly. Active listening is a communication technique that the listener listens to the speaker carefully and attentively. The listener also ask questions for confirming or showing his/her concern about what

© Springer International Publishing AG, part of Springer Nature 2018
G. Meiselwitz (Ed.): SCSM 2018, LNCS 10914, pp. 330–341, 2018.
https://doi.org/10.1007/978-3-319-91485-5_25

the speaker said. This kind of support helps to make the elderly feel cared and to relieve their anxiety and loneliness. However, due to the lack of the number of volunteers comparing to that of the elderly who are living alone, the volunteers may not be always available when they are needed. In order to improve the results, always-available and trustable conversational partners in sufficient number are demanded.

The ultimate goal of this study is the development of a computer graphics animated virtual listener who can engage active listening to serve elderly users at a level close to human listeners. In order to conduct successful active listening, it is considered essential for the listener to establish the rapport from the speaker (elderly users). Rapport is a mood which a person feels the connection and harmony with another person when (s)he is engaged in a pleasant relationship with him/her, and it helps to keep long-term relationships [8, 10]. In order to achieve this, like a human listener, the virtual listener has to observe the speaker's behaviors, to estimate how well the speaker is engaging the conversation [15], and then decides how it should respond to the user.

The estimation of the level of the speaker's engagement in the active listening conversation is therefore one of the essential functions of the active listener agent. In the context of active listening, the level of the speaker's engagement in the conversation can be considered to be expressed by his/her attitude toward the listener. The utterances of the speaker are obvious cues for the estimation of the speaker's engagement. However, due to the nature of active listening conversation, the speaker may utter in arbitrary contexts, It is difficult to utilize this information. Non-verbal behaviors are considered more general and more robust (less user-dependent). On the other hand, benefits from the advance of sensor device technology, machine learning models from multimodal sensory information has been proving to be effective in estimation human communication behaviors [2].

This paper reports our progress in developing the estimation model of speaker's engagement for active listener agents with non-verbal features based on a multimodal corpus of active listening conversation. Since there are only three situations in dyadic active listening, the speaker is speaking, the listener is speaking, neither of the two participants are speaking, we developed the model for each situation because the available features are different in each situation. The non-verbal features include head movements, facial expressions, prosodic information, and speech turn information.

2 Related Works

The research works on making robots and agents to be the communication partners of the elderly and dementia patients have been getting popularity. One of the methods to mitigate the progression of dementia is "coimagination" which was proposed by Otake et al. [13]. It is a method by using pictures as the references for the topics in group conversation. All participants who are elderly people have equal chance to listen, to talk, to ask questions, and to answer questions. It is reported that the elderly who participated in this activity talked and smiled more frequently than before. However, this method has the limitation that all of the participants have to meet at one single place which may be difficult in practical.

Bickmore et al. [3] proposed a companion agent to ease the anxiousness of elderly inpatients. Huang et al. [10] developed a rapport agent which reacts to facial expressions, backchannel feedbacks, and eye gazes of the user. The agent is designed to show behaviors which are supposed to elicit rapport. However, it does not try to estimate and react to the user's internal state or "mood". For example, when the user looks in bad mood, showing the agent's concern on the user by saying "Are you OK?" like human do. The SEMAINE project [12, 14] was launched to build a Sensitive Artificial Listener (SAL). SAL is a multimodal dialogue system with the social interaction skills needed for a sustained conversation with the user. They focused on realizing "really natural language processing [4] " which aims to allow users to talk with machines as they would talk with other people.

These works were developed base on the subject studies in the U.S. or in other western countries where the subjects' communication style may diverge from that of Japanese ones [7]. In this study, we collected an active listening corpus of Japanese subjects and analyzed Japanese style verbal/non-verbal behaviors which potentially improve the effectiveness of a listener.

3   Active Listening Corpus

This work shares the same data corpus with our previous work [9]. The corpus is collected in a human-human teleconferencing experiment which is conducted to imitate the situation where a virtual listener which is displayed on a 2D surface. The collected video data were evaluated and annotated in the aspect of how positive the speaker's engagement attitude by both of the participants and a third person from an objective viewpoint. The annotated scenes are then extracted for the development of automatic estimation method.

3.1   Experiment Setup

Due to the difficulty in recruiting elderly participants, and the experiment procedure explained in the following sections may be difficult the elderly, college students are recruited in this experiment. Furthermore, we considered the implicit criteria in judging communication partner's attitude to be positive or negative should not vary largely between younger and senior generations.

Eight pairs of participants in the same gender were recruited as the experiment participants. All of them were college students and native Japanese speakers. The two participants of each pair were recruited with the condition that they are close friends. This is because close friends were assumed easier to talk with each other in the limited experiment period. In order to simulate the situation of talking with a 2D graphical agent, the participants of each pair were separated into two rooms (Figs. 1 and 2) and talked with each other via Skype teleconferencing software. In each session, one participant played the role of speaker, and the other one played the role of the listener.

They were instructed to sit on a chair so that the move of their lower bodies can be controlled within a limited range. Each room was equipped with two video cameras. One was used for recording the participant from the front. The other one was used for logging the Skype window which was duplicated on another monitor. The speaker talked with the listener who was projected on a large screen near life-size. The height of the projected image was adjusted so that the speaker can see the listener's eyes roughly at the level for eye contact. Natural head movements and eye gazes shifts can be further analyzed.

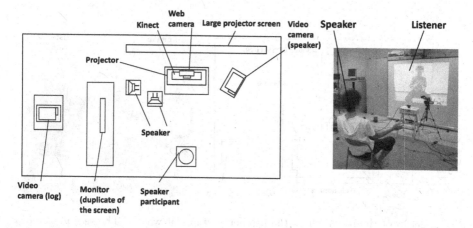

Fig. 1. Setup of the room where the speaker participant was in. The listener was projected roughly as life-size and the second monitor was used for video logging

3.2 Experiment Procedure

Speaker participant initiates the session and talks to the listener about his/her family. The topics of the conversation were "pleasant experience with family" and "unpleasant experience with family." These topics were chosen because they are common for almost everyone including the young experiment participants and the elderly. Listener participant was instructed to try their best to be a good active listener. That is, listen to the speaker carefully and attentively, follow the speaker's talk with questions or other feedbacks like nods or laugh as possible as the participant can.

They interchanged their roles in the sessions and started to talk from the pleasant experience at first because it should be easier to do (Table 1). The duration of one session was set to be seven minutes because it is considered long enough for the participants to start to talk something meaningful and keep the

Table 1. Topic and subject assignment of each active listening session

Session	Topic	Speaker	Listener
1	Pleasant experience with family	A	B
2		B	A
3	Unpleasant experience with family	A	B
4		B	A

whole experiment within a reasonable time period. During the experiment, the experimenters were outside of the two rooms without intervening the participants.

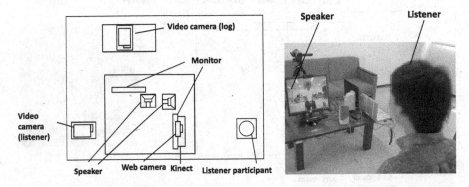

Fig. 2. Setup of the room where the listener participant was in. The second monitor was used for video logging

3.3   Evaluation of Speaker's Attitude

After the experiment, four evaluators (two male college students and two female college students) were recruited to evaluate the attitude of the participants by annotating on the recorded video corpus. The recruitment was done in the condition that the evaluators neither participated the active listening experiment, nor have close relationship with the participants to ensure objective evaluation results. How the attitude of the speaker is supposed to be perceived by the listener, and how he/she perceived the listener's attitude were evaluated in 7-scale measure from value 1 (negative) to 7 (positive). The evaluators were instructed to evaluate the participants's conversation base on their observation and perception on both verbal and nonverbal behaviors of them. Appropriate back-channel feedbacks like nods, questions, silence, agreeing opinions, smiles, or laughs were provided as positive examples in the instruction to the evaluators. In order to prevent gender bias, the video data of each participant pair were annotated by one male evaluator and one female evaluator. The video annotation tool,

ELAN [11] was adopted for this purpose. In order to align the granularity and label positions among different evaluators, they were instructed to annotate their evaluation with the four following rules:

1. The whole time line has to be labeled without blank segments
2. Starting and ending positions of the labels should be aligned to utterance boundaries
3. One label can extend to multiple utterances
4. The duration of one individual label is at most 10 s

Figure 3 shows an example segment after the evaluation annotation. Although the label values are discrete, the label values are immediately successive to one another, that is, there is always a label value at any time point.

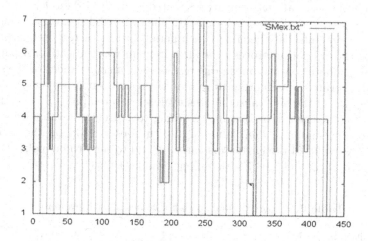

Fig. 3. Conceptual diagram of evaluation value sampling. The horizontal axis shows time (ms) and the vertical axis shows evaluation values

After the evaluation annotation, totally 3,644 labels were gathered from the four evaluators. Table 2 shows the number of labels in each value. From the distribution of the data values, varieties can be observed among the evaluators.

Table 2. Number of the 1 to 7 value labels gathered from all four evaluators

Evaluator	1	2	3	4	5	6	7
A	15	39	80	235	253	88	46
B	26	50	92	188	345	55	23
C	37	98	193	413	276	78	34
D	22	54	113	302	284	138	67

4   Automatic Estimation of Speaker's Engagement

4.1   Estimation Target Values

In order to realize automatic estimation of speaker's engagement attitude, there are two issues have to be solved. The first one is when to generate estimation results, and the second one is what should be the target value of the estimation. For a virtual listener agent to function properly, it has to show its behaviors to react to speaker's behaviors in reasonably short intervals to allow the speaker to feel that it is responsive and life-like. The period of one utterance which is usually within several seconds serves to be a candidate. In a dyadic conversation, there are three possible situations regarding to utterances, the speaker is speaking, the listener is speaking, and no one is speaking. Since the available information are not equal in these situations, they are estimated by dedicated models that we call *speaker*, *listener*, and *silences*, respectively. The estimated speaker's engagement attitude are generated at the end of each period by the corresponding model.

During the evaluation process, the measurement was subjective. A particular value of the annotation may mean different degree of engagement of the speaker to individual evaluators. In order to sum up the results from different evaluators, the elimination of the bias caused by the evaluators is required. The label values are then standardized with Z score regarding to individual evaluators. Table 3 shows these normalized label values. From the observation on the table, the average (Z score: 0) of the evaluators can be found to tend to be higher than the middle value (4) of the 1 to 7 scale except evaluator C.

Table 3. Z score normalized label values of each evaluator

Evaluator	1	2	3	4	5	6	7
A	−0.781	−0.562	−0.343	−0.124	0.096	0.315	0.534
B	−0.802	−0.603	−0.405	−0.207	−0.009	0.190	0.388
C	−0.750	−0.499	−0.249	0.002	0.252	0.503	0.753
D	−0.772	−0.544	−0.316	−0.089	0.140	0.367	0.595

The original annotation contained only the integer values between 1 and 7. After the Z score transformation, the label values were transformed to various real number values roughly ranging from −0.8 to 0.8. Target values of automatic estimation are then set to be the average values of the two evaluators annotation label values. Since the label boundaries are not aligned to the time periods of automatic estimation, label values are weighted regarding to time intervals. The data of 13 participants (nine were male and four were female, 26 experiment sessions) were used for the model training. Table 4 shows the distribution of the target values of automatic estimation for each model. They are used in the machine learning phase, one individual period is used as one instance for training or validation.

Table 4. Distribution of the data sets used in the training of speaker, listener, silence models

	Instances	Num per session	Max	Min	Average	Stdev
Speaker	3,099	119.2	0.844	−0.775	−0.020	0.226
Listener	524	20.2	0.662	−0.737	−0.013	0.213
Silence	3,648	140.3	0.844	−0.775	−0.019	0.226

4.2 Feature Sets

The speaker's behaviors during the duration of individual label were considered to affect the judgement of the evaluators and were used as the cues for the estimation of the speaker's attitude (or the level of engagement in the conversation with the listener). All behaviors including verbal ones and nonverbal ones of the speaker may affect the evaluators' perception. Regarding to the evaluation of active listening conversation, we selected the low-level communicational facial expressions, head movements, prosody and speech turn for the estimation. These features were selected base on the following hypotheses. Smiles may imply that the speaker is in his/her pleasant mood. Nods may imply that the speaker agrees to what the listener is talking about or shows his/her willing to listen to the listener. The speaking frequency of the speaker may imply his/her willing to talk with the listener. Prosodic information in the voice of speakers is also considered to propagate the emotional state of the speaker. The followings are the list of all 98 features used in the estimation.

- Facial expression (F): features related to facial expressions. Most features are extracted with face recognition software tool, visage|SDK[1] at 30 fps. Action units of facial expressions are extracted according to CANDIDE model [1] which is designed for face tracking rather than well-known FACS [5] in psychology field (59 features).
  - Average intensity of the following action units in current estimation period: nose winker, jaw push z/x, jaw drop, upper lip raiser, lip stretcher left/right, lip corner depressor, left/right outer brow raiser, left/right inner brow raiser, left/right brow lowerer, left/right eye closed, rotate eyes left, rotate eyes down, and lower lip x push
  - Intensity values of the same action unit set immediately before the estimation (average of the frames in last 10 ms)
  - average intensity of the same action unit set up to now
  - number of smiles per second in the current estimation period (hand labeled)
  - number of laughs per second in the current estimation period (hand labeled)
- Head movements (H): features related to head movements. Most features extracted with visage|SDK at 30 fps (19 features).

---

[1] http://visagetechnologies.com/products-and-services/visagesdk/.

- Average three-dimension head position (x, y, z) of current utterance
- Average three-axis head rotation ($roll$, $pitch$, yaw) of current utterance
- Three-dimension head position immediately before the estimation (average of the frames in last 10 ms)
- Three-axis head rotation immediately before the estimation (average of the frames in last 10 ms)
- Average three-dimension head position up to now
- Average three-axis head rotation up to now
- Number of nods per second during current utterance (hand labeled)

- Prosody (P): the prosodic information when the speaker-role participant is speaking. Praat[2] was used to compute the following prosodic features of the the current utterances. Since the raw values can have a large diversity among the speakers, relative values are used here (11 features).
  - Ratio of average pitch/intensity of current utterance comparing to the average pitch up to now
  - Maximum pitch/intensity of current utterance comparing to the average of current utterance
  - Maximum pitch/intensity of current utterance comparing to the average of all utterances up to now
  - Minimum pitch/intensity of current utterance comparing to the average of current utterance
  - Minimum pitch/intensity of current utterance comparing to the average of all utterances up to now
  - Speech rate of current utterance
- Speech turn (T): the features related to speech turns (nine features).
  - Ratio of the time period when no one was speaking (silent) regarding to the total session time up to now
  - Average number of silent periods regarding to the total session time up to now
  - Average duration of silent periods up to now
  - Ratio of the time period when the speaker was speaking (speaking) regarding to the total session time up to now
  - Average number of speaker's speaking periods regarding to the total session time up to now
  - Average duration of speaking periods
  - Duration of speaker's last utterance
  - Speaker of last utterance preceding this one
  - Three-gram pattern of the speaker of last three utterances

4.3   Support Vector Regression Models

Support vector regression models are trained for the three periods, speaker is speaking (speaker model), listener is speaking (listener model), neither speaker nor listener is speaking (silence model). The SMO Regression implementation

---

[2] http://www.fon.hum.uva.nl/praat/.

of Weka [6] toolkit is used in data mining experiments. All four feature sets are used in training speaker model while prosody feature set was absent in the training of listener and silence models. PUK kernel [16] was adopted because it achieves best results in the experiments based on our data corpus. The complexity parameter were explored from one to 10 (10 steps) and the combinations of normalization/standardization were explored to find optimal results. Correlation and R^2 (coefficient of determination) are used as the metrics in evaluation model performance. From the definitions, correlation is computed with equal weights regardless of data values while the errors at extreme values has larger impacts in the computation of R^2.

Figures 4 and 5 show the 10-fold cross validation results of each estimation model regarding to all combinations of feature sets. Training parameters are aligned to the values which achieve highest correlation with full feature set of each model (TPFH for speaker and TFH for listener/silence models). The order of the performance of the three models is silence, speaker, and listener. The fact that silence model performed best in spite of less available features may imply the influence of verbal information in the judgement of speaker's engagement in speaker and listener models. Generally but not always, better performance is achieved with more feature sets. Speaker model has best performance (correlation: 0.55, R^2: 0.35) with feature set TFH, listener model has best performance (correlation: 0.41, R^2: 0.18) with feature set H, silence model has best performance (correlation: 0.59, R^2: 0.39) with feature set FH. Facial expression features and head movement features have larger impact on the results while the prosodic information is least effective.

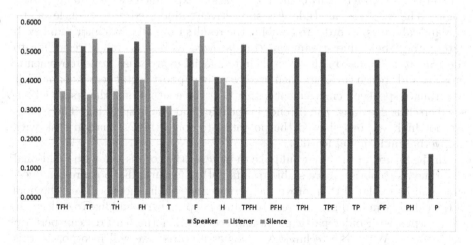

Fig. 4. Correlation results of the estimation models, speaker, listener, silence regarding to feature sets

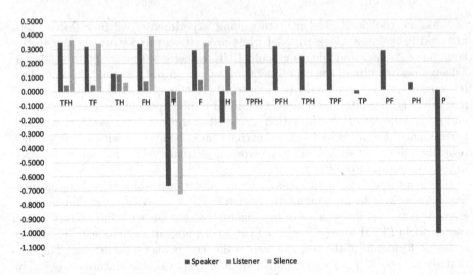

Fig. 5. R^2 results of the estimation models, speaker, listener, silence regarding to feature sets

5   Conclusion

In order to realize an active listener agent for the elderly, the ability for the agent to evaluate the engagement level in the conversation is essential. In this work, we conducted an active listening conversation experiment with human participants. The speakers' attitude was subjectively annotated into seven levels by human evaluators. In order to combine the results from all participants and evaluators, the label value stream was standardized to Z-score values regarding to the evaluators. Nonverbal features including facial expressions, head movements, prosody and speech turns are then used to train support vector regression models to estimate speaker's engagement attitude in three situations, the speaker himself/herself is speaking, the listener is speaking, no one is speaking. The results are not high yet but showed the potential to solve this estimation task with non-verbal multimodal features.

In the future, at first we would like to increase the corpus size with additional experiments to obtain more stable results. Other features like postures and gestures can be explored to improve the performance of the models. The behaviors of the elderly can be quite different to young peoples, we would like to collect the data corpus with older participants and see whether the features can perform well enough. When the technology becomes matured, we will incorporate this function to a fully working listener agent and evaluate it with the elderly for long period.

References

1. Ahlberg, J.: Candide-3 – an updated parameterised face. Technical report. LiTH-ISY-R-2326, Department of Electrical Engineering, Linkoping University, January 2001
2. Baltrusaitis, T., Ahuja, C., Morency, L.: Multimodal machine learning: a survey and taxonomy. CoRR abs/1705.09406 (2017). http://arxiv.org/abs/1705.09406
3. Bickmore, T., Bukhari, L., Vardoulakis, L.P., Paasche-Orlow, M., Shanahan, C.: Hospital buddy: a persistent emotional support companion agent for hospital patients. In: Nakano, Y., Neff, M., Paiva, A., Walker, M. (eds.) IVA 2012. LNCS (LNAI), vol. 7502, pp. 492–495. Springer, Heidelberg (2012). https://doi.org/10.1007/978-3-642-33197-8_56
4. Cowie, R., Schröder, M.: Piecing together the emotion jigsaw. In: Bengio, S., Bourlard, H. (eds.) MLMI 2004. LNCS, vol. 3361, pp. 305–317. Springer, Heidelberg (2005). https://doi.org/10.1007/978-3-540-30568-2_26
5. Ekman, P., Friesen, W.: Facial Action Coding System. Consulting Psychologists Press, Palo Alto (1978)
6. Frank, E., Hall, M.A., Witten, I.H.: The weka workbench. In: Data Mining: Practical Machine Learning Tools and Techniques, 4th edn. Morgan Kaufmann (2016)
7. Hofstede, G., Hofstede, G.J., Minkov, M.: Cultures and Organizations: Software of the Mind, 3rd edn. McGraw-Hill, New York (2010)
8. Huang, H.H., Matsushita, H., Kawagoe, K., Sakai, Y., Nonaka, Y., Nakano, Y., Yasuda, K.: Toward a memory assistant companion for the individuals with mild memory impairment. In: 11th IEEE International Conference on Cognitive Informatics & Cognitive Computing (ICCI*CC 2012), Kyoto, pp. 295–299, August 2012
9. Huang, H.H., Shibusawa, S., Hayashi, Y., Kawagoe, K.: Toward a virtual companion for the elderly: an investigation on the interaction between the attitude and mood of the participants during active listening. In: 1st International Conference on Human-Agent Interaction (iHAI 2013), Sapporo, Japan, August 2013
10. Huang, L., Morency, L.-P., Gratch, J.: Virtual rapport 2.0. In: Vilhjálmsson, H.H., Kopp, S., Marsella, S., Thórisson, K.R. (eds.) IVA 2011. LNCS (LNAI), vol. 6895, pp. 68–79. Springer, Heidelberg (2011). https://doi.org/10.1007/978-3-642-23974-8_8
11. Lausberg, H., Sloetjes, H.: Coding gestural behavior with the NEUROGES-ELAN system. Behav. Res. Methods **41**(3), 841–849 (2009)
12. McKeown, G., Valstar, M.F., Cowie, R., Pantic, M.: The SEMAINE corpus of emotionally coloured character interactions. In: IEEE International Conference Multimedia and Expo, pp. 1079–1084 (2011)
13. Otake, M., Kato, M., Takagi, T., Asama, H.: Coimagination method: supporting interactive conversation for activation of episodic memory, division of attention, planning function and its evaluation via conversation interactivity measuring method. In: International Symposium on Early Detection and Rehabilitation Technology of Dementia, pp. 167–170 (2009)
14. Pammi, S., Schro, M.: Annotating meaning of listener vocalizations for speech synthesis. In: 3rd International Conference on Affective Computing and Intelligent Interaction (ACII 2009), pp. 1–6 (2009)
15. Tickle-Degnen, L., Rosenthal, R.: The nature of rapport and its nonverbal correlates. Psychol. Inq. **1**(4), 285–293 (1990)
16. Ustun, B., Melssen, W., Buydens, L.: Facilitating the application of support vector regression by using a universal pearson vii function based kernel. Chemometr. Intell. Lab. Syst. **81**(1), 29–40 (2006)

Agents, Models and Algorithms in Social Media

Designing SADD: A Social Media Agent for the Detection of the Deceased

James Braman[1]([✉]), Alfreda Dudley[2], and Giovanni Vincenti[3]

[1] School of Technology,
Art and Design Computer Science/Information Technology,
Community College of Baltimore County,
7201 Rossville Boulevard, Rosedale, MD 21237, USA
jbraman@ccbcmd.edu
[2] Department of Computer and Information Sciences, Towson University,
8000 York Rd, Towson, MD 21252, USA
adudley@towson.edu
[3] Division of Science, Information Arts and Technologies,
University of Baltimore, 1420 N. Charles St, Baltimore, MD 21201, USA
gvincenti@ubalt.edu

Abstract. As the number of profiles and user generated content online continues to grow, many accounts in time will inevitably belong to those that are deceased. Questions arise as what to do with content on these profiles when it becomes known that the owner of the profile has died. Additionally, many users have several profiles spread across various social networking sites. This project describes our efforts to create SADD - A Social Media Agent for the Detection of the Deceased. This tool aims to systematically identify social media accounts that may belong to a deceased user based on related posts from other users as well as changes in profile interaction. As part of this project, we present a basic framework and prototype for the program and include survey results about the perception of this type of tool which examine how users may wish to preserve online content. This project is a continuation of previous research related to virtual memorialization of web based and virtual world content and interaction.

Keywords: SADD · Online memorials · Thanatechnology · Digital legacy
Social networks

1 Introduction

Social media use has grown exponentially over the last several years, with no signs of reversing. As more individuals turn to social media of all kinds, the number of accounts and profiles online continue to increase. The culmination of video, photos, text and other content over time is what makes up a user's profile and timeline on a particular social media site. As social media is used in many aspects of our lives, the number of users, content and time spent with these profiles continues to expand and evolve. Despite the pervasiveness of social media technologies, few studies have been conducted on the impact on those connected to the departed. Even though death is a major component of one's life, current technology design does not yet fully consider the

© Springer International Publishing AG, part of Springer Nature 2018
G. Meiselwitz (Ed.): SCSM 2018, LNCS 10914, pp. 345–356, 2018.
https://doi.org/10.1007/978-3-319-91485-5_26

inevitable death of the user [1]. Death is part of life no matter how much we try to avoid the concept. In preparation we prepare wills and set aside documents in the event of our death and pay for life insurance to help compensate for end of life expenses for surviving family members. However, this preparation and protection of our digital assets are not considered as often. Our digital assets contain a wealth of information, knowledge and digital artifacts that we may wish to pass over to members of our families, or in some way preserve. Our social media accounts contain information and representations of who we were in life, thus transforming into a digital memorial of our lives (even inadvertently). In this paper, we continue to add onto our previous work examining the implications of social networking sites and virtual worlds regarding death [2–4]. This project however, concentrates on the development of a program which aims to systematically identify accounts of users that are deceased on social networking sites. Therefore, we describe our efforts to create SADD - A Social Media Agent for the Detection of the Deceased.

The automatic detection of accounts of deceased users is inherently problematic. There are many factors to explore and hurdles to overcome to automatically flag these profiles. Through the implementation of SADD we intend to expand our research on social media related to deceased user profiles and explore how other users interact with the content. SADD will also allow us to examine which types of digital content may become a person's online memorial (purposefully or inadvertently). How does one determine if an account has just become inactive rather than an account of someone who is deceased? There are many elements within the page that can be examined such as titles, posts, "liked" websites, images, videos and other connected/friended accounts. There is a plethora of content in a social media site to examine for a wide variety of purposes. Boyd and Ellison [5] describe a web-based social networking site as a service that allows users to: "(1) Construct a public or semipublic profile within a bounded system (2) articulate a list of other users with whom they share a connection, and (3) view and traverse their list of connections and those made by others within the system" (pg. 211). From the user's profile the connections, content and other activity can yield important patterns.

As the death of the user is an often under considered event in the protection of one's content, more research and focus on this topic is needed. Massimi and Charise [1] introduced the concept of thanatosensitivtiy, as a humanistically grounded approach to HCI concerning mortality, dying and death in the creation of interactive systems. We see a need for a multidisciplinary approach to these problems between computing and thanatology. In this paper, we expand our previous research as it is related to SADD by presenting a rationale for the program, followed by additional survey data and a discussion. Our future work section describes the next phase of the project.

2   Rationale

As part of our ongoing research, we have been examining the impact of social media and virtual worlds for the past several years. One of the overarching themes of these projects is the idea of virtual/online memorials and how users interact with profiles of the deceased. Additionally, we have proposed many questions as to how these memorials should be designed or maintained and what to do with content. As we investigate how information can be preserved, presented and interacted with, we must also find a way to discover accounts of those that are deceased. Thus the introduction of the SADD Model. There are several benefits of the implementation of SADD such as:

1. Potentially assisting families and friends in discovering unknown accounts of loved ones that have died.
2. Assisting social networking providers with an automated process to detect deceased user's accounts.
3. Identifying these accounts can help memorialize and save content.
4. Identifying accounts of deceased users can be beneficial for security reasons.

The detection of these profile can allow family members to preserve content such as images, posts and videos posted or stored within the deceased's account. In previous studies, some users have expressed the desire to have content memorialized, either their own content when they are deceased or to protect a loved one's content [4]. However, the detection of posthumous profiles brings up many related issues. These issues increase in complexity if the wishes of the deceased are not known or ambiguous. Some questions to consider include:

- Who should maintain the profiles of the deceased?
- Is it known and verified that the original owner is indeed deceased?
- What was the final wishes of the deceased?
- What are the Terms of Service related to the accounts being considered?
- What content remains in the account?
- Is the content public or private information?

As many of these elements are unknown, particularly at the time of death, one should consider that their social media presence at any point may become their digital representation (digital legacy). Some of our previous research has proposed several questions that one can use to assess content as they are posting or storing content online or on a social networking site which include [3, 4]:

1. Is this content something I'm alright with if it becomes part of my digital legacy?
2. Is this content something that should be protected if something were to happen to me?
3. If this content should be protected, how can it be protected?

The main motivation in developing SADD to identify profiles of the deceased it to help those grieving and to also protect the final wishes of the deceased (if known). As more people turn to creating and maintaining a presence on multiple social networking sites, the potential for large number of profiles of the deceased will also rise. Being able to identify and protect this content is important. For some, being able to visit the profile of the deceased friend or family member can provide comfort and time for reflection. Others may not want to see the content as it may illicit various emotions reminded them of the loss. Having profiles memorialized can also be used to have "conversations" with the deceased through posting messages on one's page or timeline. One can also explore the deceased's timeline to better understand aspects of who they were, important events and feelings or just to view other people's comments. However, it is possible that over time some users befriend other users in which they do not know in real life. Therefore, it may not be known if they are deceased in the real world. We are also interested in knowing more about how the manner of death may affect how individuals interact with SNS and virtual memorial design [4]. Using the NASH categories (Natural, Accidental, Suicide, Homicide) used for some classification of death, we may be able to capture important patterns of interaction, however, this information would need to be known of the deceased. Another interesting phenomenon related to interacting with the profiles of the deceased is RIP trolling. This is where offensive and inflammatory posts and/or images are purposefully left to cause offense. RIP trolling or sometimes referred to as memorial trolling may be an increasing problem on social media as offensive posts and fictitious profile accounts can be easily created. Phillips [6] notes that often these individuals "scour the site for the most sensitive people and the most sensitive subjects". Some social networking site reporting tools, however, can limit some of these occurrences in a limited degree. For those effected, it can be quite upsetting. If these accounts could be automatically detected as belonging to a deceased person, actions could be taken to memorial or deactivate that page which could prevent some of these activities.

3   Framework and Preliminary Design

The preliminary version of SADD was programmed using the Python programming language (version 2.7), due to its flexibility and the large number of available libraries which make connecting to social media sites easier. Additionally, there are numerous text analysis tools available. The SADD agent is set to run at various intermittent time intervals on public Twitter posts. Twitter was chosen as our preliminary social networking site to test the program at identifying user accounts of probable deceased users. Currently, the agent is in a preliminary stage and performs searches for keywords related to death to identify deceased user profiles. For example, it may search for the term "#RIP" to see which twitter user has posted that term in relation to another user's profile. The web component of SADD is essentially a web-crawler which mines for specific words within a subset of social networking sites.

Several features can be obtained from crawled pages as part of various indexing strategies [7, 8] such as: page content, descriptions, hyperlink structure, hyperlink text, keywords, page titles and text with different fonts. This additional structural information could provide additional information depending on the type of profile being examined and its interconnectedness to other profiles. Some information however would need to be ignored. A common word filter is used to filter out commonly used words that occur in a relatively high frequency as part of the English language. In general, these common words will have little or no importance for the agent. The filtering of common words is a useful technique that can be used for many diverse purposes. Natural languages typically contain common words that serve to connect pieces of sentences, prepositions and other common elements. These elements are often referred to as stop-words, and are typically words like "the", "and", "of", and "in". The removal of stop-words is often associated with balancing the processing speed versus retrieval quality [9]. It is argued that processing speed will be improved even though some quality may be lost by the removal of these terms [10]. The selection of stop-words selected for this project was based on Leech et al. [11] and from a partially compiled list from the Onix Text Retrieval Toolkit [12] as well as added words through experimentation. Removal of additional terms also cuts down on the amount of information stored in the database. The ability to store all information retrieved from Twitter posts is limited.

Figure 1 illustrates a preliminary design model of the SADD agent, with several components still to be developed. After a profile is examined, text content and structural elements are interpreted by the agent. In development for SADD, is the Grief Counselor Component and Sentiment Analysis, which is based on derivation of our previous work in affective computing [13]. Moreover, sentiment analysis of posts combined with discovered variations in a user's timeline could yield useful results. The Grief Counselor Component is not meant to serve as a real grief counselor, but instead is the main element in determining if the profile under investigation is indeed deceased. Recommendations made by the agent included derived information about the confidence of the user's status in addition to analysis patterns.

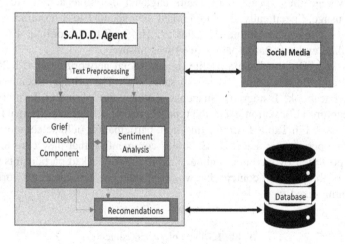

Fig. 1. SADD Agent Model

There are many difficulties to overcome in SADD's implementation, some of which will be addressed in the future work section. First, the agent is limited in the social networks it can currently access at this time (i.e., Twitter). Also, the ability to perform social network analysis on current connections and interactions with other users as well as the ability to establish patterns in Twitter posting is under development. SADD is also susceptible to "fake news" and hoax content. For instance, it once picked up a previous hoax of a suicide of a celebrity due to numerous reposting and instances of "#RIP" in relation to the account. We also must be careful about the number of requests sent as not to have the program banned, but in doing so may cause some information to be missed. Additionally, there are limits to how many messages can be examined at one time. The database also has needed to be redesigned several times for efficiency. SADD currently is using a MySQL database. As previously stated, there is much research that is needed to most effectively extract targeted profiles. To better understand how people interact with a feel about deceased user profiles a survey was administered and described in the next section.

4 Feedback

Using similar, but expanded questions from a previous survey, we asked a series of questions to undergraduate students at a large community college [4]. The survey was distributed electronically at the end of the Fall 2017 academic semester. Participants were selected from a limited number of technology courses such as introductory courses and various levels of computer programming. Limitations of the survey include low response rate and high level of technology use of those participating. Twenty-eight participants responded, which consisted of 16 males (57.1%) and 12 (42.9%) females with an average age of 23.3 years old. The majority were majoring in computer science, information technology, computer engineering and/or general studies. Participants noted on average they spend many hours engaged in online activity reveling high technology usage. Specifically, 2 (7.1%) noted they spend 1 to 3 h online, 15 (53.6%) spent 4 to 6 h engaged in online activities, 6 (21.4%) 7 to 9 h online, 4 (14.3) more than 9 h online and one participant (3.6%) was unsure.

When asked "Do you have a profile on any social networking site?", 25 (89.3%) said "Yes", and the remaining 3 (10.7%) said "No". The primary social network of choice was Facebook, Instagram, Snapchat, Twitter, Google+ and Twitter. We also wanted to capture information about the type of content posted to participant's profiles which is described in Table 1. In this question, participants could select more than one response. One person that selected "E. None" commenting that they consume content rather than post their own content online. Next the survey asked participants to rate the importance of the personal content that was contained on their social networking sites, which is summarized in Table 2.

Table 1. Type of posted content

Choice	Total (n = 28)
A. Pictures	20
B. Text based posts	19
C. Video	18
D. Music or other audio	7
E. None	3

Following these introductory questions, the survey became more focused on death related questions. The next question asked, "Do you know anyone that has died, but their social networking profile is still present after their death?". Eighteen (64.3%) responded "Yes" and 10 (35.7%) responded "No". Participants were also asked "Would you want your social networking page to remain active after your death?".

Table 2. Content ratings

Rating	Total (n = 28)
A. Not at all important	6
B. A Little important	4
C. Somewhat important	14
D. Very important	2
E. Highly important	2

Only 5 (17.9%) said "Yes", whereas the remaining 23 (82.1%) said "No". Related to this question targeted which content participants would want protected after their death through a multiple-choice question, where more than one answer could be selected. A summary can be found in Table 3.

Table 3. Digital assets that need protecting

Category	Total (n = 28)
A. Photos	19
B. Documents	13
C. Music	5
D. Video	13
E. Intellectual property (i.e. things that you or others have created)	11
F. Personal Information (i.e. tax documents, addresses, financial data etc.)	19
G. No Response	1

Participants were also asked what they would want to happen to their digital content after their death. From twenty-seven participants that responded 13 (48.2%) reported that their wishes would be to have the content "deleted", 8 (29.6%) reported "preserved with restrictions" and 6 (22.2%) reported that their final wishes would be for their site to remain the same as it currently is. Next the survey question asked "If a family member or friend died and had various social media accounts, would you want the ability to find out what social media accounts they had? (Assuming you do not know what social media accounts they had)". The results were mixed 10 (35.7%) responding "Yes", 8 (28.6%) responding "No" and 10 (35.7%) responding with "Maybe". Tied to this question, participants were also asked "How would knowing the deceased person's social media accounts help you? (Or not). Twenty participants responded to this question. Five (25%) of the participants responded that they would not want such information. One person noted that such private information should not be made public in any way after the person's death. Another participant responded:

"Knowing the deceased person's social media accounts would just upset me and make me sad seeing the individual the way that they were while they were alive. Scrolling through their social media account would just make me reminisce and have more of a difficult time with getting over their death and moving on."

Ten (50%) of the other responses were more positive in nature for having such information. It was noted that this would allow for insight into the deceased, and that being reminded of their life could be helpful in dealing with the loss. The other five (25%) commented on how the information could be used to identify the accounts to protect the content of the deceased, either for memorialization purposes or to close the accounts. One person brought up an interesting point regarding the manner of death and the role knowing the deceased's accounts. For instance, in the case of a homicide or suicide, the person's timeline and connections may be of importance. Next participants were asked "If you died and your family and friends did not know what social media accounts you had, would you want them to be able to find these accounts?". Fifteen (53.6%) of the participants said "Yes" that they would want their social media accounts known. Nine (32.1%) responded "No" and 3 (10.7%) responded as "Maybe". One (3.6%) participant skipped the question and did not respond. Some related comments include:

- If my family and friends did not know about my social media accounts when I was alive then I would not want them to know about them when I am dead.
- They would get to know another side of me
- Because its personal, it would make me or them uncomfortable
- I would want them to see all memories about me

The survey then shifted the focus of the questions over to virtual memorials. The next question asked participants if they had ever seen or encountered an online (social networking based) or virtual memorial. From the responses 9 (32.1%) said "Yes", 15 (53.6%) said "No" and 4 (14.3%) said "Maybe". Two individuals commented: "Facebook page of a deceased was updated by a family member informing people of the death of the individual's page" and "I have seen many online memorials for famous individuals who have passed away, but never for anyone that I have known personally". Next the survey asked, "Do you currently have any documentation dictating your final wishes for your online content?". Only 2 (7.1%) said "Yes" and the remaining 26 (92.9%) said "No". Next the survey asked, "Would you feel better about your own death if you knew there would be an online memorial dedicated to you?". From this question 13 (46.4%) said "No", 8 (28.6%) said "Yes" and 7 (25%) said "Maybe" or "Unsure". The survey also questioned if they had ever received a friend request or suggestion from someone on a social media site that you know was deceased? Three responded "Yes" (10.7%) and the remaining 25 (89.3%) reported "No". No participant selected "Unsure". Lastly, participants were asked "Have you ever received a friend request or suggestion from someone you suspected was using a falsified/fake account?". Twenty-two (78.6%) responded "Yes" they believe they have had fake/falsified friend requests. Five (17.8%) responded "No" and one (3.6%) was "Unsure".

5 Discussion

Although the number of participants in this survey was limited, the results will be helpful for future design considerations. The responses from this survey showed consistency with previous survey results [4]. In this particular survey, however, there were several additional questions about attitudes toward the automatic detection of content. The focus of these questions was to include the assessment of younger participants who typically have more of a high degree of technical knowledge and use social media extensively. Feedback from this demographic is important as they will be likely to continue using social media in the future, which may provide key insights. Maciel and Pereira [14] note the importance of exploring the concept of death within this age group as it applies to the web due to their high use of the Internet and social media in general. Participants from the survey responded that they do post a range of content on their social media profiles, primarily consisting of pictures, test and video. However, many responses varied in the importance level of such posts, with the majority (50%) saying they only viewed their content as "somewhat important" overall. This is contrast with the type of content they viewed as needing protection in the event of their death. When participants were asked overall if they would want their digital assets (which would include social media content and presence) to be deleted, preserved or to leave it as it was at the time of their death, results were mixed with only 48.2% wanted the contented deleted.

When asked about wanting to know the social media accounts of a friend or family members after their death, particularly, if such accounts were not known, result also varied. Where 10 (35.7%) responded "Yes", 8 (28.6%) responding "No" and 10 (35.7%) responded with "Maybe". Some viewed knowing such information as an invasion of privacy, where others viewed knowing such information as helpful. As related to SADD, the identification of the accounts of deceased users to both preserve and memorialize the page or to shut down the profile is important. On some social networking sites, any active account can still appear in search results or become listed as potential "friends". As noted by Goldberg [15], receiving a "friendship request" from someone that you know is deceased can be unsettling. From the survey results, though limited in occurrence (three users or 10.7%) had said yes, they have had this happen. As the number of users increase, this trend may also rise. From a security standpoint these accounts should be limited (memorialized) or deleted to make sure that active accounts are not the target for hacking attempts. If no one is monitoring the account, it could be used by a hacker to send unsolicited content to other users or control could be gained over stored content. If the username and content is compromised on one account, it may be a similar password and username on other accounts which could then also be hacked. Security concerns were also noted by several participants in the survey as a reason to know about such accounts.

Having your final wishes known and the ability for someone to carry out those final wishes is important for users of social media and online content to consider. Not only is knowing what online services you were participating is needed, but also passwords and other account information. However, as discussed by Massimi and Charise, relying on passwords alone may not be enough in the case of systems requiring biometric authentication [1]. These types of systems would prove to be problematic in the case of a deceased user unless backup measures are in place. Often, the final wishes of the deceased are not known in addition to not knowing their account information, including passwords. As asked in the survey, participants typically did not have any documentation dictating their final wishes for online content. Only two (7.1%) noted they did have such documentation whereas the remaining 26 (92.9%) did not. There has been some research proposed that would allow for self-destructing data as proposed through *Vanish* [16]. Although the aim of *Vanish* is privacy and security of archived data, this idea/approach would be helpful in the case of a user's death.

6 Summary and Future Work

In future research we intend to experimentally test SADD in a variety of situations. There is a great need for additional features to be added to the program such as social network analysis tools, additional keyword combinations and the ability to filter out erroneous information. We also intend to expand its functionality for several social networking sites. Additionally, we plan for more specific surveys asking users additional questions. For instance, would users be willing to have their account monitored systematically to see if they are still alive to protect their content and to make sure that their final wishes are carried out? This would require preplanning and monitoring of their activities online. Should SADD like components be designed and added automatically to social networks?

One limitation of this study was the small sample size. This is reflected in our analysis in the application of the SADD framework. In our next phase, our aim is to combine data collected from several previous studies for a more in-depth analysis to better understand implications of online memorials (both social networking based and virtual). This will help to better understand how users interact with such data to better design SADD. A better understanding of what and how users comment and interact with online memorials can also lead to better text based analysis of posts on pages of the deceased. It may also be possible to examine text posted by living users who are having suicidal ideation which could theoretical be identified for intervention. "Suicidal ideation is defined as thinking about, considering, or planning suicide" [17, p. 226). Worldwide there are close to 800,000 people that die by suicide annually [18]. However, this may be difficult to detect as social network posts may be in contradiction to the actual person's thoughts or posted content. It may also be difficult to know the actual person's information or location to provide help.

Through an interdisciplinary approach between computer science and thanatology we may be able to design a working program to help identify profiles of deceased users. The goal is to help protect the final wishes of the deceased and to assist the bereaved. In this paper we have described our project SADD - A Social Media Agent for the Detection of the Deceased, in its preliminary form. As the amount of digital content belonging to those that have died increases, the need for automatic detection and content protection will become progressively important.

References

1. Massimi, M., Charise, A.: Dying, death, and mortality: towards thanatosensitivity in HCI. In: CHI 2009 Extended Abstracts on Human Factors in Computing Systems, pp. 2459–2468. ACM (2009)
2. Braman, J., Dudley, A., Vincenti, G.: Death, social networks and virtual worlds: a look into the digital afterlife. In: Proceedings of the 9th International Conference on Software Engineering Research, Management and Applications. Baltimore, MD, USA (2011)
3. Braman, J., Vincenti, G., Dudley, A., Wang, Y., Rodgers, K., Thomas, U.: Teaching about the impacts of social networks: An end of life perspective. In: Proceedings of the 15th HCI International Conference on Online Communities and Social Computing, Las Vegas, Nevada (2013)
4. Braman, J., Dudley, A., Vincenti, G.: Memorializing the deceased using virtual worlds: A preliminary study. In: Proceedings of the 19th HCI International Conference on Online Communities and Social Computing, 9–14 July 2017, Vancouver, Canada (2017)
5. Boyd, D., Ellison, N.: Social network sites: Definition, history, and scholarship. J. Comput.-Mediated Commun. **13**, 210–230 (2008)
6. Phillips, W.: LOLing at tragedy: Facebook trolls, memorial pages and resistance to grief online. First Monday **16**(12) (2011)
7. Hu, W., Yeh, J.: World wide web search engines. In: Si, S.N., Murthy, V.K. (eds.) Architectural Issues of Web-Enabled Electronic Business. Idea Group Publishing (2003)
8. Google Inc.: Search engine optimization starter guide (2010). Accessed. http://static. googleusercontent.com/media/www.google.com/en//webmasters/docs/search-engine-optimization-starter-guide.pdf
9. Hiemstra, D.: Term-specific smoothing for the language modeling approach to information retrieval: the importance of a query term. In: Proceedings of the 25th Annual International ACM SIGIR Conference on Research and Development in Information Retrieval, pp. 35–41 (2002)
10. Blok, H.E., Hiemstra, D., Choenni, S., de Jong, F., Blanken, H.M., Apers, P.M.: Predicting the cost-quality trade-off for information retrieval queries: facilitating database design and query optimization. In: Proceedings of the Tenth International Conference on Information and Knowledge Management, pp. 207–214. ACM (2001)
11. Leech, G., Rayson, P., Wilson, A.: Word Frequencies in Written and Spoken English. Harlow (2001)
12. Lextek International: Onix text retrieval toolkit API reference - Stop word list 1 (2011). Accessed. http://www.lextek.com/manuals/onix/stopwords1.html
13. Vincenti, G., Braman, J., Trajkovski, G.: Hybrid emotionally aware mediated agent architecture for human-assistive technologies. In: 2008 Spring AAAI Symposium on Emotion, Personality and Social Behavior, Stanford, CA. USA (2008)
14. Maciel, C., Pereira, V.C.: Social network users' religiosity and the design of post mortem aspects. In: Kotzé, P., Marsden, G., Lindgaard, G., Wesson, J., Winckler, M. (eds.) INTERACT 2013. LNCS, vol. 8119, pp. 640–657. Springer, Heidelberg (2013). https://doi.org/10.1007/978-3-642-40477-1_43
15. Goldberg, L.: Posthumous profiles on virtual social networks: death and language. [Blog post] (2017). Accessed. https://www.leonardogoldberg.com/single-post/2017/06/07/Posthumous-profiles-on-virtual-social-networks-death-and-language
16. Geambasu, R., Kohno, T., Levy, A.A., Levy, H.M.: Vanish: increasing data privacy with self-destructing data. In: USENIX Security Symposium, vol. 316 (2009)

17. Ordaz, S., Goyer, M., Ho, T., Singh, M., Gotlib, I.: Network basis of suicidal ideation in depressed adolescents. J. Affect. Disord. **226**, 92–99 (2018). ISSN 0165-0327, https://doi.org/10.1016/j.jad.2017.09.021
18. WHO: Suicide Rates. Global Health Observatory (2017). http://www.who.int/gho/mental_health/suicide_rates/en/. Accessed 12 Oct 2017

Human Factors in the Age of Algorithms. Understanding the Human-in-the-loop Using Agent-Based Modeling

André Calero Valdez[✉] and Martina Ziefle

Human-Computer Interaction Center,
RWTH Aachen University, Campus-Boulevard 57, Aachen, Germany
{calero-valdez,ziefle}@comm.rwth-aachen.de

Abstract. The complex interaction of humans with digitized technology has far reaching consequences, many of which are still completely opaque in the present. Technology like social networks, artificial intelligence and automation impacts life at work, at home, and in the political sphere. When work is supported by decision support systems and self-optimization, human interaction with such systems is reduced to key decision making aspects using increasingly complex interfaces. Both, algorithms and human operators become linchpins in the opaque workings of the complex socio-technical system. Similarly, when looking at human communication flows in social media, algorithms in the background control the flow of information using recommender systems. The users react to this filtered flow of information, starting a feedback-loop between users and algorithm—the filter bubble. Both scenarios share a common feature: complex human-algorithm interaction. Both scenarios lack a deep understanding of how this interaction must be properly designed. We propose the use of agent-based modeling to address the human-in-the-loop as a part of the complex socio-technical system by comparing several methods of modeling and investigating their applicability.

Keywords: Agent-based modeling · Social simulation
Cognitive simulation · Opinion forming · Industrie 4.0
Internet of things · Internet of production

1 Introduction

Our world is becoming increasingly complex, but how?. According to Moore's law [1] the power of modern computers increases with an exponential rate. As with any exponential growth curve, when a certain tipping point is reached, growth becomes fast so quickly that it seems unbounded. This tipping has been reached for computing power. Thanks to the inventions like mobile Internet, cloud storage, cloud computing [2], and with the rise of AI technologies such as deep learning [3], much of this newly available power can finally be used

© Springer International Publishing AG, part of Springer Nature 2018
G. Meiselwitz (Ed.): SCSM 2018, LNCS 10914, pp. 357–371, 2018.
https://doi.org/10.1007/978-3-319-91485-5_27

by algorithms that utilize Big Data using GPUs to find solutions to previously seemingly intractable problems (e.g., image recognition, face detection, speech recognition, etc.).

However, it seems that many problems still remain unfeasible to be solved by algorithms alone. When it comes to questions of ethics, responsibility, or intimacy [4] humans and their elaborated and cognizant decision making are required. Similarly, for many other problems the combination of skills of humans and computers is considered to be the optimal strategy—the human-in-the-loop helps the computer to overcome its limitations, or vice versa [5]. The challenge for such settings is designing the human-computer interface to reduce "friction-loss" at the seam. For many problems efficient interfaces exist. The interaction of algorithm and human is already so pervasive that many users are even unaware of the algorithmic procedure and complexity running behind the scenes. A user that uses Google to find search results does not need to know about the *Page-Rank* algorithm or the personalization of search results based on complex *tensor decomposition* techniques.

The interaction of humans and algorithms thus brings visible and invisible complexity to the table that has deep consequences for almost all areas of human life. For example, it affects work by drastically changing requirements in jobs. Will future jobs be only available to workers with an understanding of data that can focus on monitoring complex data? It affects the private life by changing what is made available by social media to the users and by creeping into all economic decisions. If a financial scoring algorithm detects patterns of established social injustice and recommends not giving a poor family credit, does it perpetrate prejudice [6]? It affects political life by introducing algorithmic decisions into the political discourse on social media, in search, and in political ad targeting. They who have the data, have the voters? Some of these questions oversimplify the underlying problems of biased data, biased learning, and biased decision-making.

But what exactly is this new complexity? What are its reasons? And how can human factors research address these?

2   Complex Systems and Agent-Based Modeling

To understand where complexity comes from it is necessary to look at the underlying systemic structure of phenomena. In this article we investigate different levels of complexity, how they bring rise to emergent phenomena and look into models of understanding such phenomena. For this purpose we investigate two fields of application, that are heavily influences by algorithms. The first aspect is *opinion forming* in social media and the second aspect is the interaction of humans and algorithms in working environments.

2.1   Complexity and Chaotic Systems

It is necessary to understand where complexity arises, what causes it, and how it can be addressed. But first, we must differentiate complex from complicated. Both concepts relate to a system of components, but the terms address different aspects of the systems. The term complicated is inherently linked to human understanding. It refers to being hard to understand. A long differential equation is a good example for something that is complicated. However, it is not necessarily complex, as it may have only few little parts. Complexity, in contrast, refers to the number of sub-components and their interactions that lead to system behavior that is hard to predict. A complex system does not need to be complicated, and a complicated system does not need to be complex.

Complexity is often associated with both the number of the constituent parts of a system and the intricate interactions of those parts [7,8]. The latter component yields another requirement for complexity—complex systems are dynamic. A static system with many parts is never truly complex, as it can be extensively studied. But, what makes complex problems so hard to solve?

Emergent Phenomena. When we study complex systems, it is not necessarily true that we will be able to derive the system behavior from an understanding of the parts of the system. The whole is more than the sum of its parts. By merely observing the behavior of its parts, we remain ignorant to the behavior of the system.

One complex system that can easily be observed in the real world is the strategy of a colony of ants finding food. Each ant shows only a very simple behavior. Each ant has no model of the world, yet it explores the world search for food, dropping pheromones while walking. When returning to the ant hill with food, it follows the pheromones dropped earlier, but this time increasing the amount of pheromone dropped. Every ant follows pheromone traces and thus reinforces the trail, when it carries food back to the hill, dropping pheromones along the way. The wind carries away all those pheromones that are not continuously reinforced. This process optimizes the ants' path to food sources without any individual knowledge of the world and without the need for communication or control between ants. The relatively efficient food finding that results from this behavior, emerges from the interaction of ants, pheromones, food, and wind. This systemic behavior is utilized in ant colony algorithms [9] to solve complex problems with only local knowledge.

But, when looking at the inside of the "internal program" of the ant, only someone who is familiar with ant colony algorithms, would guess this emergent behavior; although it occurs consistently, robustly, and relatively error-free. While we can observe and understand the individual behavior quite well, it is this non-observability of emergence from the individual that makes understanding of such system hard.

Non-Linear Behavior. A linear system is a system that can be described using linear equations. The input into such a system linearly influences the output of the system. Typical descriptive sentences about linear systems are sentences such as "X is proportional to Y". The more you add of "X" the more you get of "Y".

The methods to study such systems typically are also *linear models*, such as multiple-linear regression models and correlation models. Such models are relatively easy to grasp, as they follow "the more of this, the more of that" logic. But, many systems are not as easily described. Some show non-linear behavior as for example interest rates in a bank account. An account with an interest rate of 10% is easily predicted over the course of 1 year—it increases about 10%. However, most humans fail in predicting and mentally comparing different interest rates (i.e. exponential growth) over longer periods of time [10]. How large is the difference of an account with 10% interest rate to an account with 12% interest rate? If you put 1,000$ into an account and let it sit over the course of 50 years, the first will yield approx. 100,000$ the latter almost 300,000$.

Systems as these can be described using *non-linear equations* (e.g. polynomials, exponential-functions). In non-linear systems small changes can have large effects, and vice versa, large changes can result in small effects. And while such systems are harder to understand, it gets even worse when feedback loops are part of the system.

Systems that include feedback loops, exhibit behavior that affects the behavior of the system itself. To adequately describe such systems it is necessary to utilize differential equations. Differential equations are equations that include the derivative of a function. For example, when calculating how long it would take for an object to fall: One must consider the acceleration due to gravity, the longer it falls, the faster it becomes. But, air-friction depends on the speed of the object. The faster it is, the stronger the counter-acting force of air resistance. So how long does it take for a sky-diver to fall from 500 meters? In this case this equation can be solved explicitly. But, when one increases the distance to 5,000 meters, the differences in air-density makes this problem intractable to analytical processes. In such cases numeric simulation procedures are required to estimate adequate solutions.

Chaotic Behavior. A system is said to behave chaotically when even only a slight difference in the initial conditions leads to completely different behavior [11], often described as the butterfly effect [12].[1] This does not mean that the system behaves randomly, it can be very deterministic, but still the outcome is hard to predict from similar inputs. Some of these systems show very different behavior for similar initial conditions, but predicting the outcome of two such states in the future is intractable. If the state remains in some close boundaries but is hard to predict, we call these systems strange attractors—most famously the Lorenz-Attractor.

---

[1] The butterfly effect refers to the phenomenon that a single strike of the wing of a butterfly in Europe could potentially cause a thunderstorm in China through cascading effects.

A famous example for a chaotic system is a double-pendulum, where it is very hard to predict the position of the 2nd weight from its origin position (see Fig. 1). Not all complex systems behave chaotically and not all chaotic systems are complex—as the double-pendulum. Some complex systems have parameter configurations that behave chaotically, as system that seems to behave relatively stable can be destabilized by small changes if it an underlying chaotic system is assumed [13].

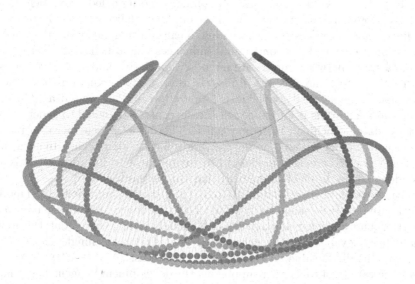

Fig. 1. Trajectory of the 2nd weight of a simulated double pendulum

Self Similarity and Levels of Observation. Complex systems often form from self-regulating processes [14]. The system produces new parts to compensate for changes in the environment, and all other parts react by adapting functions and interactions with other parts. This process leads to structures of self-similarity, where the whole reflects on the structure of its parts. Depending on the level of observation (micro-, meso-, macro-scale) other features are prominent, even though they are governed by the same few rules on the micro level. A consequence of this fractal nature are power-law distributions. Often such systems can not be described using normally distributed patterns, but require power-law distributions, as very few components play very important roles, and very many components play smaller roles in such systems. This makes understanding of such systems harder, as typical measures of describing data are meaningless in this context. A mean and a standard deviation have little meaning in a power-law distributions. Such effects are observed in species differentiation [15], network topologies [16], language construction [17] and many others.

2.2 Agent-Based Modeling

Since such systems are not easily described or understood analytically due to the aforementioned factors, one approach to understanding complex systems is simulation. When the rules of behavior of the individual parts are relatively simple, and the amount of parts and their interactions is still tractable, simulating the parts can provide means to understanding system behavior.

One approach for such simulations is agent-based modeling. In agent-based modeling, the individuals and the environment are modeled, their behavior is predicted from model parameters and output is visualized to facilitate understanding. The idea is that each individual agent autonomously decides what to do next and no centralized control unit influences macro-behavior. Such models provide a space between empirical methods, that derive knowledge from observation (induction), and theoretical models, that derive predictions from theory (deduction). Agent-based modeling is situated on middle ground as it utilizes theory and transforms it into a model, then executes the model to generate new empirical data, which can then be studied. The model is generative in nature. Furthermore, an agent-based model may allow for learning agents. In such cases each agent tries to create a model of the environment to improve its decisions. This adds another layer of complexity that can be modeled [18].

Because the data is generated "ab-initio" it is crucial that agent-based models are validated—either using verification of the model-code or using real-world data [19]. The challenge for agent-based models is to ensure that the model actually has relevance for real world settings and is not a meaningless abstraction of theory, with little value for real-world applications.

A large variety of tools exist supporting the development of agent-based models. The Wikipedia-page on agent-based modeling compares about 20 different frameworks. Most famous, due to its approachability is Netlogo [20]. Netlogo is a tool that allows creating agent-based models using a dialect of the Logo programming language, which was itself designed to teach kids the basics of programming. It adds both user interface components, visualization components, and analytical tools to the modeling process, allowing reproducible research that is easy to understand inside Netlogo (see Fig. 2). It further allows batch-running experiments to investigate not just how a single model behaves—which could be unstable—but also how a model behaves when changing parameters. One benefit of Netlogo is that it comes with a model library pre-installed, so that interested users can immediately start their own experiments.

3 Fields of Application and Model Types

Agent-based models come in as many varieties as one might imagine, depending on their level of sophistication, field of application, and their disciplinary background. When we want to understand the interaction of algorithms and human users, it is necessary to consider these options carefully. For this purpose we look at two different fields of application, where complexity arises in them and how agent-based modeling can be used to gain a deeper understanding of these fields.

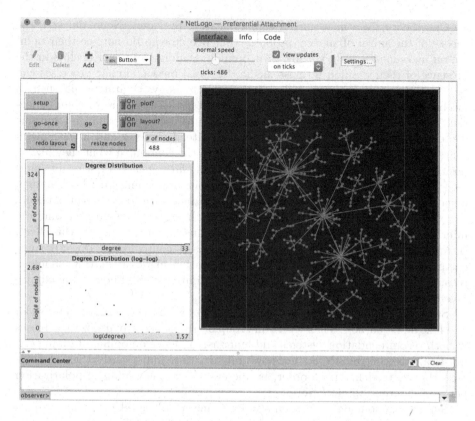

Fig. 2. Preferential attachment model in Netlogo

3.1 Social Phenomena and Social Simulation

The first field of application is the field of social simulation. Social simulations try to increase the understanding of theories and data from sociology covering a broad range from questions such as friendship relations, inequality, cooperation behavior as well as questions from opinion forming in the political sphere. Here, we will focus on the latter.

Research on opinions and how they are spread has been conducted since the 1960 s investigating questions of opinion leadership [21]. In this context, theory differentiates between "leaders" and "followers". Opinion leaders are the people in society that others seek out for the validation of their own opinions. Opinion leadership is topic dependent [22] and can be identified using standardized questionnaires [23]. The opinions of others are often changed in the process of *deliberation* [24], where the interaction with opinion heterogeneity affects opinion forming [25] and increases political awareness. When such deliberation processes are conducted online, as in social media, only little evidence points towards similar changes [26]. How is the communication in digital media different?

One theory from mass media communication by Noelle-Neumann [27], the theory of the spiral of silence, proposes that people who do not see their opinion on mass media, stop voicing their opinion, leading to further decline in the presence of their opinion in mass media. In social media, similar processes can be witnessed [28]. However, this is no simple analogy. Digital media such as facebook has algorithmic components that affect the structure of communication. For example, *recommender systems* [29] are used to recommend suitable content for users and keep them on the website. Users see and interact with content they like, and the recommender adjusts the content to what they like. A positive feedback loop that Pariser [30] called the *filter bubble*. First results indicate that the filter bubble impacts political opinion forming [31]. Both human behavior and algorithmic behavior lead to a reduction in diversity of exposure to opinions [32]. Although modern recommender systems try to overcome the filter bubble problematic [33], it is unclear how users compensate for this effect. In this context, many questions arise: Do all users compensate? Do only some compensate? How does this affect recommendation algorithms? How should a recommendation engine behave? Should it maximize profit? Should it maximize diversity?

Such questions become increasingly important with the rise of social bots [34] and fake news [35, 36]. When a certain group of people actively utilizes the biases in both recommendation systems and humans, influence on opinion forming can drastically be centralized to a few powerful few. More importantly, these settings pose threats to democratic order, by increasing the perception of polarization between different opinion groups.

Similar phenomena have been studied using agent-based modeling. In particular, phenomena such as homophily, network effects [37], polarization [38] forming and social differentiation [39] have all their counterpart agent-based models that help understand these phenomena. However, an integration of the algorithmic part is still missing and further no model exists that covers both micro and macro scales.

To fully understand the complexity of the interaction of human and algorithmic participants in digital communication, a model that unifies both social and information theoretic aspects is needed. Such a model would include both human participants, the underlying network structure, and a model of the algorithms governing digital communication (i.e. recommender system agents). The underlying *network structure* can also be generated from the agents using preferential attachment (as seen in Fig. 2 and demonstrated in [40]).

Modeling the User. For *the user model*, different model types can be considered, depending on what part of interaction is to be modeled. If the aim is to understand the spread of rational discourse in networks, a BDI-Model (Belief, Desire, Intent) of agents may be sufficient to model behavior of users [41, 42] and their personality. However, when the topic becomes irrational as in hate-speech, affect-based models of communication must be integrated into the agents. For this purpose different existing theories must be empirically evaluated and an approach to convert these to agent-based descriptions is needed. A typical theoretical description of such systems are structural equation models, where latent constructs can be depicted as internal state variables and paths can be modeled as probabilistic state transformations. However, in this aspect only very little previous work exists, and model transformation techniques are needed.

Modeling the Algorithm. The *algorithm model* is seemingly simpler. The most direct approach would be to use the actual underlying algorithm to simulate the algorithmic participant. However this is, for most cases, not very practical. Real world data would be needed. It is more fruitful to analyze the behavior of such algorithms and replicate behavior on a larger scale. A collaborative filtering algorithm could for example be replicated by generating artificial documents with a limited set of artificial tags, and then randomly assign tag preferences to user agents. To cosine-similarity could be used to model preferences for users and documents, and users are then repeatedly exposed to top-ten recommendations. The abstraction of the model comes from the limitation of articles and the artificial assignment of topics as tags. For such a model, the algorithm can rely on ground truth and deviations from optimal recommendations can be observed. Similar models could be generated for other algorithms such as content-based models, hybrid models and matrix or tensor-based approaches.

3.2 Cognitive Modeling for Human Factors Research

Another field of complex interaction of humans and algorithms is at work [43]. For example, when users interact with computers the underlying algorithms of computer programs are used implicitly and rarely with big complications. However, with the rise of artificial intelligence, Big Data and machine learning, an increasing amount of opaque algorithms are used in software and hardware solutions. Opaque in this case means that the algorithm adapts to the user and the environment, without the user necessarily noticing this adaptation. It is particularly opaque, if the systems adjusts to the behavior of other users as well and utilizes this knowledge for all users.

Decision Support Systems and Human Factors. One such setting is e.g. a decision support system that is used to analyze large amounts of data and guide the user to choose optimal solutions. Such systems are in use in the medical field, in logistics, in production, and even in programming environments. However, what if the decision that is suggested is not optimal? How do users cope with such situations [44]? Such settings arise, when uncommon data is present. Typically, rare events may lead to situations that the decision support system might be completely unaware of and thus false suggestions are made. The human user, who is not limited to data form previous settings, could extrapolate (or ask a colleague) what could cause such a situation, if s/he notices the deflective solution. The risk of blindly following suggested solutions is increased when the suggestion process becomes increasingly opaque. Further, the better the algorithms become the less trained human operators will become in solving these increasingly rare situations [45].

The interaction of such systems and human factors in this regard is often investigated using serious gaming environments [46]. Here, factors such as the decision correctness, the impact of UI interfaces, or environmental behavior can be experimentally modified to test the influence of these factors on human decision making [47]. Problematic for such experiments is that each experiment takes relatively long time, needs to be trained adequately by participants, and only few factors can be modified at a time. If a large amount of factors should be modified, no human sample could ever run multiple trials in all settings.

For such cases agent-based modeling can be used to simulate human operators—the human-in-the-loop. So-called cognitive agents derive their internal model from models of the cognitive sciences [48]. The models try to replicate mental organs and their interactions to produce output that is similar in nature as human cognition, learning, and behavior. The different cognitive architectures that are used in cognitive agent modeling are based on different models from cognitive neuroscience.

Cognitive Architectures. The most famous models in this regard are the ACT-R, SOAR and Clarion models, which differ in their approach. They are considered so-called cognitive architectures, which are both a model for human intelligence and the simulation of this intelligence.

The *ACT-R* architecture is based on declarative memory and procedural memory. Procedural rules are derived from declarative memory and perceptual input and encoded as symbolic representations [49]. The core architecture uses different buffers for goals, memory retrieval, visual perception, motor-procedures, and a central executive unit. A plethora of articles exist that utilize ACT-R to simulate perception, attention, memory, and learning [50]. These simulations match the behavior of human subjects to a large extent, so that even the evaluation of user interfaces using ACT-R has been successfully implemented.

The *SOAR* architecture [51] was developed to model general intelligence. SOAR-Agents try to behave as "rational" as possible, all knowledge that is applicable to a given situation is evaluated and utilized. For this purpose the *SOAR* architecture contains models of procedural memory and of working memory. Knowledge is procedurally defined as simple rules. For prediction SOAR uses a problem-space computational model. This means that it tries to model actions as transformations of problem-state. A problem state is the representation of the world and the evaluation of this state towards a certain goal. By applying actions (or rules) to this state, the agent can evaluate whether the action brings the agent closer to the goal or not. With each transformation deeper levels of "thinking-ahead" can be realized for each agent. Soar agents can be defined as having mutual knowledge and individual knowledge aspects, to simulate team behavior and cooperation. This aspect has helped SOAR to be used in many game-based AI scenarios. Opponents in computer games or military simulations can cooperate as teams using team strategies but nevertheless require an individual perception and action [52].

The *Clarion* architecture was designed to reflect on two modes of cognition [53]. It has both explicit cognition, such as planning how to solve a problem, and implicit cognition, such as assessing the danger in a situation holistically. Behavior is then generated from four subsystems. The action centered subsystem stores knowledge about internal and external actions. The Non-action centered subsystem stores knowledge about the world in general. Both systems have explicit and implicit representations. The motivational subsystem derives actions from basic and higher needs (similar to Maslow), which can be formulated as rules. From these needs the evaluation of actions (internal and external) is conducted and selected in the meta-cognitive subsystem. The latter is used to adapt goals in accordance with basic and derived higher needs with knowledge of the world. Clarion is unique, as it allows agents to modify their own goals with respect to their needs. Clarion has successfully been used to model the acquisition of cognitive skills and complex sequential decision making [54].

4   Discussion

We have seen that complexity arises from the interaction of human and algorithmic participants in various fields of application. In social interactions complexity arises from the underlying social network structure and increases when algorithmic participants are included in the picture. In working environments complexity arises from the application of "smart technologies" such as AI, machine learning, etc., in working processes. Here, the complexity of the interaction of humans and algorithms stems from the opacity of the inner-workings of the algorithms—a lack of understandable AI. Although one could argue, that it rather stems from the opacity of the inner-workings of human decision making as well. In working scenarios, AI is typically used to simplify hard problems. While this is successful most of the time, this approach becomes more complex, when the AI fails or produces non-optimal output. The interaction of humans and algorithms have in both cases not been studied extensively to ensure a deep understanding of the interaction.

For both settings, the utilization of agent-based modeling seems to provide access to a simulative validation of theories of interaction between humans and algorithms. The selection of the right modeling approach and validation with empirical data is critical for an effective use of agent-based models. Depending on the level of study, different tools and modeling techniques are required. Future work will have to focus on connecting micro-, meso- and macro-levels, while at the same time keeping computational performance at a reasonable level.

Finally, one must keep in mind that modeling a system with chaotic behavior can be modeled using agent-based modeling. However, the insights from these models are hard to translate to real-world applications. It could be the case that a models' only insight is that the underlying system is chaotic in nature. Every slight change in configuration would lead to drastically different outcomes. The benefit of such insight would rather be in evaluating traditional approaches of understanding such a system. Modeling such a system using traditional static linear models can be considered futile, if inherent chaotic properties are discovered.

The core benefit of simulation and agent-based modeling in human factors research is to identify configurations that are probabilistic in nature, that have stable states, and that have stabilizing environmental influences. Knowing such influences can be informative for policy making, political decision making and the design of "smart technologies". This knowledge should ensure that the intended goals are at least probabilistically in line with the expected outcomes of the system design.

Acknowledgments. This work was funded in part by the State of North Rhine-Westphalia under the grant number 005-1709-0006, project "Digitale Mündigkeit" and project-number 1706dgn017. The authors also thank the German Research Council DFG for the friendly support of the research in the excellence cluster "Integrative Production Technology in High Wage Countries".

References

1. Schaller, R.R.: Moore's law: past, present and future. IEEE Spectr. **34**(6), 52–59 (1997)
2. Alagöz, F., Valdez, A.C., Wilkowska, W., Ziefle, M., Dorner, S., Holzinger, A.: From cloud computing to mobile internet, from user focus to culture and hedonism: the crucible of mobile health care and wellness applications. In: 2010 5th International Conference on Pervasive Computing and Applications (ICPCA), pp. 38–45. IEEE (2010)
3. Schmidhuber, J.: Deep learning in neural networks: an overview. Neural Netw. **61**, 85–117 (2015)
4. Kraemer, F., Van Overveld, K., Peterson, M.: Is there an ethics of algorithms? Ethics Inf. Technol. **13**(3), 251–260 (2011)
5. Holzinger, A.: Interactive machine learning for health informatics: when do we need the human-in-the-loop? Brain Inform. **3**(2), 119–131 (2016)
6. Kamishima, T., Akaho, S., Sakuma, J.: Fairness-aware learning through regularization approach. In: 2011 IEEE 11th International Conference on Data Mining Workshops (ICDMW), pp. 643–650. IEEE (2011)
7. Waldrop, M.M., Gleick, J.: Complexity: The Emerging Science at the Edge of Order and Chaos. Viking, Info London (1992)
8. Byrne, D.S.: Complexity Theory and the Social Sciences: An Introduction. Psychology Press (1998)
9. Dorigo, M., Birattari, M.: Ant colony optimization. In: Encyclopedia of Machine Learning, pp. 36–39. Springer, New York (2011)
10. Bertrand, M., Mullainathan, S., Shafir, E.: Behavioral economics and marketing in aid of decision making among the poor. J. Public Policy Market. **25**(1), 8–23 (2006)
11. Hasselblatt, B., Katok, A.: A First Course in Dynamics: With a Panorama of Recent Developments. Cambridge University Press, Cambridge (2003)
12. Lorenz, E.: Predictability: does the flap of a butterfly's wing in Brazil set off a tornado in Texas? (1972)

13. Rickles, D., Hawe, P., Shiell, A.: A simple guide to chaos and complexity. J. Epidemiol. Commun. Health **61**(11), 933–937 (2007)
14. Brown, J.H., Gupta, V.K., Li, B.L., Milne, B.T., Restrepo, C., West, G.B.: The fractal nature of nature: power laws, ecological complexity and biodiversity. Philos. Trans. Roy. Soc. B: Biol. Sci. **357**(1421), 619–626 (2002)
15. Harte, J., Kinzig, A., Green, J.: Self-similarity in the distribution and abundance of species. Science **284**(5412), 334–336 (1999)
16. Song, C., Havlin, S., Makse, H.A.: Self-similarity of complex networks. Nature **433**(7024), 392 (2005)
17. Li, W.: Random texts exhibit zipf's-law-like word frequency distribution. IEEE Trans. Inf. Theory **38**(6), 1842–1845 (1992)
18. Bonabeau, E.: Agent-based modeling: Methods and techniques for simulating human systems. Proc. Natl. Acad. Sci. **99**(suppl 3), 7280–7287 (2002)
19. Sargent, R.G.: Verification, validation, and accreditation: verification, validation, and accreditation of simulation models. In: Proceedings of the 32nd Conference on Winter Simulation, Society for Computer Simulation International, pp. 50–59 (2000)
20. Wilensky, U.: Netlogo. Center for Connected Learning and Computer Based Modeling. Northwestern University (1999). http://ccl.northwestern.edu/netlogo
21. Rogers, E.M., Cartano, D.G.: Public Opin. Q, pp. 435–441. Methods of measuring opinion leadership, D.G. (1962)
22. Myers, J.H., Robertson, T.S.: Dimensions of opinion leadership. J. Market. Res., 41–46 (1972)
23. Childers, T.L.: Assessment of the psychometric properties of an opinion leadership scale. J. Market. Res., 184–188 (1986)
24. Suiter, J., Farrell, D.M., O'Malley, E.: When do deliberative citizens change their opinions? evidence from the irish citizens' assembly. Int. Polit. Sci. Rev. **37**(2), 198–212 (2016)
25. Zaller, J.: Political awareness, elite opinion leadership, and the mass survey response. Soc. Cogn. **8**(1), 125–153 (1990)
26. Dimitrova, D.V., Shehata, A., Strömbäck, J., Nord, L.W.: The effects of digital media on political knowledge and participation in election campaigns: evidence from panel data. Commun. Res. **41**(1), 95–118 (2014)
27. Noelle-Neumann, E.: Die Schweigespirale. Riper [ie Piper] (1980)
28. Gearhart, S., Zhang, W.: 'Was it something i said?' no, it was something you posted!' a study of the spiral of silence theory in social media contexts. Cyberpsychol. Behav. Soc. Network. **18**(4), 208–213 (2015)
29. Resnick, P., Varian, H.R.: Recommender systems. Commun. ACM **40**(3), 56–58 (1997)
30. Pariser, E.: The Filter Bubble: What the Internet is Hiding From You. Penguin, UK (2011)
31. Dylko, I., Dolgov, I., Hoffman, W., Eckhart, N., Molina, M., Aaziz, O.: The dark side of technology: an experimental investigation of the influence of customizability technology on online political selective exposure. Comput. Hum. Behav. **73**, 181–190 (2017)
32. Nguyen, T., Hui, P.M., Harper, F., Terveen, L., Konstan, J.: Exploring the filter bubble: the effect of using recommender systems on content diversity, pp. 677–686 (2014)
33. Calero Valdez, A., Ziefle, M., Verbert, K.: HCI for recommender systems: the past, the present and the future. In: Proceedings of the 10th ACM Conference on Recommender Systems, pp. 123–126. ACM (2016)

34. Ferrara, E., Varol, O., Davis, C., Menczer, F., Flammini, A.: The rise of social bots. Commun. ACM **59**(7), 96–104 (2016)
35. Allcott, H., Gentzkow, M.: Social media and fake news in the 2016 election. Technical report, National Bureau of Economic Research (2017)
36. Woolley, S.: Automating power: Social bot interference in global politics. First Monday **21**(4) (2016)
37. Clemm von Hohenberg, B., Maes, M., Pradelski, B.S.: Micro influence and macro dynamics of opinions. SSRN (2017)
38. Macy, M.W., Kitts, J.A., Flache, A., Benard, S.: Polarization in dynamic networks: a hopfield model of emergent structure. In: Dynamic Social Network Modeling and Analysis, pp. 162–173 (2003)
39. Mark, N.: Beyond individual differences: social differentiation from first principles. Am. Soc. Rev., 309–330 (1998)
40. Leskovec, J., Backstrom, L., Kumar, R., Tomkins, A.: Microscopic evolution of social networks, pp. 462–470 (2008). cited By 331
41. Kosinski, M., Stillwell, D., Graepel, T.: Private traits and attributes are predictable from digital records of human behavior. PNAS **110**(15), 5802–5805 (2013)
42. Bachrach, Y., Kosinski, M., Graepel, T., Kohli, P., Stillwell, D.: Personality and patterns of facebook usage. In: Proceedings of the ACM Web Science Conference, pp. 36–44. ACM New York (2012)
43. Valdez, A.C., Brauner, P., Ziefle, M.: Preparing production systems for the internet of things the potential of socio-technical approaches in dealing with complexity (2016)
44. Brauner, P., Calero Valdez, A., Philipsen, R., Ziefle, M.: Defective still deflective – how correctness of decision support systems influences user's performance in production environments. In: Nah, F.F.-H.F.-H., Tan, C.-H. (eds.) HCIBGO 2016. LNCS, vol. 9752, pp. 16–27. Springer, Cham (2016). https://doi.org/10.1007/978-3-319-39399-5_2
45. Arnold, V., Sutton, S.: The theory of technology dominance: Understanding the impact of intelligent decision aids on decision maker's judgments. Adv. Account. Behav. Res. **1**(3), 175–194 (1998)
46. Brauner, P., Calero Valdez, A., Philipsen, R., Ziefle, M.: How correct and defect decision support systems influence trust, compliance, and performance in supply chain and quality management. In: Nah, F.F.-H., Tan, C.-H. (eds.) HCIBGO 2017. LNCS, vol. 10294, pp. 333–348. Springer, Cham (2017). https://doi.org/10.1007/978-3-319-58484-3_26
47. Brauner, P., Valdez, A.C., Philipsen, R., Ziefle, M.: On studying human factors in complex cyber-physical systems. In: Mensch und Computer 2016-Workshopband (2016)
48. Sun, R.: Cognition and Multi-agent Interaction: From Cognitive Modeling to Social Simulation. Cambridge University Press, Cambridge (2006)
49. Anderson, J.R., Bothell, D., Byrne, M.D., Douglass, S., Lebiere, C., Qin, Y.: An integrated theory of the mind. Psychol. Rev. **111**(4), 1036 (2004)
50. ACT-R Research Group: ACT-R Publications. Carnegie Melon University (2018). http://act-r.psy.cmu.edu/publication/
51. SOAR Research Group: Soar Website. University of Michigan (2018). https://soar.eecs.umich.edu/
52. Laird, J.E.: It knows what you're going to do: adding anticipation to a quakebot. In: Proceedings of the Fifth International Conference on Autonomous Agents, pp. 385–392. ACM (2001)

53. Sun, R.: Duality of the Mind: A Bottom Up Approach to Cognition. Lawrence, Mahew (2002)
54. Sun, R., Merrill, E., Peterson, T.: From implicit skills to explicit knowledge: a bottom-up model of skill learning. Cognit. Sci. **25**(2), 203–244 (2001)

Personalized Emotion-Aware Video Streaming for the Elderly

Yi Dong[1](✉), Han Hu[2], Yonggang Wen[2], Han Yu[1], and Chunyan Miao[1,2]

[1] NTU-UBC Research Center of Excellence in Active Living for the Elderly, IGS,
Nanyang Technological University, Singapore, Singapore
{ydong004,ascymiao,han.yu}@ntu.edu.sg
[2] School of Computer Science and Engineering,
Nanyang Technological University, Singapore, Singapore
{hhu,ygwen}@ntu.edu.sg

Abstract. We consider the problem of video therapy services for the elderly based on their current emotional status. Given long hours watching TV in the elder population, most of the existing TV services are not geared for them. The elderly cannot tolerate complexity and negativity due to decline in cognitive abilities. In addition, the program is not adapted to the user's current emotional status. As a result, existing TV services can not achieve optimal performance across a broad set of user types and context. To provide content tailored to individual needs, and interests of the elderly, caregivers have to select an appropriate program manually. However, this can not scale well due to shortage of caregivers and high monetary cost. We present the personalized emotion-aware video streaming system, a redesign of conventional TV system to provide appropriate program flexibly, efficiently and responsively. Our proposed architecture adds video affective profiling, real-time emotion detection and Markov decision process based video program generation to the streaming service to this end. We present a complete implementation of our design. Trace-driven simulation has shown the effectiveness of our system.

1 Introduction

Watching television is a common leisure activity for older people. Retirees spend more than half their leisure time watching television according to American time use survey [15]. Many of them follow the broadcast schedule passively, which may contribute to the development of cognitive impairment. However, watching television can benefit the elderly in multiple ways if the program is selected carefully. Many studies have shown that watching TV can be mentally stimulating, thus benefit the elderly by anxiety and stress reduction, improved cooperative behavior and lowered use of psychoactive medications. Therefore, this type of non-pharmacological therapy video service can serve as multi-sensory stimulation and therapeutic use of music or video, which is recommended by many guidelines for the elderly, especially dementia caring [2].

The goal of this paper is to develop a non-pharmacological therapy video service. Building such a service is challenging on two fronts. First, video contents should be cognitively congruent to the viewer's abilities, needs, and interests. Second, the viewer's needs changes dynamically. Thus, we need a system that is

1. **personalized** enough to meet the needs for each individual viewer.
2. capable of selecting video contents based on viewer's emotional status **in near real-time.**

Table 1. A summary of existing proposal and how they fall short of our requirements.

	Personalized	Adaptive in real-time	cost-efficient
Broadcast schedule	✗	✗	✓
VoD services	✓	✗	✓
Carehome staff guided	✓	✓	✗
Our approach	✓	✓	✓

As Table 1 shows, existing proposals fail on one or more of these requirements. For instance, broadcast schedules are not tailored for individual needs and adaptive to viewer's current emotional status. Similarly, Video on Demand(VoD) services provide control of the contents but they are not adapted as well. To address the aforementioned problems, traditional care-home relies on staffs to stay with residents so that they can choose the appropriate programs wisely and create a stimulus for interaction and conversation. Therefore, staffs themselves may need to monitor the emotional status of the elderly and find the signals when programs agitate the elderly. By long time observation, they may summarize the personal preference of each elderly and provide more accurate control. However, this kind of service cannot scale well. The shortage of care giver is becoming a big challenge and the monetary cost is a heavy burden. Thus, potential benefits promised by non-pharmacological video services remain unrealized.

We address these challenges and present the design and implementation of personalized emotion-aware video streaming system for the elderly. The design of our system is composed of accurate emotion detection using user equipment and the online personalized video program selection.

The user's emotion response can be measured accurately while watching the video by front cameras and wearable devices. We jointly employ the visual cues (facial expressions) and physiological signals like ECG to infer the emotional status of the viewer. Collectively, they can capture both the outer expressions as well as the inner feelings. We assume that users watch videos on devices with front cameras. Therefore, their frontal image with facial expressions can be easily captured. As for the ECG signals, a wrist band like Apple watch or Microsoft Band can measure accurately. Therefore, the emotion can be recognized and represented as a 2D emotion model [11] whose axes are valence and arousal.

To adapt video programs efficiently, we select video clips based on user's current emotional status. Our solution to content adaptation lies in the observation that the content characteristics are highly related to the emotion aroused [17]. Previous studies have developed affective models for video clips based on its content characteristics. Therefore, we can choose the next clip with appropriate affective properties based on user's current emotional status, optimizing the overall experience.

Our goal is to select appropriate video based on user's current emotion. Specifically, the following goals need to be achieved.

- Stable arousal. Read the signals when current video agitates someone and avoid this type of videos.
- High valence. Find out what the viewer really likes, and achieve the maximum variance of content and emotional experience.
- Exploration and exploitation tradeoff. To avoid boredom, the viewer should be exposed to new video clips with different ideas, art and culture. However, the affective effects of new clips are not verified.

The reward function should balance the three components.

In this paper, we demonstrate that, for such decision making, Markov decision process (MDP) can derive the optimal policy if the emotion-video model is known.

We implemented a prototype of non-pharmacological therapy video service. In real trace-driven evaluation, we assess our proposed algorithms based on publicly available emotional response dataset DEAP. The experiments show that our system is usable and robust.

We summarize our key contributions as follows.

- Design of system architecture for the personalized emotion aware video streaming system (Sect. 3).
- Formulating the video program adaptation as an Markovian Decision Process (Sect. 4).
- Real-world implementation of the system and trace-driven evaluation. (Sect. 5).

2   Related Works

Our system is built based on previous research on affective video content modeling, multimodal user emotion detection, adaptive video streaming. However, our system differs from previous work from multiple perspectives.

Adaptive video streaming: As a killer application of the Internet, video streaming has received tremendous attention in both industry and academia [12]. Nowadays, adaptive video streaming services are usually hosted by cloud computing platforms where video clips has been transcoded into different bitrate versions [5–7]. Existing client-side video players employ adaptive bitrate (ABR) algorithms to optimize user quality of experience (QoE). ABR algorithms dynamically choose the bitrate for each video segment based on user's network conditions or client buffer occupation [8]. This line of research focus on the resource allocation and optimization based on network resource and device usage [9]. In contrast, our proposal is the first effort on video adaptation based on viewer's current emotional status.

Video services for the elderly: The Gerontological Society of America published Video Respite[TM], which are video tapes developed for the persons with dementia. Early research shows the tapes to be calming for the persons with dementia [13]. Recent studies suggest that such type of video services has considerable promise as an individual and group activity in long-term setting [1]. Memory-Lane.tv[1] has developed an interactive and multi-sensory media collection for the memory impaired. It supports modern delivery methods such as tablets, computers and Apple TV. However, these video services require staffs to operate following guidelines to achieve promising results. Our system can release the burden of the caregivers and can be deployed easily.

3   System Architecture

In this section, we present our proposed system architecture for personalized emotion-aware video streaming systems. It is possible to adapt the program based on user's current emotional status. Specifically, we give detailed illustration

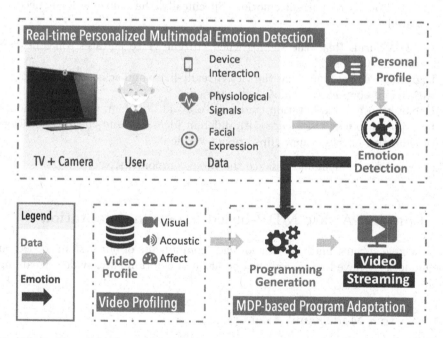

Fig. 1. An overview of the personalized emotion aware video streaming system. The system captures user's emotional status and adapts the program accordingly in a real-time manner.

---

[1] http://memory-lane.tv/.

on the three components: real-time personalized multimodal emotion detection, video profiling and MDP-based program adaptation (Fig. 1).

3.1 Emotion Detection

Emotion response while watching video is being sensed by ECG (Electrocardiography) signals and facial expressions. These signals may complement each other. Facial expressions are captured using front cameras which are ubiquitous in mobile devices while ECG signals are sensed by smart wrist or extracted from facial images.

3.2 Video Affective Profiling

This module can analyze the emotional impact of a video content will have on viewers, in terms of valence and arousal.

3.3 Video Program Selection

A set of videos that can provide a variance of video play experience is stored in our server. The system should work in the similar way that staffs in care giving institutions do to optimize the user's experience. Our goal is to select appropriate video based on user's current emotion. Specifically, the following goals need to be achieved.

- Stable arousal. Read the signals when current video agitates someone and avoid this type of videos.
- High valence. Find out what the viewer really likes, and achieve the maximum variance of content and emotional experience.
- Exploration and exploitation tradeoff. To avoid boredom, the viewer should be exposed to new video clips with different ideas, art and culture. However, the affective effects of new clips are not verified.

The reward function should balance the three components.

4   Emotion-Aware MDP-based Program Adaptation

we now give specific math models for each components of our system. Based on the models, we have formulated the operation of emotion aware video streaming as an MDP.

4.1 System Models and Assumptions

In this subsection, we present specific mathematical models for the emotion-aware content enhancement system, including user emotional status model, video response model and content caching model. For ease of reference, we summarized the notations in the Table 2.

Table 2. Definitions of Notations

Notation	Definition
s, s'	States
a	Action (The video to play)
r	Reward
$\mathcal{S}$	Set of all nonterminal states
$\mathcal{A}(s)$	Set of all actions possible in state s
$\mathcal{R}$	Set of all possible rewards
t	Discrete time step
T	Final time step of an episode
S_t	State at time t
A_t	Action at time t
R_t	Reward at time t
π	Policy, decision-making rule
$\pi(s)$	Action taken in state s under deterministic policy π
$\pi(a\vert s)$	Probability of taking action a in state s under stochastic policy π
$p(s'\vert s, a)$	Probability of transition to state s', from state s taking action a
γ	Discouont-rate parameter
$v_\pi(s)$	Value of state s under policy π (expected return)
$v_*(s)$	Value of state s under the optimal policy
V, V_t	Array estimates of state-value function v_π or v_*

Emotional Status Model. As shown in Fig. 2, we capture a person's emotions from both outward expressions and inner feelings. Specifically, we develop a machine learning based model to infer user's emotional status from physiological signals (eg. ECG) and facial expressions. We adopt a 2D emotional model [11] whose axes are valence (pleasure to displeasure) and arousal (high to low).

Video Response Model. Videos can render the experience necessary to arouse an emotion. The dynamics of emotion is the hall mark of a successful video. Emotions are created in the film's text. Since emotion is a short-term experience that peaks and burns rapidly, we can use a transitional matrix $\mathcal{T}$ to capture the dynamics of emotions. In this matrix, $P_{sa} = P(s, s', a) = \Pr(s_{t+1} = s' \vert s_t = s, a_t = a)$ denotes the probability that video a in state s at time t will lead to state s' at time $t + 1$.

Accumulated Rewards Model. The viewers are satisfied by the videos. We define the immediate reward received after the transition to a new emotion s' with video a as $R_{sa} = R(s, s', a)$. We assume that viewers gain more

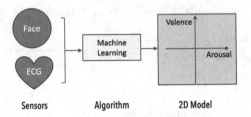

Fig. 2. The viewer's emotion can be extracted jointly from his/her facial expression and ECG signals. The Arousal-Valence model is used to model the emotion.

rewards while watching more videos. Moreover, we consider the law of diminishing returns. The more we experience something, the less effect it has. Therefore, we put a discount factor $0 \leq \gamma < 1$ to capture this characteristic. Therefore, the accumulated reward over infinite horizon is $\sum_{t=0}^{\infty} \gamma^t R(s_t, s_{t+1}, \pi(s_t))$. Please note that if the status is in termination, the reward will be 0.

4.2 MDP Formulation

As shown in Fig. 3, the viewer and TV interact in a sequence of discrete time steps, $t = 0, 1, 2, 3, \ldots, T$. At each time step t, the mobile device receives the representation of viewer's emotion state, $S_t \in \mathcal{S}$, where $\mathcal{S}$ is the set of possible emotions and on that basis selects an *video*, $A_t \in \mathcal{A}(S_t)$, where $\mathcal{A}(S_t)$ is the set of possible states available in state S_t. On time step later, in part as the impact of the video viewed, the viewer's emotional needs are fulfilled and can be quantified as a numerical reward, $R_{t+1} \in \mathcal{R} \subset \mathbb{R}$ and find oneself in a new emotional state, S_{t+1}. The figure above diagrams the viewer-TV interaction.

At each time step, the mobile device implements a mapping from states to probabilities of selecting each possible videos. This mapping is called policy and is denoted π_t, where $\pi_t(a|s)$ is the probability that $A_t = a$ if $S_t = s$.

In a real system deployment, we can model this problem as a Markov decision problem (MDP). An MDP is a tuple $(\mathcal{S}, \mathcal{A}, \mathcal{P}_{sa}, \mathcal{R}_{sa}, \gamma)$ as defined in the previous subsection. The solution of the MDP is a policy π that tells us how to make decision, i.e., choose an action when a particular state is observed. There are many possible policies but the goal is to derive the optimal policy π^*, which can be formally given as:

$$\pi^*(s, t) = \underset{a \in \mathcal{A}}{\operatorname{argmax}} \left[\sum_{s' \in \mathcal{S}} \mathcal{P}(s, a, s', t) [\mathcal{R}(s, a, s', t) + V^*(s', t+1)] \right], \quad (1)$$

where V denotes the value function. The optimal policy can be developed by dynamic programming algorithms [16].

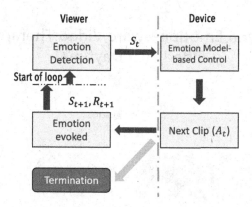

Fig. 3. The viewer-TV interaction in Markov Decision Process based program generation. The loop starts from the emotion detection.

5   System Implementation

We have built a prototype of personalized emotion-aware video therapy service that closely follows the architecture described in the previous sections. This prototype uses Microsoft band

5.1   Hardware Settings

For the client side, our system support most of the main stream mobile devices (e.g. Android/iOS) or PC that connected with a webcam. Users can log in our web portal (Fig. 4) to check the status of the sensor connection. Once the user has authorized the web browser to use webcam are connected, the video session is ready to start.

In our current implementation, the client side captures the frontal face image via front cameras and transmit the compressed image to the server side for emotion detection. Under good lighting condition, both facial expressions and ECG signals are extracted from the webcam image. Instead of relying on a webcam to capture ECG signals, We recommend users to wear ECG sensors if the lighting condition is not satisfactory.

Multimodal emotion detection and video program adaptation is deployed in the server side, given high computational demands. Our server is with 2 Intel® Xeon® E5-2603, 2 Nvidia K80 graphic cards and 64 GB memory.

5.2   Software Implementation

The user interface of the client front-end is shown in Fig. 4. We have implemented a web service for video streaming and affective profile visualization. The video player on the left can enter full screen mode to provide an immersive experience. The affective profile is visualized in the line chart on the right side. A vertical dotted line is synchronized with the player to indicate the current playing time.

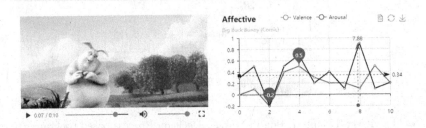

Fig. 4. Screenshot of the front end of personalized emotion-aware video service.

We had implemented the affective video profiling and multimodal emotion recognition by machine learning algorithms. Specifically, we trained SVR (Support Vector Regression) models from DEAP dataset [10], using LibSVM [3]. Audio features have been extracted by openSmile toolbox [4]. In addition, general video features were extracted using LIRE library [14]. Beyond that, we have extracted CNN features using Matlab neural network toolbox. The models in this part were developed for proof of concept. We believe that the accuracy of the two models will increase if more advanced deep learning algorithms are employed.

For personalized emotion-aware video program adaptation, we implement the proposed MDP model using Markov Decision Process (MDP) Toolbox for Python[2].

5.3   Limitations

We acknowledge three potential limitations in our study.

- **Lighting condition and user's head position.** Accurate emotion detection requires clear frontal facial image. The motion of viewer's head will cause motion blur which is the noise for our emotion detection algorithm. Therefore, our system only works under good light condition where viewer's head seldom moves.
- **Delay due to computation and communication.** Given high computation demands and the large size of media files, the delay of emotion detection is noticeable. In real-world deployment, we should strike a good balance between accuracy and delay.
- **Generalization of the emotion detection model.** In this paper, we have designed multimodal algorithms based on DEAP dataset. We have not tested the generalization of the emotion detection models in other settings. We plan to investigate this problem in our future work.

---

[2]   https://github.com/sawcordwell/pymdptoolbox.

6   Conclusions

In this work, we designed and implemented a personalized emotion-aware video therapy system for the elderly. Unlike conventional TV systems that use fixed program, we adapt the program based on user's current emotional status and the video affective attributes. We have formulated the emotion-aware video program adaptation as an MDP problem which can be solved by dynamic programming algorithms. We have deployed the emotion detection, video profiling and video program adaptation on a server with Nvidia K80 GPU. The system can effectively adapt the video program based on user's current emotional status. Trace-driven simulation validates that it is a robust and usable system.

Acknowledgments. This research is supported by the National Research Foundation, Prime Minister's Office, Singapore under its IDM Futures Funding Initiative; Nanyang Technological University, Nanyang Assistant Professorship and the Interdisciplinary Graduate School, Nanyang Technological University, Singapore.

References

1. Caserta, M.S., Lund, D.A.: Video respite® in an alzheimer's care center: group versus solitary viewing. Activities, Adapt. Aging **27**(1), 13–28 (2003)
2. National Collaborating Centre: Dementia: A nice-scie guideline on supporting people with dementia and their carers in health and social care. British Psychological Society (2007)
3. Chang, C.C., Lin, C.J.: Libsvm: a library for support vector machines. ACM Trans. Intell. Syst. Technol. (TIST) **2**(3), 27 (2011)
4. Eyben, F., Weninger, F., Gross, F., Schuller, B.: Recent developments in opensmile, the munich open-source multimedia feature extractor. In: Proceedings of the 21st ACM International Conference on Multimedia, MM 2013, pp. 835–838. ACM, New York (2013)
5. Gao, G., Hu, H., Wen, Y., Westphal, C.: Resource provisioning and profit maximization for transcoding in clouds: a two-timescale approach. IEEE Trans. Multimed. **19**(4), 836–848 (2017)
6. Gao, G., Wen, Y., Hu, H.: Qdlcoding: Qos-differentiated low-cost video encoding scheme for online video service. In: IEEE INFOCOM 2017 - IEEE Conference on Computer Communications, pp. 1–9, May 2017
7. Gao, G., Zhang, W., Wen, Y., Wang, Z., Zhu, W.: Towards cost-efficient video transcoding in media cloud: insights learned from user viewing patterns. IEEE Trans. Multimed. **17**(8), 1286–1296 (2015)
8. Hu, H., Wen, Y., Niyato, D.: Spectrum allocation and bitrate adjustment for mobile social video sharing: potential game with online qos learning approach. IEEE J. Sel. Areas Commun. **35**(4), 935–948 (2017)
9. Hu, H., Jin, Y., Wen, Y., Chua, T.S., Li, X.: Toward a biometric-aware cloud service engine for multi-screen video applications. In: Proceedings of the 2014 ACM Conference on SIGCOMM. SIGCOMM 2014, pp. 581–582. ACM, New York (2014)
10. Koelstra, S., Muhl, C., Soleymani, M., Lee, J.S., Yazdani, A., Ebrahimi, T., Pun, T., Nijholt, A., Patras, I.: Deap: a database for emotion analysis; using physiological signals. IEEE Trans. Affect. Comput. **3**(1), 18–31 (2012)

11. Lang, P.J.: The emotion probe: studies of motivation and attention. Am. Psychol. **50**(5), 372 (1995)
12. Li, B., Wang, Z., Liu, J., Zhu, W.: Two decades of internet video streaming: a retrospective view. ACM Trans. Multimedi. Comput., Commun. Appl. (TOMM) **9**(1s), 33 (2013)
13. Lund, D.A., Hill, R.D., Caserta, M.S., Wright, S.D.: Video respite: an innovative resource for family, professional caregivers, and persons with dementia. Gerontologist **35**(5), 683–687 (1995)
14. Lux, M., Chatzichristofis, S.A.: Lire: lucene image retrieval: an extensible java CBIR library. In: Proceedings of the 16th ACM International Conference on Multimedia, pp. 1085–1088. ACM (2008)
15. Robinson, J., Godbey, G.: Time for Life: The Surprising Ways Americans Use Their Time. Penn State Press (2010)
16. Sutton, R.S., Barto, A.G.: Reinforcement Learning: An Introduction, 2nd edn. MIT Press, Cambridge, MA (2017)
17. Zhu, Y., Hanjalic, A., Redi, J.A.: QoE prediction for enriched assessment of individual video viewing experience. In: Proceedings of the 2016 ACM on Multimedia Conference, pp. 801–810. ACM (2016)

Automatically Generating Head Nods with Linguistic Information

Ryo Ishii, Ryuichiro Higashinaka[(⊠)], Kyosuke Nishida[(⊠)],
Taichi Katayama[(⊠)], Nozomi Kobayashi[(⊠)], and Junji Tomita[(⊠)]

NTT Media Intelligence Laboratories, NTT Corporation, 1-1 Hikari-no-oka,
Yokosuka-shi, Kanagawa, Japan
{ishii.ryo,higashinaka.ryuichiro,nishida.kyosuke,katayama.taichi,
kobayashi.nozomi,tomita.junji}@lab.ntt.co.jp

Abstract. In addition to verbal behavior, nonverbal behavior is an important aspect for an embodied dialogue system to be able to conduct a smooth conversation with the user. Researchers have focused on automatically generating nonverbal behavior from speech and language information of dialogue systems. We propose a model to generate head nods accompanying an utterance from natural language. To the best of our knowledge, previous studies generated nods from the final words at the end of an utterance, i.e. bag of words. In this study, we focused on various text analyzed using linguistic information such as dialog act, part of speech, a large-scale Japanese thesaurus, and word position in a sentence. First, we compiled a Japanese corpus of 24 dialogues including utterance and nod information. Next, using the corpus, we created a model that generates nod during a phrase by using dialog act, part of speech, a large-scale Japanese thesaurus, word position in a sentence in addition to bag of words. The results indicate that our model outperformed a model using only bag of words and chance level. The results indicate that dialog act, part of speech, the large-scale Japanese thesaurus, and word position are useful to generate nods. Moreover, the model using all types of linguistic information had the highest performance. This result indicates that several types of linguistic information have the potential to be strong predictors with which to generate nods automatically.

Keywords: Nod · Generation · Japanese dialogue
Linguistic information

1 Introduction

Nonverbal behavior in human communication has important functions of transmitting emotions and intentions in addition to verbal behavior [2]. This means that an embodied dialogue system should be able to express nonverbal behavior according to the utterance to communicate smoothly with the user [10,28,35]. Against such a background, researchers have focused on constructing automatic

© Springer International Publishing AG, part of Springer Nature 2018
G. Meiselwitz (Ed.): SCSM 2018, LNCS 10914, pp. 383–391, 2018.
https://doi.org/10.1007/978-3-319-91485-5_29

generation models of nonverbal behavior from speech and linguistic information. Among nonverbal behaviors, nodding of the head is very important for emphasizing speech, giving and receiving speech authority, giving feedback, expressing conversational engagement, and intention of starting to speak [12,14,31,33,34]. It has been shown that nodding improves the naturalness of avatars and dialog systems and promotes conversation.

Nodding accompanying an utterance has the effect of strengthening the persuasive power of speech and making it easier for the conversational partner to understand the content of the utterance [27]. Researchers have tried to generate nods during speaking from speech and natural language. In particular, they used several acoustic features, such as sound pressure and prosody, for generating nods [1,4,6,10,24,25,37]. However, it has been difficult to accurately generate nods at an appropriate time according to an utterance from only speech.

A few studies have tackled the problem of generating nods from natural language. These studies focused on the final word in the phrase of an utterance and analyzed the co-occurrences with nods. They found that morphemes related to the interjections, feedback, questionnaire, and conjunctions appearing in turn-keeping [8,9] tend to co-occur with nods. On the basis of this information, a simple automatic nod-generation model was proposed [10,32]. It was found that the behavior of humanoid robots and avatars that generated nods with the model gave a better impression of naturalness. It is thought that if a model that can generate nodding more accurately is constructed, it will lead to smoother communication between the dialog system and user. Therefore, a more accurate nod-generation model should be constructed by clarifying the relevance of more detailed language information and nodding. It is also known that the relevance of a speech feature to nodding and vice versa depends on the language; for instance this is weaker in Japanese [8]. A detailed examination of a nod-generation model using language information is thus considered important.

In this research, we constructed a highly accurate head-nod-generation model using natural Japanese language by focusing on the various text analyzed linguistic information such as dialog act, part of speech, a large-scale Japanese thesaurus, and word position in a sentence, which has not been investigated. A dialogue act is information indicating the intention of a speaker in a whole utterance, and it is believed that the occurrences of nods change in accordance with the intention. We hypothesized that words in phrases other than the final phrase and lexicons of utterances had strong relationships with head nodding.

We collected a corpus consisting of 24 Japanese dialogues including utterances and head-nod information. Next, we used the corpus to create our model that generates a nod during a phrase by using bag of words, dialog act, part of speech, a large-scale Japanese thesaurus, and word position in a sentence in addition to the bag of words. The results indicate that our model using dialog act, part of speech, the large-scale Japanese thesaurus, and word position outperformed a model using only bag of words and chance level. The results indicate that dialog act, part of speech, the large-scale Japanese thesaurus, and word position are useful to generate nods. Moreover, the model using all types of language information had the highest performance. This result indicates that several types of linguistic information have the potential to be strong predictors with which to generate nods automatically.

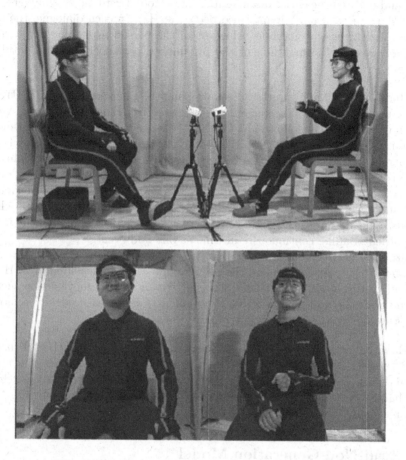

Fig. 1. Photograph of two participants having dialogue

2   Corpus

To collect a Japanese conversation corpus including verbal and nonverbal behaviors for generating nods in dialogue, we recorded 24 face-to-face two-person conversations (12 groups of two different people). The participants were Japanese males and females in their 20s to 50s who had never met before. They sat facing each other (Fig. 1). To gather more data on nodding accompanying utterances, we adopted the explanation of an animation participants have not seen as the conversational content. Before the dialogue, they watched a famous popular cartoon called "'Tom & Jerry" in which the characters do not speak. In each dialogue, one participant explained the content of the cartoon to the conversational partner within ten minutes. At any time during this period, the partner could freely ask questions about the content.

We recorded the participants' voices with a pin microphone attached to the chest and videoed the entire discussion. We also took bust (chest, shoulders, and head) shots of each participant (recorded at 30 Hz). In each dialogue, the data on the utterances and nodding behaviors of the person explaining the cartoon were collected in the first half of the ten-minute period (120 min in total) as follows.

- Utterances: We built an utterance unit using the inter-pausal unit (IPU) [26]. The utterance interval was manually extracted from the speech wave. A portion of an utterance followed by 200 ms of silence was used as the unit of one utterance. We collected 2965 IPUs. Moreover, we used J-tag [5] which is a general morphological analysis tool for Japanese to divide IPU into phrases. We collected 11877 phrases in total.
- Head nod: A head nod is a gesture in which the head is tilted in alternating up and down arcs along the sagittal plane. A skilled annotator annotated the nods by using bust/head and overhead views in each frame of the videos. We regarded nodding continuously in time as one nod event.
- Gaze: The participants wore a glass-type eye tracker (Tobii Glass2). The gaze target of the participants and the pupil diameter were measured at 30 Hz.
- Hand gesture and body posture: The participants' body movements, such as hand gestures, upper body, and leg movements, were measured with a motion capture device (Xsens MVN) at 240 Hz.

All verbal and nonverbal behavior data were integrated at 30 Hz for display using the ELAN viewer [36]. This viewer enabled us to annotate the multimodal data frame-by-frame and observe the data intuitively. In this research, we only handled utterance and head-nod data in the corpus we constructed. Nods occurred in 1601 out of the 2965 IPUs.

3   Head-Nod-Generation Model

The goal of our research was to demonstrate that bag-of-words, dialog acts, parts of speech, a large-scale Japanese thesaurus, and word position in a sentence is useful for generating nods. We evaluated our proposed model for generating nods from several types of linguistic information and the previously constructed estimation model using only the final word at the end of an utterance [8,9]. We constructed another estimation model using all types of linguistic information to evaluate the effectiveness of this fusion (All model). The feature values of linguistic information for each phrase were as follows.

- Length of phrase (LP): Number of characters in a phrase.
- Word position (WP): Word position in a sentence.

- Bag of words (BW): The word injunctions related to feedback (e.g., "en", "ee", "aa", "hi", etc.) and particles related to questioning and turn-keeping (e.g., "de", "kara", "kedo", "'kana", "janai", etc.) co-occurring with the nod is used for estimation of noding the previous studies [8,9]. To deal with more generic word information in addition to them, we examined the number of occurrences of all words, not some morphemes. We used J-tag [5], a general morphological analysis tool for Japanese.
- Dialogue act (DA): A dialogue act was extracted using an estimation technique for Japanese [7,29]. The technique can estimate a dialogue act using the word N-grams, semantic categories (obtained from a Japanese thesaurus Goi-Taikei), and character N-grams. The dialog acts and number of IPUs are listed in Table 1.
- Part of speech (PS): Number of occurrences of parts of speech of words in a phrase. We used J-tag [5] to extract part-of-speech information.
- Large-scale Japanese thesaurus (LT): Large-scale Japanese thesaurus is a large lexical database of Japanese. Nouns, verbs, adjectives and adverbs are grouped into sets of cognitive synonyms (synsets), each expressing a distinct concept. Synsets are interlinked by means of conceptual-semantic and lexical relations.

We constructed the nod-generation models by using J.48 [], which implements a decision tree in Weka [3] and evaluated the accuracy of the models and the effectiveness of each type of linguistic information. The class was a binary value as to whether a nod occurred.

We used 24-fold cross validation using a leave-one-person-out technique with the data for the 24 participants. We evaluated how well a participant's nods could be estimated with an estimator generated only from data of other people. As shown in Table 2, the performance of the model using only LP, WP, DA, PS, or LT was higher than the chance level. However, the performance of the model using BW with an F-score of 0.423 was lower than chance level. This suggests that BW, which was used in a previous study [8,9], was not useful for generating nods in our experiment. The model using LT had the highest performance among the models using only LP, WP, BM, DA, PS, or LT, with an F-score of 0.588. This suggests that LT is most useful in generating nods. In addition, the performance of mode using all information was higher than that using LT. This suggests that using several types of language information is useful to generate nods.

4   Discussion

The experimental results indicate that BW is not useful to generate nods. The reason is that the amount of data is not large, and it is thought that learning cannot be done well because the frequency of each word included in the learning data is too small. Because it is costly and difficult to collect a massive amount of multimodal data, BW is not effective. On the other hand, LT is most effective since such information is super classified rather than the word; therefore, the possibility that it could be learned well even with a relatively small amount of

Table 1. Dialogue act labels

Label	Dialogue Act	Label	Dialogue Act
DA0	Greeting	DA15	Question (habit)
DA1	Provision	DA16	Question
DA2	Self-disclosure (fact)		(desire)
DA3	Self-disclosure	DA17	Question
	(experience)		(plan)
DA4	Self-disclosure (habit)	DA18	Question
DA5	Self-disclosure		(evaluation)
	(positive preference)	DA19	Question
DA6	Self-disclosure		(other)
	(negative preference)	DA20	Question
DA7	Self-disclosure		(Yourself)
	(neutral preference)	DA21	Sympathy
DA8	Self-disclosure	DA22	Non-sympathy
	(desire)	DA23	Confirmation
DA9	Self-disclosure (plan)	DA24	Proposal
DA10	Self-disclosure	DA25	Repeat
	(other)	DA26	Paraphrase
DA11	Acknowledgment	DA27	Approval
DA12	Question	DA28	Thanks
	(information)	DA29	Apology
DA13	Question (fact)	DA30	Filler
DA14	Question	DA31	Admiration
	(experience)	DA32	Other

data can be considered. All linguistic information is useful to generate nods. This suggests that using several type of language information has the potential to generate nonverbal behaviors.

In this research, we used language information extracted from a unit of phrase and tried to determine whether nodding occurs in the phrase. We did not consider

Table 2. Evaluation result of generation models.

Used feature values	Precision	Recall	F-score
Chance level	0.500	0.500	0.500
LP	0.561	0.556	0.558
WP	0.526	0.528	0.527
BW	0.353	0.527	0.423
DA	0.513	0.533	0.523
PS	0.521	0.528	0.524
LT	0.614	0.538	0.578
ALL	0.578	0.599	0.590

the time-sequential information as a feature. We plan to focus on time-sequential linguistic information to generate nods. Furthermore, we would like to work on constructing a model that can generate the detailed parameters of nods such as number and depth.

5   Conclusion

We constructed a highly accurate head-nod-generation model using natural Japanese language. In this research, we focused on various text analyzed linguistic information such as dialog acts, parts of speech, a large-scale Japanese, and word positions in sentences. In an experiment, we found that our estimation model these types of information outperformed that using bag-of-words information alone. We also found that a model using all types of linguistic information is most useful to generate nods. These results indicate that several types of linguistic information have the potential to be strong predictors to generate nods automatically.

In the future, we will focus on time-sequential linguistic information to generate nods. We would like to work on constructing a model that can generate the detailed parameters of nods such as number and depth. Furthermore, we plan to construct a model for generating the occurrence timing of nods within an utterance and a model for generating nonverbal behaviors such as gaze, which is important for turn management [11–13,19–22], expression of conversational engagement [15–18,30], and body posture.

References

1. Beskow, J., Granstrom, B., House, D.: Visual correlates to prominence in several expressive modes. In: Proceedings of INTERSPEECH (2006)
2. BirdWhistell, R.L.: Kinesics and Context. University of Pennsylvania Press, Philadelphia (1970)
3. Bouckaert, R.R., Frank, E., Hall, M.A., Holmes, G., Pfahringer, B., Reutemann, P., Witten, I.H.: WEKA-experiences with a java open-source project. J. Mach. Learn. Res. 11, 2533–2541 (2010)
4. Busso, C., Deng, Z., Grimm, M., Neumann, U., Narayanan, S.: Rigid head motion in expressive speech animation: analysis and synthesis. In: IEEE Transactions on Audio, Speech, and Language Processing, pp. 1075–1086 (2007)
5. Fuchi, T., Takagi, S.: Japanese morphological analyzer using word cooccurrence -Jtag. In: Proceedings of International Conference on Computational Linguistics, pp. 409–413 (1998)
6. Graf, H.P., Cosatto, E., Strom, V., Huang, F.J.: Visual prosody: facial movements accompanying speech. In: Proceedings of IEEE International Conference on Automatic Face and Gesture Recognition, pp. 381–386 (2002)
7. Higashinaka, R., Imamura, K., Meguro, T., Miyazaki, C., Kobayashi, N., Sugiyama, H., Hirano, T., Makino, T., Matsuo, Y.: Towards an open-domain conversational system fully based on natural language processing. In: Proceedings of International Conference on Computational Linguistics, pp. 928–939 (2014)

8. Ishi, C.T., Haas, J., Wilbers, F.P., Ishiguro, H., Hagita, N.: Analysis of head motions and speech, and head motion control in an android. In: Proceedings of IEEE/RSJ International Conference on Intelligent Robots and Systems, pp. 548–553 (2007)
9. Ishi, C.T., Ishiguro, H., Hagita, N.: Analysis of prosodic and linguistic cues of phrase finals for turn-taking and dialog acts. In: Proceedings of International Conference of Speech and Language, pp. 2006–2009 (2006)
10. Ishi, C.T., Ishiguro, H., Hagita, N.: Head motion during dialogue speech and nod timing control in humanoid robots. In: Proceedings of ACM/IEEE International Conference on Human-Robot Interaction, pp. 293–300 (2010)
11. Ishii, R., Kumano, S., Otsuka, K.: Multimodal fusion using respiration and gaze behavior for predicting next speaker in multi-party meetings. In: Proceedings of the International Conference on Multimodal Interaction (ICMI 2015), pp. 99–106 (2015)
12. Ishii, R., Kumano, S., Otsuka, K.: Predicting next speaker using head movement in multi-party meetings. In: Proceedings of the International Conference on Acoustics, Speech, and Signal Processing (ICASSP 2015), pp. 2319–2323 (2015)
13. Ishii, R., Kumano, S., Otsuka, K.: Analyzing gaze behavior during turn-taking for estimating empathy skill level. In: Proceedings of the 19th ACM International Conference on Multimodal Interaction (ICMI 2017), pp. 365–373 (2017)
14. Ishii, R., Kumano, S., Otsuka, K.: Prediction of next-utterance timing using head movement in multi-party meetings. In: Proceedings of the 5th International Conference on Human Agent Interaction (HAI 2017), pp. 181–187 (2017)
15. Ishii, R., Miyajima, T., Fujita, K., Nakano, Y.: Avatar's Gaze Control to Facilitate Conversational Turn-Taking in Virtual-Space Multi-user Voice Chat System. In: Gratch, J., Young, M., Aylett, R., Ballin, D., Olivier, P. (eds.) IVA 2006. LNCS (LNAI), vol. 4133, p. 458. Springer, Heidelberg (2006). https://doi.org/10.1007/11821830_47
16. Ishii, R., Nakano, Y.I.: Estimating user's conversational engagement based on gaze behaviors. In: Prendinger, H., Lester, J., Ishizuka, M. (eds.) IVA 2008. LNCS (LNAI), vol. 5208, pp. 200–207. Springer, Heidelberg (2008). https://doi.org/10.1007/978-3-540-85483-8_20
17. Ishii, R., Nakano, Y.I.: An empirical study of eye-gaze behaviors: towards the estimation of conversational engagement in human-agent communication. In: Proceedings of the 2010 Workshop on Eye Gaze in Intelligent Human Machine Interaction (EGIHMI 2010), pp. 33–40 (2010)
18. Ishii, R., Nakano, Y.I., Nishida, T.: Gaze awareness in conversational agents: estimating a user's conversational engagement from eye gaze. ACM Trans. Interact. Intell. Syst. 3(2), 1–25 (2013). Article No. 11
19. Ishii, R., Otsuka, K., Kumano, S., Yamamoto, J.: Predicting of who will be the next speaker and when using gaze behavior in multiparty meetings. ACM Trans. Interact. Intell. Syst. 6(1), 4 (2016)
20. Ishii, R., Otsuka, K., Kumano, S., Yamamoto, J.: Using respiration to predict who will speak next and when in multiparty meetings. ACM Trans. Interact. Intell. Syst. 6(2), 20 (2016)
21. Ishii, R., Otsuka, K., Kumano, S., Matsuda, M., Yamato, J.: Predicting next speaker and timing from gaze transition patterns in multi-party meetings. In: Proceedings of the International Conference on Multimodal Interaction (ICMI 2013), pp. 79–86 (2013)

22. Ishii, R., Otsuka, K., Kumano, S., J., Yamato, S.: Analysis and modeling of next speaking start timing based on gaze behavior in multi-party meetings. In: Proceedings of the International Conference on Acoustics, Speech, and Signal Processing, pp. 694–698 (2014)
23. Ishii, R., Shinohara, Y., Nakano, Y., Nishida, T.: Combining multiple types of eye-gaze information to predict user's conversational engagement. In: Proceedings of the 2011 Workshop on Eye Gaze in Intelligent Human Machine Interaction (EGIHMI 2011) (2011)
24. Iwano, Y., Kageyama, S., Morikawa, E., Nakazato, S., Shirai, K.: Analysis of head movements and its role in spoken dialogue. In: Proceedings of International Conference on Spoken Language, pp. 2167–2170 (1996)
25. Munhall, K.G., Jones, J.A., Callan, D.E., Kuratate, T., Vatikiotis-Bateson, E.: Visual prosody and speech intelligibility: head movement improves auditory speech perception. Psychol. Sci. **15**(2), 133–137 (2004)
26. Koiso, H., Horiuchi, Y., Tutiya, S., Ichikawa, A., Den, Y.: An analysis of turn-taking and backchannels based on prosodic and syntactic features in Japanese map task dialogs. Lang. Speech **41**, 295–321 (1998)
27. Lohse, M., Rothuis, R., Gallego-Pérez, J., Karreman, D.E., Evers, V.: Robot gestures make difficult tasks easier: the impact of gestures on perceived workload and task performance. In: Proceedings of the SIGCHI Conference on Human Factors in Computing Systems (CHI 2014), pp. 1459–1466 (2014)
28. McBreen, H.M., Jack, M.A.: Evaluating humanoid synthetic agents in e-retail applications. IEEE Trans. Syst. Man Cybern. A Syst. Hum. **31**, 5 (2001)
29. Meguro, T., Higashinaka, R., Minami, Y., Dohsaka, K.: Controlling listening-oriented dialogue using partially observable Markov decision processes. In: Proceedings of International Conference on Computational Linguistics, pp. 761–769 (2010)
30. Nakano, Y.I., Ishii, R.: Estimating user's engagement from eye-gaze behaviors in human-agent conversations. In: Proceedings of the 15th International Conference on Intelligent User Interfaces (IUI 2010), pp. 139–148 (2010)
31. Ooko, R., Ishii, R., Nakano, Y.I.: Estimating a user's conversational engagement based on head pose information. In: Vilhjálmsson, H.H., Kopp, S., Marsella, S., Thórisson, K.R. (eds.) IVA 2011. LNCS (LNAI), vol. 6895, pp. 262–268. Springer, Heidelberg (2011). https://doi.org/10.1007/978-3-642-23974-8_29
32. Sakai, K., Ishi, C.T., Minato, T., Ishiguro, H.: Online speechdriven head motion generating system and evaluation on a tele-operated robot. In: Proceedings of IEEE International Symposium on Robot and Human Interactive Communication, pp. 529–534 (2015)
33. Maynard, S.: Interactional functions of a nonverbal sign: head movement in Japanese dyadic casual conversation. J. Pragmat. **11**, 589–606 (1987)
34. Maynard, S.: Japanese Conversation: Self-contextualization Through Structure and Interactional Management. Ablex Publishing Corporation, Norwood (1989)
35. Watanabe, T., Danbara, R., Okubo, M.: Effects of a speech-driven embodied interactive actor interactor on talker's speech characteristics. In: Proceedings of IEEE International Workshop on Robot-Human Interactive Communication, pp. 211–216 (2003)
36. Wittenburg, P., Brugman, H., Russel, A., Klassmann, A., Sloetjes, H.: ELAN: a professional framework for multimodality research. In: Proceedings of International Conference on Language Resources and Evaluation (2006)
37. Yehia, H.C., Kuratate, T., Vatikiotis-Bateson, E.: Linking facial animation, head motion and speech acoustics. J. Phonetics **30**(3), 555–568 (2002)

Reducing Interactions in Social Media: A Mathematical Approach

Erick López Ornelas[(⊠)]

Information Technology Department,
Universidad Autónoma Metropolitana – Cuajimalpa, Mexico City, Mexico
elopez@correo.cua.uam.mx

Abstract. Networks arise in many diverse contexts, ranging from web pages and their links, computer networks, and social networks interactions. The modelling and mining of these large-scale, self-organizing systems is a broad effort spanning many disciplines. This article proposes the use of morphological operators, based on Mathematical Morphology, to simplify a set of interactions in Twitter, that can be considered as a complex network. By applying these techniques, it is possible to simplify the social network and thus identify important interactions, communities and actors in the network. Reducing interactions is then, a crucial step for simplify and understand the networks. An analysis based on the visualization of the simplification was carried out to verify the pertinence of the proposed technique.

Keywords: Morphological operators · Social media network
Network simplification · Twitter · Information extraction

1 Introduction

Now a day, social media networks such as Twitter, LinkedIn and Facebook, are provide a cheaper way for user to share ideas, exchange information and stay connected with people. The use of social media applications on mobile devices achieves rapid growth in social media network users and leads to generate vast amount of user generated content.

This large user base and their discussions produces huge amount of user generated data. Such social media data comprises rich source of information which is able to provide tremendous opportunities for companies to effectively reach out to a large number of audience.

With the current popularity of these Social media networks (SMN), there is an increasing interest in their measurement and modelling. In addition to other complex networks properties, SMN exhibit shrinking distances over time, increasing average degree, and bad spectral expansion.

Unlike other complex networks such as the web graph, models for SMN are relatively new and lesser known. In this kind of networks, models may help detect, simplify and classify communities, and better clarify how news and gossip is spread in social networks.

© Springer International Publishing AG, part of Springer Nature 2018
G. Meiselwitz (Ed.): SCSM 2018, LNCS 10914, pp. 392–402, 2018.
https://doi.org/10.1007/978-3-319-91485-5_30

Network simplification can provide benefits to applications of various domains and for suggesting like-minded people to user which are still unknown to him/her.

An important practical problem in social networks is to simplify the network of users based on their shared content and relationship with other users.

In other hand, Mathematical morphology is generally studied as an aspect of image processing [1]. As digital images are usually two-dimensional arrangements of pixels, where spatial relationships between elements of the image are essential features.

Mathematical Morphology is a theory that studies the decomposition of lattice operators in terms of some families of elementary lattice operators [2]. When the lattices are considered as a multidimensional graph (e.g. Social Media Network), the elementary operators can be characterized by structuring functions. The representation of structuring functions by neighborhood graphs is a powerful model for the construction of morphological operators.

This article proposes the use of morphological operators, based on Mathematical Morphology, to simplify a set of interactions in a complex social network. By applying these morphological operators, it is possible to simplify the social network and thus execute important queries in the network.

The structure of the article is as follows. In Sect. 2 similar work is showed, then in Sect. 3 we explain the essentials of mathematical morphology and network representation. The morphological operators are then explained in Sect. 4. An example of this simplification is carry out in this section. Then in Sect. 5 a query modeling is explained. The conclusions and directions for further work are given in the final section.

2   Related Work

Similar work has been conducted on simplifying networks. In [3], the authors developed 3 different algorithms. The first decomposes a large network into some smaller sub-networks, generally overlapped. The remaining two carry out simplification based on commute times within the network. The algorithms produce a multilayered representation. All three algorithms use their simplified representations to perform matching between the input network.

In [4], the author uses a simplification algorithm to generate simplified network for input into their network layout algorithms. The network is not visualized and presented to the user as a way to help them better understand the network. Instead, a series of progressively simplified networks are used to guide the positioning of the nodes in the network.

Additional simplification algorithms have been proposed to assist in robot path planning [5], classifying the topology of surfaces [6], and improving the computational complexity and memory requirements of dense graph processing algorithms [7].

Simplification may also be accomplished through the clustering technique. In [8], authors define one such clustering algorithm for better visualizing the community structure of network graphs. The kind of graphs that they are targeting with this technique are those that would have a naturally occurring community structure.

Some authors present another approach to visually simplifying large scale graphs [9]. They have developed different methods for randomly sampling a network and using that sampling to construct the visual representation of the complex network.

In [10], the authors focus on providing metrics for simplifying graphs that represent specific network topologies. The goal of this work is to simplify and visualize complex network graphs while maintaining their semantic structures. Although network topologies are certainly reasonable candidates for these visualization techniques, there are other sets of graph data that could greatly benefit from simplification techniques for visualization. Their general approach is similar to ours in that they are using characteristics of the graphs that frequently occur, using Morphological operators some physical characteristics are kept and some nodes "grow" and some "shrink" in order to obtain a better simplification. This process will be explained in next section.

3 Mathematical Morphology and Networks

The study on Mathematical morphology, started at the end of the Sixties and was proposed by Matheron and Serra [11]. Mathematical morphology rests in the study of the geometry and forms; the principal characteristic of the morphological operations is image segmentation and conservation of the principal features forms of the nodes [1].

Despite its origin, it was recognized that the roots of this theory were in algebraic theory, notably the framework of complete lattices [12]. This allows the theory to be completely adaptable to non-continuous spaces, such as graphs and networks. For a survey of the state of the art in mathematical morphology, we recommend [13].

The algebraic basis of Mathematical morphology is the lattice structure and the morphological operators act on lattices [2]. In other words, the morphological operators map the elements of a first lattice to the elements of second one (which is not always the same as the first one). A lattice is a partially ordered set such that for any family of elements, we can always find a least upper bound and a greatest lower bound (called a supremum and an infimum). The supremum (resp., infimum) of a family of elements is then the smallest (greatest) element among all elements greater (smaller) than every element in the considered family.

The supremum is given by the union and the infimum by the intersection. A morphological operator is then a mapping that associates to any subset of nodes another subset of nodes. Similarly, given a graph, one can consider the lattice of all sub sets of vertices [14] and the lattice of all subsets of edges. The supremum and infimum in these lattices are also the union and intersection. In some cases, it also interesting to consider a lattice whose elements are graphs, so that the inputs and outputs of the operators are graphs.

The algebraic framework of morphology relies mostly on a relation between operators called adjunction [2]. This relation is particularly interesting, because it extends single operators to a whole family of other interesting operators: having a dilation (resp., an erosion), an (adjunct) erosion (resp., a dilation) can always be derived, then by applying successively these two adjunct operators a closing and an opening are obtained in turn (depending which of the two operators is first applied), and finally composing this opening and closing leads to alternating filters.

Firstly, they are all increasing, meaning that if we have two ordered elements, then the results of the operator applied to these elements are also ordered, so the morphological operators preserve order. Additionally, the following important properties hold true:

- the dilation (resp., erosion) commutes under supremum (resp., infimum);
- the opening, closing and alternating filters are indeed morphological filters, which means that they are both increasing and idempotent (after applying a filter to an element of the lattice, applying it again does not change the result);
- the closing (resp., opening) is extensive (resp., antiextensive), which means that the result of the operator is always larger (resp., smaller) than the initial object;

In the graph $G(V, E)$, if the vertex (V_i) of the graph constitutes the digital grid and its neighbors their interactions, then the process compares and affects the interaction value of v_i on the graph constructed using the morphological transformations. These transformations are the core of the simplification.

The principle of the growing/shrinking the graph consists in transform the $G(v)$ value by affecting the nearest interaction value $val(v_i)$ present among the v neighbor's nodes. The new graph v_n is then the result of the fusion of nodes. To carry out this transformation, the morphological operations on the graph are applied and a loop is generated until the reach of one threshold parameter.

Let us assume that we have a flat structuring element that corresponds to the neighbor's nodes Structuring Element $(SE \equiv NE(v))$. Then the eroded graph $\varepsilon(G(v))$ is defined by the infimum of the values of the function in the neighborhood [15]:

$$\varepsilon(G(v)) = \{\wedge\, G(v_i), v_i \in N_E(v) \cup \{v\}\}$$

Dilation $\delta(G(v))$ is similarly defined by the supremum of the neighboring values and the value of $G(v)$ as

$$\delta(G(v)) = \{\vee\, G(v_i), v_i \in N_E(v) \cup \{v\}\}$$

Classically, opening γ is defined as the result of erosion followed by dilation using the same SE

$$\gamma(G(v)) = \delta(\varepsilon(G(v)))$$

Similarly, closing φ is defined as the result of dilation followed by erosion with the same SE

$$\varphi(G(v)) = \varepsilon(\delta(G(v)))$$

The geometrical action of the openings and closings transformations, $\gamma(G(v))$ and $\varphi(G(v))$ respectively, produce a growing/shrinking of the graph. Of course, this fusion process can be regulated using parameters for the opening and closing, but also we can regulate the growing depending on the information that we need to compare. The graph has to be updated to keep aggregating the different nodes always applying the morphological transformations of $\gamma(G(v))$ and $\varphi(G(v))$ until their parameter value is reached. In Figs. 1 and 2, some morphological operations are shown. We can see the difference applying different morphological operators on the same graph.

Fig. 1. a. Random graph selected. b. Eroded $\varepsilon(G(v))$. c. Dilated *graph $\delta(G(v))$*.

Fig. 2. a. Random graph selected. b. Opening transformation $\gamma(G(v))$. c. Closing transformation $\varphi(G(v))$.

4 Social Media Simplification

In this article, as an experiment, we show a set of interactions on Twitter. This information was extracted from Twitter and explores a trend topic appeared in México. The hashtag is *#noalaeropuerto*, and it was arising from the corruption scandals generated in construction of the new airport in Mexico City in September 2017. Among the multiple elements of analysis, we decided to use morphological operators in order to simplify the original network, the study was made taking into account the characteristics of each node for their simplification.

The information concentrated in Fig. 3 corresponds to the extraction carried out on September 5, 2017. In the network are represented 3399 nodes (tweeters) and 5502 arcs (interactions) that were made between them. This is a complex interaction network, so it is important to simplify it in order to perform a better analysis of the interactions that were generated in the social network.

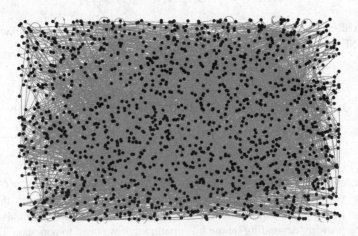

Fig. 3. Complete random graph selected with 3399 nodes and 5502 interactions.

4.1   Twitter Network Representation

At the most abstract level, given a Social Media network $G = (V, E)$, where G stands for the whole network, V stands for the set of all vertices and E for the set of all edges, each Social Media interaction can be defined as a subgraph of the network comprising a set $V_C \subseteq V$ of Social Media entities that are associated with a common element of interest.

This element can be as varied as a topic, a real-world person, a place, an event, an activity or a cause.

For instance, in the case of Twitter network, one can consider the set of vertices V to comprise the users, mentions, tweet content, tweet favorites and retweets, i.e. $V = \{U, M, Tc, Tf, Rt\}$. The edges in such an application would comprise the set of followed, followers, tweet number, image profile and location, $E = \{Fd, Fs, Tn, Ip, L\}$.

Even if we can use all these characteristics to apply morphological operators, we have decided to only use 3 node elements to carry out the simplification. These elements are the mentions, number of favorites and retweets represented by $V = (M, Tf, Rt)$.

4.2   Nodes Reduction

The principle of the union of nodes consists in transform the $G(v)$ value by affecting the nearest Tf value $val(v_i)$ present among the v neighbors, and the grouping process is the union of nodes $(v_i \cup v_j = v_n)$. The new node v_n is then the result of the fusion of nodes. To carry out this transformation, the morphological operations on the graph are applied.

Let us assume that we have a flat structuring element that corresponds to the neighborhood Structuring Element $(SE \equiv N_E(v))$. Then the eroded graph $\varepsilon(G(v))$ is defined by the infimum of the values of the function in the neighborhood and represents the minimum value found on the neighbors [16]:

$$\varepsilon(G(v)) = min\{G(v_i), v_i \in N_E(v) \cup \{v\}\}$$

Dilation $\delta(G(v))$ is similarly defined by the supremum of the neighboring values and the value of $G(v)$ and it is represented by the maximum value found on the neighbors as

$$\delta(G(v)) = max\{G(v_i), v_i \in N_E(v) \cup \{v\}\}$$

Classically, opening γ is defined as the result of erosion followed by dilation using the same SE

$$\gamma(G(v)) = \delta(\varepsilon(G(v)))$$

Similarly, closing φ is defined as the result of dilation followed by erosion with the same SE

$$\varphi(G(v)) = \varepsilon(\delta(G(v)))$$

The geometrical action of the openings and closings transformations, $\gamma(G(v))$ and $\varphi(G(v))$ respectively, produce a growing or shrinking of the selected graph. Of course, this fusion process can be regulated using parameters for the opening and closing, but also we can regulate the fusion depending on the mentions, tweet favorites or retweets. The graph has to be updated to keep aggregating the different nodes always applying the morphological transformations of $\gamma(G(v))$ and $\varphi(G(v))$ until their parameter value is reached.

For merging two adjacent nodes in a graph, certain V conditions should be verified. We can define some mention parameters that condition the difference between these values of two adjacent nodes that can be aggregate at the opening and closing operations. These parameters are called the minimal mention parameter d_1 and the maximal mention threshold d_2. To use them, we should calculate, in a first time, the mention differences in the graph. So we calculate $d_1(G(V_i), max(G(V)))$, the difference between the maximum value of mentions in the neighboring nodes, and $d_2(G(V_i), min(G(V)))$ the minimal difference. If the maximal mention parameter is higher than d_1, the opening operation $\gamma(G(v_i))$ does not merge nodes. Also, if the minimal mention parameter is higher than d_2, the closing operation $\varphi(G(V_i))$ does not merge nodes. A loop is he required to perform all the necessary aggregations for the simplification of the graph. In Figs. 4, 5, 6, 7, 8 and 9 we show the simplification process in different steps.

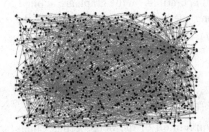 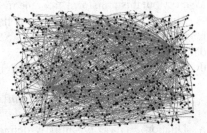

Fig. 4. Graph iterations = 5 Morphological operations applied = γ and φ, $d_1 = 50$, $d_2 = 30$,   Nodes = 1535, Interactions = 2045.

Fig. 5. Graph iterations = 20 Morphological operations applied = γ and φ, $d_1 = 50$, $d_2 = 30$,   Nodes = 1381, Interactions = 1715.

It is interesting to note that simplification is more significant in the first iterations, usually in the first 5 iterations, which is normal regarding the parameters d_1 and d_2 used. Then, the parameters do not cause so much effect and the simplification rate remains stable.

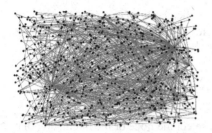

Fig. 6. Graph iterations = 35 Morphological operations applied = γ and φ, $d_1 = 50$, $d_2 = 30$, Nodes = 925, Interactions = 1251.

Fig. 7. Graph iterations = 50 Morphological operations applied = γ and φ, $d_1 = 50$, $d_2 = 30$, Nodes = 509, Interactions = 563.

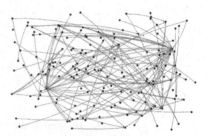

Fig. 8. Graph iterations = 65 Morphological operations applied = γ and φ, $d_1 = 50$, $d_2 = 30$, Nodes = 221, Interactions = 330.

Fig. 9. Final graph. Graph iterations = 80 Morphological operations applied = γ and φ, $d_1 = 50$, $d_2 = 30$, Nodes = 118, Interactions = 105.

5 Information Extraction

The final node characteristics is calculated using the final graph after the use of the morphological transformations of $\gamma(G(v))$ and $\varphi(G(v))$. These characteristics *{C}* are then stored separately in a database, which is useful to make meanly queries.

These features *{C}* called "metadata" [17, 18] characterizing each node are then stored and handled separately.

There are two different features extracted from the graph: (i) "node properties", that are specific to each node (user name, friends, followed, followers, tweet number, image profile and location, etc.) and (ii) "interaction characteristics", that describe the tweet (mentions, tweet content, tweet favorites and retweets).

To extract information from the final graph we decided to use *Cypher* [19] that is a declarative graph query language that allows for expressive and efficient querying and updating of the graph store. *Cypher* is a simple but powerful query language. This language allows you to focus on the domain instead of getting lost in graph database access.

Being a declarative language, *Cypher* focuses on the clarity of expressing "what" to retrieve from a graph, not on "how" to retrieve it. The query via the *Cypher* query language would be:

MATCH (*L:Node{name: 'Final-node'}*)
WHERE *tweet.likes > 200 and tweet.mentions > 6*
RETURN (*Oid*)

So to retrieve meanly information, we have to select the node or the interaction that are interested to us. We have tried different queries using *Cypher* with very interesting results. As an example we show network and the node retrieved using this Query (Fig. 10).

Fig. 10. Automatic selection using Cypher language.

6 Conclusions

Complex graphs, contains thousands of nodes of high degree, that are difficult to visualize. Displaying all of the nodes and edges of these graphs can create an incomprehensible cluttered output. We have presented a simplification algorithm that may be applied to a complex graph issue of a Social Media Network, in our case a Twitter network, in order to produce a simplify graph. This simplification was proposed by the use of morphological operators, that are based on Mathematical morphology.

We have represented the Social Media Network as a complete Lattice. In doing this, mathematical morphology has been developed in the context of a relation on a set. It has been shown that this structure is sufficient to define all the basic operations: dilation, erosion, opening and closing, and also to establish their most basic properties.

The simplification of the graph provides an approach to visualizing the fundamental structure of the graph by displaying the most important nodes, where the importance may be based on the topology of the graph and their interaction. The simplification algorithm consists in the iterative use of Opening and Closing operations that cause a growing or shrinking effect in the graph. This process generates the simplification of the network.

As can be seen from this paper, SMA have been and currently are a prominent topic in Network's analysis and simplification. With the advent of the so-called Big Data, we expect this trend to be extremely persistent [20, 21] and promising for opening novel research directions. Indeed, there is no reason to restrict the application of this simplification process the very same ideas we have described here to networks. Any kind of data can be processed with these techniques, notably, image processing.

In the proposed method based on morphological simplification, we have realized that the parameterization is a fundamental step and we must dedicate special attention to get a homogeneous simplification of nodes and interactions. This parameterization leads the process of simplification by physical characteristics of the graph, and permits to interpret in a simple way the relationship among the nodes, interactions and all characteristics associated.

Future work may be to design query-based simplification techniques that would take user's interests into account when simplifying a network. It would also be interesting to combine different network abstraction techniques with network simplification, such as a graph compression method to aggregate nodes and interactions. Also, it would be interesting to develop additional importance metrics, as well as testing and evaluating our approach with other simplification methods and on other types of graphs.

References

1. Serra, J., Soille, P.: Mathematical Morphology and Its Applications to Image Processing. Kluwer Academic Publishers (1994)
2. Serra, J.: A lattice approach to image segmentation. J. Math. Imaging Vis. **24**, 83–130 (2006)
3. Qiu, H., Hancock, E.: Graph simplification and matching using commute times. Pattern Recogn. **40**, 2874–2889 (2007)
4. Frishman, Y., Tal, A.: Multi-level graph layout on the GPU. Trans. Visual. Comput. Graphics **13**(6), 1310–1319 (2007)
5. Rizzi, S.: A genetic approach to hierarchical clustering of euclidean graphs. Pattern Récogn. **2**, 1543–1545 (1998)
6. Ban, T., Sen, D.: Graph based topological analysis of tessellated surfaces. In: Proceedings of Eighth ACM Symposium on Solid Modeling and Applications, pp. 274–279, (2003)
7. Kao, M., Occhiogrosso, N., Teng, S.: Simple and efficient graph compression schemes for dense and complement graphs. J. Comb. Optim. **2**(4), 351–359 (1998)
8. Girvan, M., Newman, M.E.J.: Community Structure in Social and Biological Networks. PNAS **99**(12), 7821–7826 (2002)
9. Davood R., Stephen, C.: Effectively visualizing large networks through sampling. In: 16th IEEE Visualization (VIS 2005), p. 48 (2005)
10. Gilbert, A., Levchenko, K.: Compressing network graphs. In: Proceedings of LinkKDD 2004 (2004)
11. Serra, J.: Image Analysis and Mathematical Morphology. Academic Press, London (1982)
12. Heijmans, H.: Morphological Image Operators. Advances in Electronics and Electron Physics Series. Academic Press, Boston (1994)
13. Najman, L., Talbot, H.: Mathematical Morphology: From Theory to Applications. ISTE-Wiley (2010)
14. Vincent, L.: Graphs and mathematical morphology. Sig. Process. **16**, 365–388 (1989)
15. Flouzat, G., Amram, O.: Segmentation d'images satelitaires par analyse morphologique spatiale et spectrale. Acta Stereologica **16**, 267–274 (1997)

16. Zanoguera, F.: Segmentation interactive d'images fixes et séquences séquences vidéo basée sur des hiérarchies de partitions, Thèse de Doctorat en Morphologie Mathématique, ENSMP, 13 décembre (2001)

17. Amous, I., Jedidi, A., Sèdes, F.: A contribution to multimedia document modeling and organizing. In: Bellahsène, Z., Patel, D., Rolland, C. (eds.) OOIS 2002. LNCS, vol. 2425, pp. 434–444. Springer, Heidelberg (2002). https://doi.org/10.1007/3-540-46102-7_45

18. Chrisment, C., Sedes, F.: Multimedia Mining, A Highway to Intelligent Multimedia Documents. Multimedia Systems and Applications Series, vol. 22, p. 245. Kluwer Academic Publisher (2002). ISBN 1-4020-7247-3

19. Holzschuher, F., Peinl, R.: Performance of graph query languages: comparison of cypher, gremlin and native access in Neo4j. In: Proceedings of the Joint EDBT/ICDT Workshops, pp. 195–204. ACM (2013)

20. Lafferty, J., Zhai, C.: Document language models, query models, and risk minimization for information retrieval. ACM SIGIR Forum **51**(2), 251–259 (2017)

21. Zwaenepoel, W.: Really big data: analytics on graphs with trillions of edges. In: LIPIcs-Leibniz International Proceedings in Informatics, vol. 70 (2017)

Pointing Estimation for Human-Robot Interaction Using Hand Pose, Verbal Cues, and Confidence Heuristics

Andrew Showers[✉] and Mei Si[✉]

Rensselaer Polytechnic Institute, Troy, NY 12180, USA
{showea,sim}@rpi.edu

Abstract. People utilize pointing directives frequently and effortlessly. Robots, therefore, will need to interpret these directives in order to understand the intention of the user. This is not a trivial task as the intended pointing direction rarely aligns with the ground truth pointing vector. Standard methods require head, arm, and hand pose estimation inhibiting more complex pointing gestures that can be found in human-human interactions. In this work, we aim to interpret these pointing directives by using the pose of the index finger in order to capture both simple and complex gestures. Furthermore, this method can act as a fall-back for when full-body pose information is not available. This paper demonstrates the ability of a robot to determine pointing direction using data collected from a Microsoft Kinect camera. The finger joints are detected in 3D-space and used in conjugation with verbal cues from the user to determine the point of interest (POI). In addition to this, confidence heuristics are provided to determine the quality of the source information, whether verbal or physical. We evaluated the performance of using these features with a support vector machine, decision tree, and a generalized model which does not rely on a learning algorithm.

Keywords: Pointing · Object detection · Object localization
Social interaction

1 Introduction

Pointing directives are commonly used in social interactions to express an object of interest. In order for social robots to have meaningful interactions with people, interactions like these will need to be implemented. Unfortunately, systematically interpreting this seemingly trivial behavior has proved to be a difficult task. The difficulty stems from the number of factors influencing the pointing vector. A common solution in minimizing this error is to detect the head pose [7,8] or the gaze direction [2]. However, this correction requires additional information which may not be available due to occlusion. Ideally the system would have a more reliable fall-back method than using the hand pose exclusively.

© Springer International Publishing AG, part of Springer Nature 2018
G. Meiselwitz (Ed.): SCSM 2018, LNCS 10914, pp. 403–412, 2018.
https://doi.org/10.1007/978-3-319-91485-5_31

In this work, we propose to integrate contextual information for helping decoding the users' intentions. In order to frame the pointing direction within a given context, we use natural language processing as well as confidence heuristics to determine the accuracy of the information sources. The verbal component can extend to a large range of descriptors, including object names, attributes, and spatial references. However, in order to use this information the system needs to be aware of the user's environment. Fortunately, with the advancement of object detection and localization, this task is becoming feasible. By using object detection to learn about the environment, we can now leverage verbal cues as a way to filter out irrelevant objects.

We claim that the proposed method which combines these channels of information can interpret pointing directives significantly better than with the hand pose alone. Separately, the hand pose is too unreliable as shown by previous studies and using language alone may lead to overly complex and unnatural descriptions. By fusing these channels together we can attain a reliable and natural way of handling this social interaction.

2   Related Work

Previous studies have attempted to use the vector from the head-hand line [9], head-finger line [10], and forearm direction [4] as the pointing direction. Unfortunately these models lack accuracy. An example of this inaccuracy can be seen in the study performed by Abidi et al. [2] where the vector formed from the elbow to the hand/finger provides unsatisfactory results when trying to guide a robot. They found that using the pointing vector as the only feature led to 38% satisfaction. Improvements were made by combining the pointing vector with the gaze direction. However this relies on additional features that may not be easy for a robot to obtain or susceptible to noise.

In order to address this, a study performed by Ueno et al. [5] created a calibration procedure where the user points to the camera directly before pointing to the intended object. Offsets are determined by the difference in the vector formed from the eyes to the fingertip and the vector from the eyes to the target object. This method achieved accuracy between 80%–90% depending on the camera position.

An alternative approach is to apply learning algorithms for finding a transformation from a set of observed spacial features to the intended pointing direction. Droeschel et al. [3] used Gaussian Process Regression (GPR) to dramatically reduce the error of the desired pointing direction angles in comparison to the source vector. Similarly, Shukla et al. [6] implemented a probabilistic model to learn the pointing direction by providing hand pose images. This approach can be extended to other subproblems as well. Pateraki et al. [7] implemented a least squares matching technique to minimize error with estimating the head pose.

3   Proposed Method

We propose a computational model that takes verbal and visual input. The verbal input has multiple sources of ambiguity which can be observed. The message could become corrupted due to a failure in the Automatic Speech Recognition (ASR), leading to a loss of keywords and potentially the intention of the user. Even when the message is accurately retrieved, the information provided can have varying levels of description. The level of detail required to resolve this ambiguity is too verbose in comparison to what is naturally practiced. Furthermore, identifying what information to search for is difficult given free expression. Similarly, visual input is susceptible to noise and may lack contextual information not captured by the hand pose. Since both inputs carry inherent ambiguity, we adjust the system's trust on them dynamically.

3.1   Preprocessing

Knowledge of the objects and their respective position within the room is required for resolving the pointing directive. This knowledge is acquired with the use of an additional camera at a known location in the room. The image captured is fed into a Single Shot MultiBox Detector (SSD) [1] for object detection and localization. The model is pre-trained on the VOC0712 dataset containing 20 classes. The localization only provides a bounding box on the input image which does not provide the required spatial information of the objects (Fig. 1).

Fig. 1. Layout of the room

While the object position can be inferred from the Kinect depth stream, we used AR markers as a simple alternative. Given the limited classes within the model, we manually encoded 3 AR tags as object type "book". To provide some attribute information to the objects, we specified the color of each object by hand (Fig. 2).

Fig. 2. Object detection and localization with a Single Shot MultiBox Detector

3.2 Capture Process

In order to begin the capture process, subjects must use explicit keywords such as "get" or "bring" in their pointing directive. Messages were captured using a microphone and converted to text using a natural language processing library. To estimate the pointing direction, we take the line formed by the base and tip joints of the index finger. This information was captured by the Kinect using the Metrilus Aiolos finger tracking library. Spatial instances were collected until a minimum sample size, in our case 10, was reached in order to minimize noise.

3.3 Feature Vector

Physical Information. The pointing direction, i.e. the line formed between the base and tip joints, is not explicitly added to the feature vector to minimize over-fitting. Instead, the single direction is abstracted to a range of possible directions using the 5 closest objects as features. This gives a general sense of the direction while being too general to attribute to a specific sample instance.

Verbal Information. From the original message, the following categories are extracted; object names, object attributes, and spacial cues. Specifically, the number of references to each object type (e.g. "book") as well as boolean values for the presence of spacial and attribute keywords. For this paper, we define spacial cues to be keywords which may provide spacial information of the desired object, such as "bottom" and "left".

Confidence Heuristics. A useful metric to provide is the accuracy of the information source. To accomplish this, confidence heuristics were implemented for both verbal input and visual inputs. For the verbal component, weights were applied to the mention of known attributes in the workspace, the number of spacial cues, and the message length and structure. For the physical component, confidence was inversely related with the variance of the capture buffer.

3.4   Model

In order to examine the importance of these features we trained a Support Vector Machine (SVM) and Decision Tree (DT) to classify the POI. The SVM used a linear kernel and the DT was limited to a max depth of 8. By tuning these models and examining the accuracy we were able to construct a simpler and more generalized model using the features provided to the learning algorithms.

4   Experiments and Results

4.1   Experiment Setup

In order to evaluate the accuracy of our approach, we conducted the experiments in an indoor environment. There are 9 test objects in the room; 3 bottles, 3 books, and 3 chairs. The subject stands in a fixed location in the room, approximately 8–10 ft away from the objects. The object positions relative to the Kinect are shown in Fig. 3. In this configuration, objects are non-unique and similar objects are spatially near one-another (e.g. the stack of books).

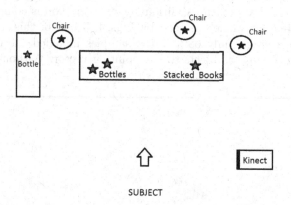

Fig. 3. Position of objects, Kinect, and the subject.

4.2   Procedure

For this experiment, we used 10 subjects, 7 male and 3 female. In order to determine the POI, the features provided are fed into the SVM and DT classifiers respectively. For the generalized model we first eliminate unrelated objects based on the verbal information provided, if any. Then by using this reduced set of objects, the POI is either assigned to the object closest to the line formed by the pointing vector. In the event of low confidence on either of the input channels, the low confidence channel is discarded. We evaluated the accuracy of this method under the following conditions:

1. Pointing Independently – not allowed to talk
2. Subject can only express the object name and some attribute while pointing.
3. Subject can only express spacial information in relation to other objects while pointing.
4. Subject can freely express any information about the object while pointing.

For condition (4), subjects were required to point to 4 objects of their choosing. This relaxed condition was used to learn what subjects naturally say while identifying objects. For each of the other conditions, subjects were given a randomly generated permutation of the objects in which to point to. Given this setup, subjects were required to give 31 pointing gestures each, providing a total of 310 samples for the entire experiment.

4.3   Results

In order to evaluate the performance of the learning algorithms, the samples were divided into two datasets for training (70%) and testing (30%) respectively. Figures 4 and 5 show the accuracy of the SVM and DT on the test dataset. Notice Fig. 6 uses the entire dataset as there is no learning algorithm involved with the generalized approach. Additionally, each method is compared to using the pointing direction alone. As seen by Figs. 4, 5 and 6, this method is very inaccurate. On average, only 68% of the samples included the desired object when taking the 5 closest objects to the pointing vector. Without any learning

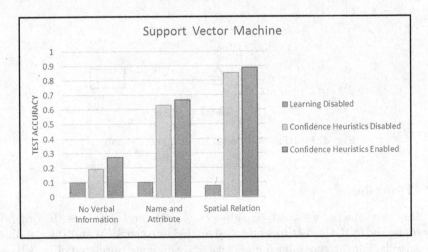

Fig. 4. Accuracy of the support vector machine on the test dataset

algorithm enabled the closest object is taken as the POI which leads to an accuracy rate no better than guessing.

The SVM was able to take advantage of additional information while managing to generalize well to new samples. In all cases the inclusion of confidence estimates for the data source led to improvements in performance. Even in the presence of no verbal information the accuracy significantly improved from 10% to 19% without heuristics. This improvement suggests patterns in the detected object positions vs their actual locations. Since the positions are detected using a combination of the SSD object localization and AR tags, these positions are susceptible to noise. Accuracy raised further to 27% with confidence heuristics enabled suggesting the existence of noise while estimating the pointing direction, such as potential dead-zones where the Kinect struggles to accurately detect the hand.

Introducing verbal information containing the object name and attribute significantly raised accuracy as would be expected. Accuracy improved to 63% and 67% with heuristics disabled and enabled respectively. Interestingly, complex verbal cues encoding spatial relations led to the highest performance in this experiment. Accuracy reached 85% with spatial cues and raised further to 89% with the inclusion of confidence metrics.

In comparison to the SVM, the DT was able to achieve similar performance for samples in which there was limited or no verbal information provided. With heuristics disabled, the DT was able to achieve accuracies of 12%, 67%, 77% for conditions (1), (2), and (3). Unfortunately decision trees are prone to overfitting, which can be seen by the decreased accuracy when introducing confidence heuristics. The tree incorporated the heuristics but the model poorly generalized to new samples. With heuristics enabled, the DT was able to achieve accuracies of 27%, 60%, 56% for conditions (1), (2), and (3). As seen in Fig. 5, as the

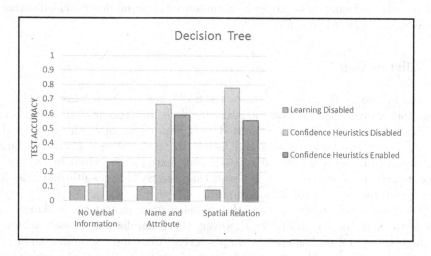

Fig. 5. Accuracy of the decision tree on the test dataset

complexity of the verbal information increased, the tendency to over-fit based on the confidence heuristics increased as well.

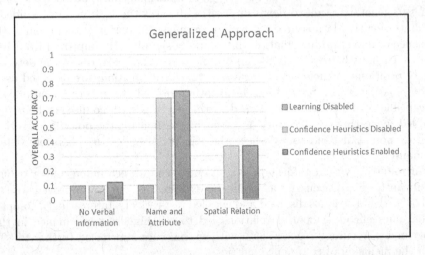

Fig. 6. Accuracy of the generalized approach on the entire dataset

The generalized approach was able to achieve accuracies of 10%, 70%, 37% for conditions (1), (2), and (3) when confidence heuristics were excluded. By including this metric, accuracies improved to 12%, 74% and 37% respectively. This model performed similarly well to the learning algorithms given simple verbal information but was unable to fully utilize more complex messages. Furthermore, the model was only able to take advantage of the heuristics to a small degree. This gives an insight to the difficulty of utilizing the abundance of information provided without the assistance of a learning algorithm.

5   Discussion

As seen by the results provided above, the SVM was able to achieve the highest overall accuracy. The DT, however, was unable to perform as well due to over-fitting. Additionally, the DT is limited to a max depth of 8 which forces the model to focus on key features and ignore some of the less significant ones. In comparison, the SVM can utilize as many features as desired so long as it benefits the classification process. Given this, the SVM is a suitable model for fusing multiple channels of information.

The generalized approach shows the challenge of extracting and making good use of the information available. However, the generalized approach achieved the highest accuracy when given a simple verbal command containing an object name and attribute. This result can be explained by the ease in which this information can be extracted, whereas spatial relations can take on many forms and is harder to analyze. Failure to resolve part of the complex message can result in the information being diminished if not entirely lost.

By implementing a suitable learning algorithm we have shown that the combination of the hand pose and varying levels of verbal information can lead to improved reliability in the absence of other body pose information required to accurately resolve pointing gestures.

6 Future Work

Future improvements can be made by eliminating the need to preprocess the room. The AR markers used here simplify capturing object positions but could be replaced by segmenting objects in the Kinect depth stream in conjunction with the object localization provided by the SSD. Such a setup may require the use of multiple cameras in order to maximize the number of objects in view.

Additionally, the generalized approach could be improved for complex verbal messages by implementing a learning algorithm that transforms the verbal input into a set of observed features. If successful, the model could reach similar accuracies as the SVM while still being applicable to dynamic environments.

References

1. Liu, W., Anguelov, D., Erhan, D., Szegedy, C., Reed, S., Fu, C.-Y., Berg, A.C.: SSD: single shot multibox detector. In: Leibe, B., Matas, J., Sebe, N., Welling, M. (eds.) ECCV 2016. LNCS, vol. 9905, pp. 21–37. Springer, Cham (2016). https://doi.org/10.1007/978-3-319-46448-0_2
2. Abidi, S., Williams, M., Johnston, B.: Human pointing as a robot directive. In: 2013 8th ACM/IEEE International Conference on Human-Robot Interaction (HRI), Tokyo, pp. 67–68 (2013)
3. Droeschel, D., Stckler, J., Behnke, S.: Learning to interpret pointing gestures with a time-of-flight camera. In: 2011 6th ACM/IEEE International Conference on Human-Robot Interaction (HRI), Lausanne, pp. 481–488 (2011)
4. Li, Z., Jarvis, R.: Visual interpretation of natural pointing gestures in 3D space for human-robot interaction. In: 2010 11th International Conference on Control Automation Robotics & Vision, Singapore, pp. 2513–2518 (2010)
5. Ueno, S., Naito, S., Chen, T.: An efficient method for human pointing estimation for robot interaction. In: 2014 IEEE International Conference on Image Processing (ICIP), Paris, pp. 1545–1549 (2014)
6. Shukla, D., Erkent, O., Piater, J.: Probabilistic detection of pointing directions for human-robot interaction. In: 2015 International Conference on Digital Image Computing: Techniques and Applications (DICTA), Adelaide, SA, pp. 1–8 (2015)
7. Pateraki, M., Baltzakis, H., Trahanias, P.: Visual estimation of pointed targets for robot guidance via fusion of face pose and hand orientation. In: 2011 IEEE International Conference on Computer Vision Workshops (ICCV Workshops), Barcelona, pp. 1060–1067 (2011)
8. Stiefelhagen, R., et al.: Enabling multimodal humanrobot interaction for the Karlsruhe humanoid robot. IEEE Trans. Rob. **23**(5), 840–851 (2007)

9. Burger, B., Ferran, I., Lerasle, F., Infantes, G.: Two-handed gesture recognition and fusion with speech to command a robot. Auton. Robots **32**(2), 129–147 (2012)
10. Yamamoto, Y., Yoda, I., Sakaue, K.: Arm-pointing gesture interface using surrounded stereo cameras system. In: Proceedings of the 17th International Conference on Pattern Recognition, ICPR 2004, vol. 4, pp. 965–970 (2004)

Opportunistic Work-Rest Scheduling
for Productive Aging

Han Yu[1,2(✉)], Chunyan Miao[1,2], Lizhen Cui[3], Yiqiang Chen[4],
Simon Fauvel[1], and Qiang Yang[5]

[1] School of Computer Science and Engineering,
Nanyang Technological University, Singapore, Singapore
{han.yu,ascymiao,sfauvel}@ntu.edu.sg
[2] Joint NTU-UBC Research Centre of Excellence in Active Living for the Elderly,
Nanyang Technological University, Singapore, Singapore
[3] School of Software Engineering, Shandong University, Jinan, China
clz@sdu.edu.cn
[4] School of Computer and Control Engineering,
University of Chinese Academy of Sciences, Beijing, China
yqchen@ict.ac.cn
[5] Department of Computer Science and Engineering,
Hong Kong University of Science and Technology, Kowloon, Hong Kong
qyang@cse.ust.hk

Abstract. Crowdsourcing platforms are interesting being leveraged by
senior citizens for productive aging activities. Algorithmic management
(AM) approaches help crowdsourcing systems leverage workers' intelligence and effort in an optimized manner at scale. However, current AM
approaches generally overlook the human aspects of crowdsourcing workers. This prevailing notion has resulted in many existing AM approaches
failing to incorporate rest-breaks into the crowdsourcing process to help
workers maintain productivity and wellbeing in the long run. To address
this problem, we extend the Affective Crowdsourcing (AC) framework
to propose the *Opportunistic Work-Rest Scheduling (OWRS)* approach.
It takes into account information on a worker's mood, current workload and desire to rest to produce dynamic work-rest schedules which
jointly minimize collective worker effort output while maximizing collective productivity. Compared to AC, OWRS is able to operate under
more diverse mood–productivity mapping functions. As it is a fully distributed approach with time complexity of $O(1)$, it can be implemented
as a personal assistant agent for workers. Extensive simulations based
on a large-scale real-world dataset demonstrate that OWRS significantly
outperforms three baseline scheduling approaches in terms of conserving
worker effort while achieving superlinear collective productivity. OWRS
establishes a framework which accounts for workers' heterogeneity to
enhance their experience and productivity.

Keywords: Mood · Productive aging · Crowdsourcing · Scheduling

© Springer International Publishing AG, part of Springer Nature 2018
G. Meiselwitz (Ed.): SCSM 2018, LNCS 10914, pp. 413–428, 2018.
https://doi.org/10.1007/978-3-319-91485-5_32

1   Introduction

Crowdsourcing over the world-wide web has been connecting large groups of people together to enable organizations to tap into their knowledge, expertise, time or resources for productive purposes [5]. Senior citizens in Japan and Singapore are beginning to turn to crowdsourcing platforms for productive aging activities including both paid work and volunteering [13,25]. As the likes of such technical platforms as Amazon's Mechanical Turk (mTurk), 99designs and Uber grow in popularity, a large amount of worker behaviour data have been accumulated to enable artificial intelligence (AI) approaches to collaborate with human workers to streamline various aspects of crowdsourcing. Increasingly, software algorithms have been designed to evaluate [10,16], optimize [4,18] and delegate [27,29–31] work among diverse populations of crowdsourcing workers. These software algorithms that assume managerial roles in order to allow crowdsourcing companies to oversee many workers in an optimized manner are referred to as *Algorithmic Management (AM)* [14].

While AM helps crowdsourcing systems leverage workers' intelligence and diverse skills effectively and efficiently, the human aspects of crowdsourcing workers are often overlooked [2]. This prevailing notion has resulted in the workers' experience during the crowdsourcing processes not being taken into consideration by existing AM approaches. The benefits of rest on worker productivity and wellbeing have been well-established in management science [3,11]. Yet, most existing AM approaches do not account for allocating time to rest for workers in their computational processes. So far, only one published AM research proposed an *Affective Crowdsourcing (AC)* framework to dynamically leverage changes in workers' mood in view of other situational factors to schedule rest-breaks in an optimized manner [32].

The AC framework is based on the assumption that changes in workers' mood result in linear changes in their productivity. However, existing empirical research from social sciences only support that mood is positively correlated to productivity [20]. Due to individual heterogeneity, it is possible that the mapping function between a worker's mood and his productivity follows non-linear forms. This assumption limits the applicability of AC in practical situations.

To address this limitation, we extend the AC framework to propose the *Opportunistic Work-Rest Scheduling (OWRS)* approach. It takes into account local information on a worker's mood, current workload and desire to rest to produce dynamic work-rest schedules. By scheduling work during favourable situations and scheduling rest during unfavourable situations, OWRS jointly minimizes crowdsourcing workers' collective effort output while maximizing their collective productivity. Compared to AC, OWRS is able to operate under more diverse mood–productivity mapping functions including *step*, *logarithmic* and *exponential* forms. As it is a distributed approach with time complexity of $O(1)$, it can be implemented as a personal assistant agent [21,24,26] responsible for scheduling activities.

Through extensive numerical experiments based on a large-scale real-world dataset released by Taobao.com, we demonstrate that OWRS outperforms three baseline scheduling approaches. It significantly conserves the collective effort output while making the smallest sacrifice in terms of task completion rates, thereby achieving superlinear collective productivity [23]. The proposed approach establishes a framework which accounts for workers' heterogeneity to enhance their user experience and productivity.

2 Related Work

In the algorithmic management for crowdsourcing literature, existing approaches concerning the human aspects mostly focus on providing incentives to motivate workers to work harder or inducing good mood among workers to enhance their productivity.

In [15], the authors studied how paid crowdsourcing workers perform compared to volunteers. They observed that per-task payment schemes induce workers to work faster but with reduced quality of work, while wage-based payment schemes cause workers to work more slowly but produce better quality results. This work laid the empirical foundation for computational incentive mechanisms for crowdsourcing which dynamically trade off precision, recall, speed, and total attention on tasks. However, it does not provide insight into how to advise workers to rest efficiently. Following [15], the incentives for crowdsourcing workers are further linked to their past performance so that their competence in the system can be used as a sanctioning mechanism to determine their future expected payoff [16]. This approach induces workers to produce higher quality results, but does not help workers determine the most opportune time to rest either.

In [17], the authors studied the relationship between crowdsourcing workers' mood and their creative output capacities. They then proposed two approaches for enhancing worker performance in creative tasks: affective priming and affective pre-screening. Their results confirmed that workers who are in a good mood exhibit enhanced creativity, and the approaches induce good mood among workers to create favourable conditions for them to work.

Designing work-rest schedules for workers is an important topic in management science as well [1]. Most of the work in this field focuses on maximizing the workers' wellbeing as a function of how long they have been working without considering how to simultaneously achieving how collective productivity. Outcomes from this branch of research are static schedules at a coarse granularity (e.g. rest breaks over a day).

[32] is the most closely related work to this research. It proposed an Affective Crowdsourcing framework to model the dynamics of crowdsourcing work while accounting for the effect of workers' mood on their productivity to produce optimal work-rest schedules. This paper extends [32] to propose new solutions to this optimization problem under more diverse general forms of mood–productivity mapping functions to deal with individual differences more effectively.

3   The OWRS Approach

In this section, we discuss the model and formulation of the optimization objective function to minimize effort output while maximizing collective productivity. Then, we provide efficient distributed solutions to this optimization problem under three mood–productivity mapping functions.

3.1   Preliminaries

In the context of defined task crowdsourcing systems such as mTurk, the system model consists of a set of workers $\mathbf{W} = \{w_1, w_2, ..., w_N\}$, a set of competence $\mathbf{C} = \{c_1, c_2, ..., c_K\}$ workers may possess, and a set of tasks $\mathbf{T} = \{\tau_1, \tau_2, ..., \tau_L\}$ at any given time slot t.

- *Competence* refers to know how in performing certain types of tasks. It is quantified as a scalar value in the range of $[0, 1]$ with 0 indicating no competence in a given type of tasks and 1 indicating completely competent in performing a given type of tasks. It can be treated as the probability for a worker to complete a task requiring c_m with quality acceptable to the crowdsourcer.
- *Workers* are associated with personal profiles. Each profile contains information on a worker i's competence in all available types of tasks, his pending workload queue $q_i(t)$, and his maximum productivity $\mu_i^{\max}$ which indicates the maximum workload he can complete in a given time slot t (the granularity of t depends on requirements by specific crowdsourcing platforms). The ground truth of the competence and the maximum productivity values may not be directly observable. With analytics tools such as Turkalytics [9], they can be tracked and approximated over time.
- *Tasks* are associated with task profiles. Each profile contains information on the competence or combinations of competence required perform a task τ_l, the reward for performing the task r_{τ_l} and the deadline, d_{τ_l} (expressed in terms of number of time slots) before which this task must be completed.

The number of workers, tasks and competence available in a crowdsourcing system at different time slots may differ.

Following well-established models of the dynamics of the pending task queue for each worker [27,28], we can express a worker i's pending workload at time slot $t + 1$ as:

$$q_i(t + 1) = \max[0, q_i(t) + \lambda_i(t) - \mu_i(t)] \tag{1}$$

where $\lambda_i(t)$ is the new workload entering i's backlog queue during time slot t; and $\mu_i(t)$ is the workload completed by i during time slot t. Based on the 'happy-productive worker' thesis [8,20], $\mu_i(t)$ can be expressed as a function of i's mood and effort output as:

$$\mu_i(t) = \mu(\xi_i(t), m_i(t)) \tag{2}$$

where $\xi_i(t) \in [0,1]$ is the normalized effort output by worker i during time slot t; and $m_i(t) \in [0,1]$ is i's mood during time slot t, where 1 denotes the most positive mood. Based on this function, the amount of work done by a worker depends on his current mood and how much effort he spends on the tasks. Following the same Lyapunov optimization [19] based derivation process explained in [32], the optimization objective function is:

Minimize:

$$\frac{1}{T} \sum_{t=0}^{T-1} \sum_{i=1}^{N} [\phi_i \xi_i(t) - q_i(t) \mu(\xi_i(t), m_i(t))] \tag{3}$$

Subject to:

$$0 \leqslant \xi_i(t) \leqslant 1, \forall i, \forall t \tag{4}$$

$$0 \leqslant \mu(\xi_i(t), m_i(t)) \leqslant \mu_i^{\max}, \forall i, \forall t \tag{5}$$

where $\mu_i^{\max}$ is worker i's maximum productivity based on historical observations tracked by the crowdsourcing system. By minimizing Eq. (3), we simultaneously minimize the time-averaged total worker effort output while maximizing the time-averaged collective task throughput (i.e. productivity).

Opportunistic Work-Rest Scheduling

In order to derive solutions for the above objective function, we require a computational model on how mood influences productivity. Existing literature generally agrees that good mood leads to higher productivity. However, due to individual heterogeneity, the mood–productivity mapping function may take various forms. In the proposed OWRS approach, we make provisions for three possible general forms of the mood–productivity mapping function which follow the "happy-productive worker" thesis. They are:

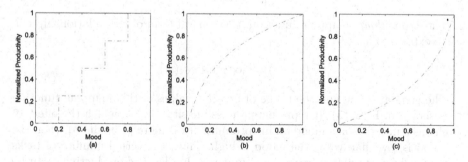

Fig. 1. General forms of the mood–productivity mapping function investigated in this paper: (a) *Step* mapping; (b) *Logarithmic* mapping; and (c) *Exponential* mapping.

1. The *Step* mapping function: the general shape and the range of possible values of this relationship is shown in Fig. 1(a). Let $f(m_i(t))$ be a mapping function between mood and productivity with 5 steps:

$$f(m_i(t)) = \begin{cases} 0 & , \text{ if } m_i(t) \in m_{vl} \\ 0.25 & , \text{ if } m_i(t) \in m_l \\ 0.5 & , \text{ if } m_i(t) \in m_m \\ 0.75 & , \text{ if } m_i(t) \in m_h \\ 1 & , \text{ if } m_i(t) \in m_{vh}. \end{cases} \quad (6)$$

There can be multiple potential ways to implement the $\mu(\xi_i(t), m_i(t))$ for this situation. In this paper, we adopt an approach which divides the mood value between 0 to 1 into 5 equal partitions labelled as "very low (vl)", "low (l)", "medium (m)", "high (h)", and "very high (vh)" denoted as $\{m_{vl}, m_l, m_m, m_h, m_{vh}\}$, respectively. In this case, changes in a worker's mood result in five different levels of productivity. It represents a possible situation in which small changes in a person's mood do not result in significant change in productivity immediately. The general trend of productivity increasing with mood follows the "happy-productive worker" thesis. As the number of steps approaches infinity, the trend will approach the linear mood–productivity mapping function. If $\xi_i(t) = 0$, $\mu(0, m_i(t)) = 0$ regardless of $\mu(0, m_i(t)$. Let $\Psi_i(t) = \phi_i - q_i(t)\mu(1, m_i(t))$, Eq. (3) can be optimized by determining the values of $\xi_i(t)$ for all workers at a given time slot as:

$$\xi_i(t) = \begin{cases} 0, & \text{if } \Psi_i(t) \geqslant 0 \text{ or } m_i(t) \in m_{vl} \\ \min\left[1, \frac{q_i(t)}{f(m_i(t))\mu_i^{\max}}\right], & \text{otherwise.} \end{cases} \quad (7)$$

Thus, $\mu(\xi_i(t), m_i(t))$ can be computed as:

$$\mu(\xi_i(t), m_i(t)) = \begin{cases} 0, & \text{if } \Psi_i(t) \geqslant 0 \text{ or } m_i(t) \in m_{vl} \\ \lfloor f(m_i(t))\mu_i^{\max} \rfloor, & \text{otherwise.} \end{cases} \quad (8)$$

2. The *Logarithmic* mapping function: the general form of this relationship can be expressed as:

$$f(m_i(t)) = \log_2(1 + m_i(t)). \quad (9)$$

The general shape and the range of possible values of this mapping function is shown in Fig. 1(b). It represents a possible situation in which the effect of changes in a person's mood on his productivity is more pronounced when the mood is low than when the mood is high. Thus, the actual number of tasks that can be completed by worker i under such a level of productivity can be expressed as:

$$\mu(\xi_i(t), m_i(t)) = \lfloor \mu_i^{\max} \log_2(1 + m_i(t)\xi_i(t)) \rfloor. \quad (10)$$

In this case, to minimize Eq. (3), we compute its first order derivative with respect to $\xi_i(t)$ for an individual worker i as:

$$\frac{d}{d\xi_i(t)}[\phi_i\xi_i(t) - q_i(t)\mu_i^{\max}\log_2(1 + m_i(t)\xi_i(t))]$$

$$= \phi_i - \frac{q_i(t)m_i(t)\mu_i^{\max}}{[1 + m_i(t)\xi_i(t)]\ln 2}. \tag{11}$$

By equating Eq. (11) to 0, we obtain the effort output $\xi_i(t)$ by worker i at time slot t which can minimize the objective function Eq. (3) as follows:

$$\xi_i(t) = \min\left[\max\left[0, \frac{q_i(t)\mu_i^{\max}}{\phi_i\ln 2} - \frac{1}{m_i(t)}\right], 1\right]. \tag{12}$$

By substituting Eq. (12) into Eq. (10), we obtain the solutions $\mu_i(t)$.

3. The *Exponential* mapping function: this general form of this relationship between mood and productivity can be expressed as:

$$f(m_i(t)) = \frac{1}{e - 1}[e^{m_i(t)} - 1]. \tag{13}$$

The general shape and the range of possible values of this mapping function is shown in Fig. 1(c). It represents a possible situation in which the effect of changes in a person's mood on his productivity is more pronounced when the mood is high than when the mood is low. Thus, the actual number of tasks that can be completed by worker i under such a level of productivity can be expressed as:

$$\mu(\xi_i(t), m_i(t)) = \left\lfloor \frac{\mu_i^{\max}}{e - 1}[e^{m_i(t)\xi_i(t)} - 1] \right\rfloor. \tag{14}$$

In this case, to minimize Eq. (3), we compute its first order derivative with respect to $\xi_i(t)$ for each worker i as:

$$\frac{d}{d\xi_i(t)}\left\{\phi_i\xi_i(t) - \frac{q_i(t)\mu_i^{\max}}{e - 1}[e^{m_i(t)\xi_i(t)} - 1]\right\}$$

$$= \phi_i - \frac{q_i(t)\mu_i^{\max}m_i(t)e^{m_i(t)\xi_i(t)}}{e - 1}. \tag{15}$$

By equating Eq. (15) to 0, we obtain the effort output $\xi_i(t)$ by each worker i at time slot t which can minimize the objective function Eq. (3) as follows:

$$\xi_i(t) = \min\left[\max\left[0, \frac{1}{m_i(t)}\ln\left(\frac{\phi_i(e - 1)}{q_i(t)\mu_i^{\max}m_i(t)}\right)\right], 1\right]. \tag{16}$$

By substituting Eq. (16) into Eq. (14), we obtain the solutions $\mu_i(t)$.

Algorithm 1. OWRS

Require: $q_i(t)$, ϕ_i, $\mu_i^{\max}$, $m_i(t)$ for a worker i at a given time t; the mood–productivity
    mapping function variable $M \in \{$Step, Logarithmic, Exponential$\}$.
1: **if** $M =$ Step **then**
2:    Compute $\mu_i(t)$ according to Eq. (8);
3: **else if** $M =$ Logarithmic **then**
4:    Compute $\mu_i(t)$ according to Eq. (10);
5: **else if** $M =$ Exponential **then**
6:    Compute $\mu_i(t)$ according to Eq. (14);
7: **end if**
8: **return** $\mu_i(t)$;

The proposed OWRS approach is summarized by Algorithm 1. It implements
computationally the intuition that the more pending tasks in a worker's backlog queue, the less emphasis placed on conserving effort output and the higher
the worker's current mood, the worker should spend more effort on completing
pending tasks during the current time slot (subject to the physical limitations
of the worker's maximum productivity). The recommendations from OWRS are
given to workers in the form of the number of tasks he should complete in a
given time slot so as to make it actionable for the worker to follow. As OWRS
only requires a worker's local information as inputs, it can be implemented as a
distributed algorithm with time complexity of $O(1)$, making it highly scalable.

4   Experimental Evaluation

To evaluate the performance of OWRS under realistic settings, we compare it
against three baseline approaches via extensive simulations. To do so, we synthe-
size a population of crowdsourcing worker agents whose performance characteristics are based on the *Tianchi* dataset[1] released by Taobao.com. This real-world
dataset allows us to construct realistic scenarios. The simulations enable us to
study the performance of OWRS under different situations.

4.1   Experiment Settings

Released by Taobao.com, this real-world dataset contains information regarding the competence and maximum productivity (as measured by the maximum
number of tasks that can be completed per time slot) of 5,547 workers. The
distributions of productivity and competence values in the dataset are shown
in Fig. 2. It can be observed that the majority of the workers are able to process 10 tasks per time slot while a relatively small percentage of them are able
to complete up to 100 tasks per time slot. The distribution of workers' competence roughly follows a Normal distribution with mean value centred around 0.5.
This dataset allows us to construct realistic simulated scenarios to facilitate the
evaluation of OWRS under different conditions.

---

[1] http://dx.doi.org/10.7303/syn7373599.

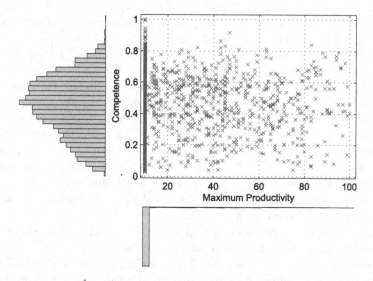

Fig. 2. The distribution of ground truth maximum productivity and competence values for 5,547 worker agents in the simulations.

The four comparison approaches included in the simulations are:

1. The *MaxEffort (ME)* approach: under this approach, a worker i always works as long as there are tasks in his backlog queue regardless of their mood. Subject to there being enough tasks in the backlog queue, worker i exerts up to $\xi_i(t) = 1$ effort at time slot t.
2. The *Mood Threshold (MT)* approach: under this approach, a worker i works whenever $m_i(t) \geqslant \theta_1$ and there are tasks in his backlog queue. $\theta_1 \in [0, 1]$ is a predetermined mood threshold used by MT. Subject to there being enough tasks in his backlog queue, worker i exerts up to $\xi_i(t) = 1$ effort at time slot t; otherwise, worker i rests.
3. The *Mood and Workload threshold (MW)* approach: this approach jointly considers a worker i's current mood and workload to determine how much effort to exert. Whenever $q_i(t)\mu(1, m_i(t)) \geqslant \mu_i^{\max}\mu(1, \theta_2)$, worker i exerts up to $\xi_i(t) = 1$ effort subject to there being enough tasks in his backlog queue; otherwise, worker i rests. $\theta_2 \in [0, 1]$ is a predetermined mood threshold used by MW.
4. The *OWRS* approach proposed in this paper.

The task allocation approach used for all comparison approaches is DRAFT [27] which dynamically distributes tasks among workers in a situation-aware manner in order to avoid over concentration of workload. At each time slot, DRAFT takes each worker i's current competence and workload as inputs to determine how many new tasks to assign to i (i.e. DRAFT determines $\lambda_i(t)$ for all i and t). The principle implemented by DRAFT is that the higher a worker i's competence and the lower his current workload, the more tasks should be assigned to him. In practice, the DRAFT approach can be replaced by other similar approaches which determine $\lambda_i(t)$.

In the simulations, the value of ϕ is varied between 5 and 100 in increments of 5. The values of θ_1 and θ_2 are varied between 0.1 and 1 in increments of 0.05. As the workload is measured in relation to the collective task throughput of the generated worker agent population, we compute the maximum collective task throughput as $\Omega = \sum_{i=1}^{N} r_i \mu_i^{max}$. In this equation, r_i is a worker agent i's competence, and $N = 5,547$ in our experiments. We adopt the concept of load factor (LF) from [30] to denote the overall workload placed on the crowdsourcing system. It is computed as the ratio between the number of tasks allocated among the worker agents during time slot t, $W_{req}(t)$, and the maximum collective productivity Ω of the system (i.e. $LF = \frac{W_{req}(t)}{\Omega}$). We vary LF between 5% to 100% in 5% increments in the simulations.

In the simulations, the mood for each worker i during time slot t, $m_i(t)$, is randomly generated in the range of $[0,1]$ following a uniform distribution. This eliminates the possibility for any of the four approaches to predict a worker agent's future mood, thereby focusing the experimental comparisons on the effectiveness of the scheduling strategies. Under each LF setting, the simulation is run for $T = 1,000$ time slots. Task deadlines are randomized following a uniform distribution. On average, a task must be completed within 3 time slots after it is assigned to a worker agent. The simulations are executed under three mood–productivity mapping functions $M \in \{Step, Logarithmic, Exponential\}$.

The performances of the four approaches are compared using the following metrics:

1. Time-averaged worker effort output:

$$\bar{\xi} = \frac{1}{TN} \sum_{t=0}^{T-1} \sum_{i=1}^{N} \xi_i(t). \tag{17}$$

The smaller the value of $\bar{\xi}$, the better the performance.
2. Time-averaged task completion rate:

$$\bar{\mu} = \frac{1}{T} \sum_{t=0}^{T-1} \frac{\sum_{i=1}^{N} \mu_i(t)}{L}. \tag{18}$$

The higher the value of $\bar{\mu}$, the better the performance.
3. Efficiency:

$$\bar{\epsilon} = \frac{\bar{\mu}}{\bar{\xi}}. \tag{19}$$

The higher the value of $\bar{\epsilon}$, the better the performance.

4.2   Results and Analysis

As worker agents adopting the ME approach consistently achieve the highest task completion rate at the expense of not resting whenever there are pending tasks, we use ME as the reference for comparing the performance of the other three approaches under different LF, ϕ, θ_1 and θ_2 settings. In sub-figures (a)–(c), (f)–(h) and (k)–(m) of Fig. 4, each data point for the OWRS approach is averaged over all ϕ values; whereas each data point for the MT and MW approaches is averaged over all θ_1 and θ_2 values, respectively. In sub-figures (d), (i) and (n) of Fig. 4, each data point is averaged over all LF values. In sub-figures (e), (j) and (o) of Fig. 4, each data point is averaged over all experimental settings. As the experiments are simulations, exact values are less important than relative performance.

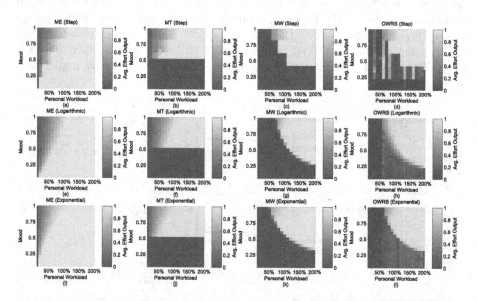

Fig. 3. The average effort output by worker agents experiencing different mood and workload conditions following each of the four comparison approaches under LF = 50%, $\theta_1 = 0.5$, $\theta_2 = 0.5$ and $\phi = 50$. (a)–(d) *Step* mood–productivity mapping; (e)–(h) *Logarithmic* mood–productivity mapping; and (i)–(l) *Exponential* mood–productivity mapping.

Figure 3 contains snapshots of the average effort output by worker agents experiencing different mood and workload conditions following each of the four comparison approaches under different mood–productivity mapping functions. The snapshots are taken under the settings $\theta_1 = 0.5$, $\theta_2 = 0.5$ and $\phi = 50$ with LF = 50%. The personal workload values in the x-axis are computed as $\frac{q_i(t)}{\mu_i^{\max}}$ for each worker agent i during each time slot t in the simulations. Under LF = 50%, DRAFT ensures that the workload in any worker agent's pending task queue never exceeds $200\% \times \mu_i^{\max}$.

Under ME (Figs. 3(a), (e) and (i)), worker agents always exert enough effort to complete as many tasks as allowed by their respective $\mu_i^{\max}$ limits. Since higher mood values results in higher productivity in general, worker agents under ME can spend less effort when their mood is high and workload is below 100%. Worker agents under MT follows the same effort output patterns as those under ME except for when their mood drops below the θ_1 which is 0.5 in this case study (Figs. 3(b), (f) and (j)). When $m_i(t) < \theta_1$, MT worker agents rest. Under MW (Figs. 3(c), (g) and (k)), worker agents jointly consider if their mood and workload pass a combined threshold before determining how much effort to exert. When the worker agents experience either high mood, high workload or both, they exert as much effort as required by the situation. For the remaining conditions, MW worker agents rest. Under OWRS (Figs. 3(d), (h) and (l)), worker agents also jointly consider their mood and workload when making effort output decisions. However, as OWRS is not a threshold-based approaches, worker agents do not rest completely. Instead, even when worker agents are experiencing low mood or low workload, OWRS worker agents still exert effort, albeit at lower levels. This allows OWRS worker agents to achieve comparatively higher task completion rates.

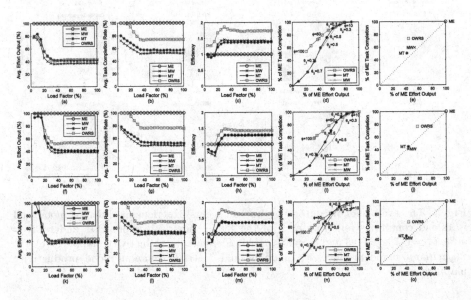

Fig. 4. Experiment results: (a)–(e) *Step* mapping; (f)–(j) *Logarithmic* mapping; and (k)–(o) *Exponential* mapping.

Figures 4(a)–(e) show the simulation results under the step mood–productivity mapping function. The time-averaged effort output $\bar{\xi}$ achieved by all four approaches are shown in Fig. 4(a). Compared to ME, MT, MW and OWRS achieve significant saving in effort as LF increases. This is partially due to the DRAFT task allocation approach used in the simulations. When LF is low, tasks are mostly concentrated on worker agents with high competence. In this case, the task backlog on individual workers who have been allocated tasks tend to be relatively high, which makes scheduling approaches allocate less time

for these workers to rest in order to meet task deadlines. As LF increases and
the workload is spread more evenly among a larger segment of the worker agent
population, there are more opportunities for scheduling approaches to slot in
opportunistic rest sessions. The $\bar{\xi}$ values achieved by MT, MW and OWRS stabilize around 40% with MW recommending the lowest worker effort output and
OWRS recommending the highest. The time-averaged task completion rate $\bar{\mu}$
achieved by all four approaches are shown in Fig. 4(b). It can be observed that
OWRS achieves the second highest $\bar{\mu}$ values which stabilize around 70% and
are 27% and 40% higher than MT and MW, respectively. Figure 4(c) shows the
performance of all four approaches in terms of efficiency. The efficiency of ME is
constant as is does not attempt to conserve worker agents' effort. OWRS achieves
the highest efficiency which stabilizes at around 50% higher than that achieved
by ME as it makes the smallest sacrifice in time-averaged task completion rate
to achieve significant saving in worker effort. The efficiency values achieved by
MT and MW under the step mood–productivity mapping function are roughly
the same. OWRS achieves over 20% higher efficiency compared to MT and MW.

Figure 4(d) shows the performance landscape of MT, MW and OWRS as a
percentage of ME under different control parameter settings. By varying the ϕ
value in OWRS, we can achieve different trade-offs between worker effort conservation and task completion rate. The performance of OWRS varies between
spending 78% of the ME effort output and achieving 99% of the ME task completion rate, to spending 20% of the ME effort output and achieving 54% of the
ME task completion rate. Under MT and MW, the performance varies between
spending 90% of the ME effort output and achieving 99% of the ME task completion rate, and spending 0% of the ME effort output and achieving 0% of
the ME task completion rate. OWRS consistently and significantly outperforms
both MT and MW, conserving significant worker effort while achieving high
task completion rates. Overall (Fig. 4(e)), OWRS achieves superlinear collective
productivity, spending 44% of effort to complete 73% of the tasks on average.
Whereas, ME, MT and MW are only able achieve linear collective productivity.

Figures 4(f)–(j) and (k)–(o) show the simulation results under the logarithmic
and exponential mood–productivity mapping functions, respectively. The overall
trend of relative performance of the four approaches are similar to those under
the step mood–productivity mapping function. The overall efficiency achieved by
OWRS under these two mapping functions are slightly lower than that under the
step mapping function. Nevertheless, OWRS is able to handle the changes in the
mapping function to achieves superlinear collective productivity and outperform
other approaches significantly.

In the cases in which θ_1 and θ_2 are set to 1, indicating that workers are
unwilling to work under any mood condition, the worker effort output and the
task completion rates achieved by both MT and MW are 0 under all experimental settings. This is expected as in MT and MW, mood is directly used
to control effort output. However, under OWRS, mood becomes one of the situational factors considered by the approach. By dynamically minimizing the
joint objective function (Eq. 3), even when ϕ is set to 100 indicating that workers place very high emphasis on effort conservation in general, OWRS is able to

maintain a long-term average effort output of between 20% to 30% taking advantage of favourable working conditions whenever possible to achieve average task completion rates of more than 50% under all three mood–productivity mapping functions investigated. By making workers sacrifice the completion of some tasks under situations which do not support high productivity, OWRS achieves significant savings in worker effort and superlinear collective productivity in view of the reduced total effort output, putting the *Parrondo's paradox* [7,22] into practice on a large scale.

5   Conclusions and Future Work

In this paper, we extended the AC framework to propose OWRS. It takes into account information on a worker's mood, current workload and desire to rest to produce dynamic work-rest schedules. By leveraging on favourable situations to work and rest when the situations do not support high productivity, OWRS minimizes crowdsourcing workers' effort output while maximizing collective productivity. Compared to AC, OWRS is able to handle more diverse mood–productivity mapping functions including *step, logarithmic* and *exponential* forms. As it is a distributed approach, the time complexity of OWRS is $O(1)$. It can be implemented as a personal scheduling agent to advise each worker on when (and how much) to work and when to rest in order to balance personal wellbeing with collective productivity goals in an optimized manner. Simulations based on a large-scale real-world dataset demonstrate that OWRS outperforms three baseline scheduling approaches. It significantly conserves the collective effort output while making the smallest sacrifice in terms of task completion rates, thereby achieving superlinear collective productivity. OWRS establishes a framework which accommodates workers' heterogeneity to enhance their experience and productivity, enabling AI-power algorithmic management to collaborate with workers in a more human-centric way.

In subsequent research, we will combine OWRS with explainable AI techniques [6,12] to provide stronger persuasion in order to enhance worker compliance to the recommended schedules.

Acknowledgements. This research is supported by the National Research Foundation, Prime Minister's Office, Singapore under its IDM Futures Funding Initiative; Nanyang Technological University, Nanyang Assistant Professorship (NAP); the Association for Crowd Science and Engineering (ACE); Shandong Province Major Scientific and Technological Special Project (2015ZDJQ01002); the Shandong Peninsular (Jinan) National Innovation Showcase Development Project; the Shandong Province Independent Innovation Special Project (2013CXC30201); National Key R&D Program (No. 2016YFB1000602); NSFC (No. 61572295); SDNSF (No. ZR2017ZB0420).

References

1. Bechtold, S.E., Sumners, D.L.: Optimal work-rest scheduling with exponential work-rate decay. Manage. Sci. **34**(4), 547–552 (1988)
2. Bigham, J.P.: What's hot in crowdsourcing and human computation. In: Proceedings of the 29th AAAI Conference on Artificial Intelligence (AAAI 2015), pp. 4318–4319 (2015)
3. Dababneh, A.J., Swanson, N., Shell, R.L.: Impact of added rest breaks on the productivity and well being of workers. Ergonomics **44**(2), 164–174 (2001)
4. Dai, P., Lin, C.H., Mausam, Weld, D.S.: POMDP-based control of workflows for crowdsourcing. Artif. Intell. **202**, 52–85 (2013)
5. Doan, A., Ramakrishnan, R., Halevy, A.Y.: Crowdsourcing systems on the world-wide web. Commun. ACM **54**(4), 86–96 (2011)
6. Fan, X., Toni, F.: On computing explanations in argumentation. In: Proceedings of the 29th AAAI Conference on Artificial Intelligence (AAAI-2015), pp. 1496–1502 (2015)
7. Harmer, G.P., Abbott, D.: Losing strategies can win by Parrondo's paradox. Nature **402**, 864 (1999)
8. Hersey, R.: Worker's emotions in the shop and home: a study of individual workers from the psychological and physiological standpoint. University of Pennsylvania Press (1932)
9. Heymann, P., Garcia-Molina, H.: Turkalytics: Analytics for human computation. In: Proceedings of the 20th International Conference on World Wide Web (WWW 2011), pp. 477–486 (2011)
10. Ho, C.J., Vaughan, J.W.: Online task assignment in crowdsourcing markets. In: Proceedings of the 26th AAAI Conference on Artificial Intelligence (AAAI-2012), pp. 45–51 (2012)
11. Iodice, P., Ferrante, C., Brunetti, L., Cabib, S., Protasi, F., Walton, M.E., Pezzulo, G.: Fatigue modulates dopamine availability and promotes flexible choice reversals during decision making. Sci. Rep. **7**, Article no. 535 (2017). https://doi.org/10.1038/s41598-017-00561-6
12. Kang, Y., Tan, A.H., Miao, C.: An adaptive computational model for personalized persuasion. In: Proceedings of the 24th International Conference on Artificial Intelligence (IJCAI 2015), pp. 61–67 (2015)
13. Kobayashi, M., Ishihara, T., Itoko, T., Takagi, H., Asakawa, C.: Age-based task specialization for crowdsourced proofreading. In: Stephanidis, C., Antona, M. (eds.) UAHCI 2013. LNCS, vol. 8010, pp. 104–112. Springer, Heidelberg (2013). https://doi.org/10.1007/978-3-642-39191-0_12
14. Lee, M.K., Kusbit, D., Metsky, E., Dabbish, L.: Working with machines: the impact of algorithmic and data-driven management on human workers. In: Proceedings of the 33rd Annual ACM Conference on Human Factors in Computing Systems (CHI 2015), pp. 1603–1612 (2015)
15. Mao, A., Kamar, E., Chen, Y., Horvitz, E., Schwamb, M.E., Lintott, C.J., Smith, A.M.: Volunteering versus work for pay: Incentives and tradeoffs in crowdsourcing. In: Proceedings of the 1st AAAI Conference on Human Computation and Crowdsourcing (HCOMP-2013), pp. 94–102 (2013)
16. Miao, C., Yu, H., Shen, Z., Leung, C.: Balancing quality and budget considerations in mobile crowdsourcing. Decis. Support Syst. **90**, 56–64 (2016)
17. Morris, R.R., Dontcheva, M., Gerber, E.M.: Priming for better performance in microtask crowdsourcing environments. IEEE Internet Comput. **16**(5), 13–19 (2012)

18. Nath, S., Narayanaswamy, B.M.: Productive output in hierarchical crowdsourcing. In: Proceedings of the 13th International Conference on Autonomous Agents and Multi-agent Systems (AAMAS 2014), pp. 469–476 (2014)

19. Neely, M.J.: Stochastic Network Optimization with Application to Communication and Queueing Systems. Morgan and Claypool Publishers, San Rafael (2010)

20. Oswald, A.J., Proto, E., Sgroi, D.: Happiness and productivity. Technical report, the Institute for the Study of Labor (IZA), Bonn, Germany (2009)

21. Pan, L., Meng, X., Shen, Z., Yu, H.: A reputation pattern for service oriented computing. In: Proceedings of the 7th International Conference on Information, Communications and Signal Processing (ICICS 2009) (2009)

22. Shu, J.J., Wang, Q.W.: Beyond parrondo's paradox. Sci. Rep. **4**(4244) (2014). https://doi.org/10.1038/srep04244

23. Sornette, D., Maillart, T., Ghezzi, G.: How much is the whole really more than the sum of its parts? 1 ⊞ 1 = 2.5: Superlinear productivity in collective group actions. PLoS ONE **9**(8) (2014). https://doi.org/10.1371/journal.pone.0103023

24. Wu, Q., Han, X., Yu, H., Shen, Z., Miao, C.: The innovative application of learning companions in virtual singapura. In: Proceedings of the 2013 International Conference on Autonomous Agents and Multi-agent Systems (AAMAS 2013), pp. 1171–1172 (2013)

25. Yu, H., Miao, C., Liu, S., Pan, Z., Khalid, N., Shen, Z., Leung, C.: Productive aging through intelligent personalized crowdsourcing. In: Proceedings of the 30th AAAI Conference on Artificial Intelligence (AAAI-2016), pp. 4405–4406 (2016)

26. Yu, H., Cai, Y., Shen, Z., Tao, X., Miao, C.: Agents as intelligent user interfaces for the net generation. In: Proceedings of the 15th International Conference on Intelligent User Interfaces (IUI 2010), pp. 429–430 (2010)

27. Yu, H., Miao, C., An, B., Leung, C., Lesser, V.R.: A reputation management model for resource constrained trustee agents. In: Proceedings of the 23rd International Joint Conference on Artificial Intelligence (IJCAI 2013), pp. 418–424 (2013)

28. Yu, H., Miao, C., An, B., Shen, Z., Leung, C.: Reputation-aware task allocation for human trustees. In: Proceedings of the 13th International Conference on Autonomous Agents and Multi-Agent Systems (AAMAS 2014), pp. 357–364 (2014)

29. Yu, H., Miao, C., Chen, Y., Fauvel, S., Li, X., Lesser, V.R.: Algorithmic management for improving collective productivity in crowdsourcing. Sci. Rep. **7**(12541) (2017) https://doi.org/10.1038/s41598-017-12757-x

30. Yu, H., Miao, C., Leung, C., Chen, Y., Fauvel, S., Lesser, V.R., Yang, Q.: Mitigating herding in hierarchical crowdsourcing networks. Sci. Rep. **6**(4) (2016) https://doi.org/10.1038/s41598-016-0011-6

31. Yu, H., Miao, C., Shen, Z., Leung, C., Chen, Y., Yang, Q.: Efficient task subdelegation for crowdsourcing. In: Proceedings of the 29th AAAI Conference on Artificial Intelligence (AAAI-2015), pp. 1305–1311 (2015)

32. Yu, H., Shen, Z., Fauvel, S., Cui, L.: Efficient scheduling in crowdsourcing based on workers' mood. In: Proceedings of the 2nd IEEE International Conference on Agents (ICA 2017), pp. 121–126 (2017)

A Model for Information Behavior Research on Social Live Streaming Services (SLSSs)

Franziska Zimmer$^{(\boxtimes)}$, Katrin Scheibe$^{(\boxtimes)}$, and Wolfgang G. Stock

Department of Information Science, Heinrich Heine University Düsseldorf,
Düsseldorf, Germany
{franziska.zimmer,katrin.scheibe}@hhu.de,
stock@phil.hhu.de

Abstract. Social live streaming services (SLSSs) are synchronous social media, which combine Live-TV with elements of Social Networking Services (SNSs) including a backchannel from the viewer to the streamer and among the viewers. Important research questions are: Why do people in their roles as producers, consumers and participants use SLSSs? What are their motives? How do they look for gratifications, and how will they obtain them? The aim of this article is to develop a heuristic theoretical model for the scientific description, analysis and explanation of users' information behavior on SLSSs in order to gain better understanding of the communication patterns in real-time social media. Our theoretical framework makes use of the classical Lasswell formula of communication, the Uses and Gratifications theory of media usage as well as the Self-Determination theory. Albeit we constructed the model for understanding user behavior on SLSSs it is (with small changes) suitable for all kinds of social media.

Keywords: Social media · Social live streaming service (SLSS)
Live video · Information behavior · Users · Lasswell formula
Uses and gratifications theory · Motivation · Self-determination theory

1 Introduction: Information Behavior on SLSSs

On social media, users act as prosumers [1], i.e. both as producers of content as well as its consumers [2]. Produsage [3] amalgamates active production and passive consumption of user-generated content. Social Networking Services (SNSs) are social media in which prosumers communicate among each other with the help of texts, images and videos. Typical examples of SNSs are Facebook[1] and Vkontakte[2] (in Russia and neighboring countries) [4]. Facebook-like SNSs are asynchronous [5], which means that the producer of the content acts at another time than the consumer of that content. There is (or, better, there can be) a closed circle of communication, if the consumer reacts to the producer's content by commenting, liking or sharing the information and if the producer gains knowledge about those acts. As the communicative acts take place in the passage of (maybe long) times, communication happens

---

[1] https://www.facebook.com/.
[2] https://vk.com/.

© Springer International Publishing AG, part of Springer Nature 2018
G. Meiselwitz (Ed.): SCSM 2018, LNCS 10914, pp. 429–448, 2018.
https://doi.org/10.1007/978-3-319-91485-5_33

slowly. With the advent of social live streaming services (SLSSs) [6], communication between all involved prosumers comes to real-time meeting.

Social live streaming services such as, for instance, Periscope[3], Ustream[4], YouNow[5], YouTube Live[6], Facebook Live[7], Instagram Live[8], niconico[9] (in Japan), Yi-ZhiBo[10], Xiandanjia[11], Yingke[12], YY Live[13] (all in China) or – for broadcasting e-sports resp. drawing – Twitch[14] and Picarto[15] are social media, which combine Live-TV with elements of Social Networking Services including a backchannel from the viewer to the streamer and among the viewers. SLSSs allow their users to broadcast their programs to everyone who wants to watch, all over the world. The streamers film either with the camera of a mobile phone or with the aid of a webcam. Some SLSSs employ elements of gamification (especially YouNow; Fig. 1) to motivate their users to continuously apply the service. The main feature of SLSSs is the simultaneity of the communication, as everything happens in real time. Summing up, SLSSs are social media platforms with the following characteristics:

- SLSSs are synchronous,
- they allow users to broadcast their own program in real-time (as in TV),
- users employ their own mobile devices (e.g., smartphones, tablets) or their PCs and webcams for broadcasting,
- the audience is able to interact with the broadcasting users and with other viewers via chats,
- some SLSSs support gamification mechanics and dynamics, and
- the audience may reward the performers with, e.g., points, badges, or money.

What information behavior do prosumers exhibit on SLSSs? In line with Bates [7] and Wilson [8] we define "information behavior" as all human behavior with relation to information and knowledge (HII: Human Information Interaction) or to information and communication technologies (HCI: Human Computer Interaction). As information behavior on SLSSs is always computer-mediated, it is subject of HCI by definition. Fisher, Erdelez and McKechnie [9, p. xix] conceptualize information behavior "as including how people need, seek, manage, give, and use information in different contexts." Similarly, Robson and Robinson [10, p. 169] propose an information

---

[3] https://www.pscp.tv/.

[4] http://www.ustream.tv/.

[5] https://www.younow.com/.

[6] https://www.youtube.com/channel/UC4R8DWoMoI7CAwX8_LjQHig.

[7] https://live.fb.com/.

[8] https://help.instagram.com/292478487812558.

[9] http://www.nicovideo.jp/.

[10] https://www.yizhibo.com/.

[11] http://www.xiandanjia.com/.

[12] https://www.inke.cn/.

[13] http://www.yy.com/.

[14] https://www.twitch.tv/.

[15] https://picarto.tv/.

Fig. 1. Live stream on YouNow (split screen of broadcaster and one participant)

behavior model that "takes into account not just the information seeker but also the communicator or information provider."

The aim of this article is to develop a heuristic model for the scientific description, analysis and explanation of prosumers' information behavior on social live streaming services in order to gain better understanding of the communication patterns in real-time social media. Why do some people broadcast live – even slices of their own lives similar to Truman Burbank (in the movie "The Truman Show" – please, have in mind that Truman was not fond of it in the end) [11]? Why do people watch such streams? And why do some people participate in the communication by giving "hearts," comments or gifts? What are the users' motivations as producers, consumers and participants? Does gamification help to motivate prosumers to use an SLSS and to lock the users to the service?

In order prepare the ground to answer these questions empirically, we are going to develop a theoretical framework for understanding information behavior on SLSSs building on the classical Lasswell formula of communication, the Uses and Gratifications theory of media usage and the psychological theory of Self-Determination.

2 The Lasswell Formula of Communication

In a first rough differentiation, we distinguish between sender-centered and audience-centered communication models as in SLSSs both aspects, namely senders (i.e., broadcasters) and viewers (i.e., audience) are equally important. One of the classical sender-centered models is the theory of Harold D. Lasswell. Lasswell [12, p. 37] introduces the following questions:

- Who
- Says What
- In Which Channel
- To Whom
- With What Effect?

This five questions lead to five sub-disciplines of communication science, which however can definitely cooperate. "Scholars who study the 'who,' the communicator, look into the factors that initiate and guide the act of communication. We call this subdivision of the field of research *control analysis*. Specialists who focus upon the 'says what' engage in *content analysis*. Those who look primarily at the radio, press, film and other channels of communication are doing *media analysis*. When the principal concern is with the persons reached by the media, we speak of *audience analysis*. If the question is the impact upon audience, the problem is *effect analysis*" [12, p. 37]. Braddock [13] adds two further questions:

- What circumstances?
- What purpose?

The extended Lasswell formula reads as follows: "WHO says WHAT to WHOM under WHAT CIRCUMSTANCES through WHAT MEDIUM for WHAT PURPOSE with WHAT EFFECT" [13, p. 88]. In terms of Braddock, the *Who* is the communicator (in SLSSs, the broadcaster); he/she acts as an individual or as a representative of a group. The *What* is the message with the two inseparable aspects of content and presentation (in SLSSs, the content and style of the broadcast). To *Whom* asks for the audience and its characteristics (in SLSSs, the viewers of the broadcast). *What Circumstances* concerning SLSSs analyzes the environment of the broadcasting act in terms of time and setting. One question Braddock asks is, "Was the communicator in a position in which he was forced or expected to say something? Was he acting as a spokesman for a group, being paid to say something, being influenced by superiors …?" [13, p. 91]. In SLSSs, for describing and analyzing influencers or micro-celebrities, for instance, is seems to be very important to realize the exact setting of the broadcast. *What medium* includes questions on the information channel. "Does it imply a mass or selected audience? … Can the audience see the communicator's expression, gestures, dress, and so on? … Does the medium require oversimplification of the message?", Braddock asks [13, p. 92]. *What Purpose* means the communicator's motives to communicate. What does the communicator want the audience to do? Interestingly, Braddock only makes mention of the motives of the communicator, but not of the audience. Concerning SLSSs, we have to study the broadcasters' motivations to produce and to perform a live stream. The last aspect *What Effect* analyzes the outcomes of the entire communication process for the audience. What are the reactions of the SLSSs' audience when they consume a live video? The entire process is a linear sequence of building blocks of the communication [14, p. 14] (Fig. 2). This representation of communication is quite similar to the signal transmission process as described by Shannon [15]; however, we have to add the component of knowledge as the content of information to Shannon's more technologically oriented model [16, p. 36].

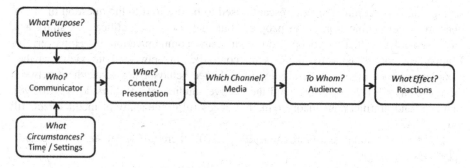

Fig. 2. The communication process according to Lasswell and Braddock

Concerning Lasswell, circuits of communication are one-way or two-way, depending upon the degree of reciprocity between communicators and audience [12]. Given that there is an audience for a live stream, on SLSSs communication is always two-way and thus reciprocal. This means that the roles between communicator and audience can change; the original communicator will become audience when the viewer reacts to his/her message (or his/her appearance), and that the original audience may communicate with the original communicator and – what is very special to SLSSs – with the rest of the audience.

The Lasswell formula found application in the study of communication via social media and user-generated content. Wenxiu [17] transferred the model from "classical" mass media (as TV) to "new media" (as Internet and its services); Jan [18] developed an analytical framework for research on enterprise social media; and Auer [19] discussed political motivated content leading to influence the audience while using social media. However, scholars were able to identify problems in Lasswell's model of the five (with Braddock's additions, seven) W-questions if applied to understand communication on social media. "Lasswell's '5W' model lacks feedback, and the role of communicator and audience is rigid, the interactivity of new media provides the communication study lots of new inspiration," Wenxiu states [17, p. 249]. Similarly, Jan questions the linear relationship in the Lasswell formula. Instead, new media "are likely to reshuffle the dynamics of existing and future communication processes" [18, p. 11]. Therefore, we turn our attention to audience-centered communication models.

3   Uses and Gratifications

In sender-centered communication models, the starting point is the active communicator, and the audience remains more or less passive. By contrast, audience-centered models place special emphasis on the receivers. In the Uses and Gratifications approach by Elihu Katz and colleagues [20, 21] "the 'needs' of the individual form the starting point" [14, p. 135]. "Audience activity is central to uses-and-gratifications research, and communication motives are key components of audience activity," Papacharissi and Rubin define [22, p. 175]. Klapper [23, p. 525] works out the difference between the Lasswell formula and Uses and Gratifications theory clearly: "We are fond of saying that mass communication research used to be directed to the question of 'What does mass communication do to people?' but that uses-and-gratifications study ask, more sensibly, 'What do people do with mass communication?'." However, for Klapper (as for us as well), there is no contradiction between both views as they complement each other. "A valid view of audience behavior lies somewhere between these extremes [of the "passive" and the "active" audience]," Rubin adds [24, p. 98]. The uses and gratifications theory remains successful in the study of media effects till today [25–27].

For Katz, Blumler, and Gurevitch [20, p. 510], there are seven steps in the audience's media usage:

- the social and psychological origins of
- needs, which generate
- expectations of
- the mass media or other sources, which lead to
- different patterns of media exposure, resulting in
- need gratifications and
- other consequences (including unintended ones).

Researchers may study the audience's needs and then uncover how they are gratified by the media. Or vice versa, we observe gratifications and look for the needs that are gratified. Of course, researchers may analyze the social and psychological origins of audience expectation and gratifications as well [20, p. 510]. It is important to realize that the need gratification and the media choice are strongly dependent on the single concrete audience member – so we have to be very careful when generalizing audience data into hypothesis or theories. All media compete with other sources of gratification, e.g. with face-to-face contacts with other people or with playing with toys [20, p. 511]. We tried to visualize the media usage steps in Fig. 3.

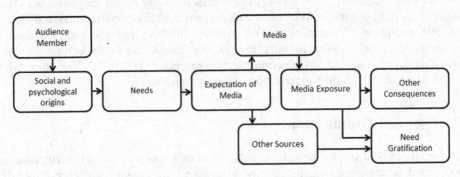

Fig. 3. The communication process according to Katz, Blumler, and Gurevitch

We have to distinguish between two aspects of gratifications. When there are needs and expectations of media, audience members seek for gratification. After media exposure, they find gratifications. According to Palmgreen et al. [28] there is a feedback loop between the gratifications sought and the gratifications obtained (Fig. 4). "Over time we would expect such feedback processes to result in a rather strong relationship between sought and obtained measures for a particular gratification as long as the seeking behavior is reinforced" [28, p. 164]. It is possible, for instance, to seek for information; however, after media exposure not obtaining the anticipated information but finding entertainment. For Palmgreen et al. [28, p. 164] it is an important research question, "are the *dimensions* of gratifications of gratification *sought* from a particular medium, content type, or program the same as the *dimensions* of gratifications perceived to be *obtained*?"

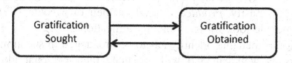

Fig. 4. Gratification sought and obtained according to Palmgreen, Wenner, and Rayburn II

What types of gratification are identified in communication science? Blumler and Katz [29] and later MacQuail [30] found four basic dimensions of gratifications:

- information,
- personal identity,
- entertainment as well as
- integration and social interaction.

Information means the motive of finding knowledge; *personal identity* is related to our motive to define our identity; *entertainment* comprises escaping from problems, relaxing, filling time, or sexual arousal; *social interaction* is the motive to interact with other people.

However, mediated social interaction is different from "normal" social interaction. Basic elements of "normal" social interaction are bodily contact, proximity, orientation, gesture, facial expression, eye-movement as well as verbal and non-verbal aspects of speech [31]. An audience member, say of a TV show and its actors, does sometimes not only passively consume the show, but he or she builds up a kind of relationship to an actor, presenter or celebrity [32]. The "media figure" is not aware of a relationship, but only the spectator. Horton and Wohl [33] name such mediated social interactions "parasocial interactions." The crucial difference between social interactions and parasocial interactions "lies in the lack of effective reciprocity," establishing an "intimacy at a distance" [33, p. 215] as bodily contact is not given as well. In mediated contexts, the fourth dimension of gratification is *parasocial interaction*.

There are other classifications of basic gratifications. It is possible to sort all motivations into the five categories of cognitive, affective, personal integrative and social integrative motivations as well as into the motive of tension release [20].

Use of social media is not the same as use of TV [34]. With Joinson [35] we can distinguish between content gratifications (gratifications based upon the content of the watched media), process gratifications, which are based on the actual experience of using the media, and – which is new on the Web – social gratifications (or gratifications in a "social environment" [36]), based upon communication and integration.

TV-oriented communication research predominantly studied the behavior of the audience. On social media, one can figure out three different roles of people [37, p. 15]:

- consumers (or lurkers),
- participators and
- producers.

For Shao [37] *consumers* only receive content and do not contribute to the communication processes. *Participators* do not initiate content communication, but they "take advantage of user-generated sites to interact with the content and other human beings" [37, p. 18]. Lastly, *producers* "produce their own contents" [37, p. 18]. All three groups of people look for and obtain gratifications.

The dimension of personal identity has to be broadened in social media. Producers and participators as well can and will articulate their personal identity. They are actively acting and presenting themselves. So we should better speak about "self-presentation" in this dimension.

Uses and Gratifications theory found and finds many diverse applications in social media research. There are numerous studies about uses and gratifications on, for instance, SNSs as Facebook [35, 38], MySpace [39] or professional SNSs [40], microblogging services as Twitter [41] or Weibo [42] and sharing services as Instagram [43] or YouTube [44]. Additionally, there are lots of papers on Uses and Gratifications concerning other Web services, e.g. messengers as WhatsApp [45] or WeChat [46].

4 Self-determination: Needs, Motivations and (Maybe) Flow

In the Uses and Gratifications theory, there is an important building block of the model called "needs." Without human needs there will be no media production or media reception. To clarify the function of needs, we turn towards Self-Determination theory, originated by Ryan and Deci [47–49]. Self-Determination theory – as a theory of human motivation [50] – seems to be an ideal psychological addition to communication science approaches as the Lasswell formula and the Uses and Gratifications theory [51].

For Ryan and Deci [47, p. 10], needs are defined as "nutrients that are essential for growth, integrity, and well-being." There are three basic needs; *autonomy* is the need to self-regulate own experiences and actions; *competence* is the need to act efficiently and to master all important life contexts; finally, *relatedness* concerns feeling socially connected, belonging to a community and feeling significant among others [47, pp. 10–11]. Needs lead to motivations. Motivations concerns "what 'moves' people to action" [47, p. 13], they "energize" and give directions to human behavior. Ryan and Deci distinguish between three regulatory styles of motivation, namely

- intrinsic motivation,
- extrinsic motivation (integration, identification, introjection, external regulation), and
- amotivation [52].

Motivations are determined either by the acting persons themselves (self-determination) or by other circumstances (nonself-determination). Those circumstances include other people as loci of causality or nonpersonal loci. There are no clear boundaries between self-determination and nonself-determination, but a continuum of the degree of (non-)self-determination of motivations. Apart from intrinsic motivations (which are always caused by internal aspects, i.e. by the acting persons' selves), motivations are caused by a combination of internal and external aspects (Fig. 5).

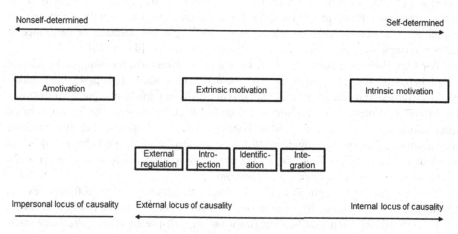

Fig. 5. Human motivations in the self-determination continuum following Ryan and Deci

Intrinsic motivation "involves people freely engaging in activities that they find interesting, that provide novelty and optimal challenge" [49, p. 235]. Intrinsically motivated human behavior is performed out of the acting person's interests, for which the primary rewards are the confirmation of one's own competence or simply enjoyment. Following Vallerand [53], there are three types of intrinsic motivations, namely to cause an activity for pure joy, to understand something new, and to arrive at an accomplishment (for the process to create something new).

While intrinsic motivations are autonomous by definition, extrinsic motivations vary "widely in the degree to which they are controlled versus autonomous" [47, p. 14]. Deci and Ryan distinguish four kinds of extrinsic motivations. *Integration* means the internalization of extrinsic causes. "When regulations are integrated people will have fully accepted them by bringing them into harmony or coherence with other aspects of their values and identity" [49, p. 236]. The external aspects of motivation are fully transformed into self-regulation resulting in self-determined extrinsic motivation. *Identification* is the adoption of external regulations for a special purpose. "For example, if people identified with the importance of exercising regularly for their own health and well-being, they would exercise more volitionally" [49, p. 236]. While integration and identification are more related to a person's self-determination, introjection and external regulation are more caused by external and nonself-determined aspects. *Introjection* entails the actor's taking in external regulations and the reaction on contingent consequences of those regulations. Prototypical examples of introjection are actions leading to the person's pride or refraining actions which could end in the person's feelings of shame or guilt. The "classic case" of extrinsic motivation is the *external regulation* "in which people's behavior is controlled by specific external contingencies" [49, p. 236]. People behave to get rewards or to avoid negative consequences – independently of their own preferences or norms.

Intrinsic as well as all types of extrinsic motivations represent personally caused actions (internally caused by the actor or externally by others). Amotivation lacks such external personal aspects. Amotivation leads to non-activity, i.e. to refrain from an action. Deci and Ryan identified three forms of amotivation, namely a felt lack of competence, a lack of interest, relevance or value, and the defiance or resistance to influence (which can also be seen as motivated nonaction) [47, p. 16].

For Max Weber, "action is 'social' insofar as its subjective meaning takes account of the behavior of others and is thereby oriented in its course" [54, p. 4]. Information behavior on social media in general and also in SLSSs in particular is partly oriented on the behavior of others. So it is social action. For social actions, there are no intrinsic motivations causing the information behavior on social media, because intrinsic motivations are autonomous and therefore not oriented towards the behavior of others as a matter of principle. Of course, all not explicitly socially oriented actions on social media may be caused by intrinsic motivations.

If we combine the sought as well as the found gratifications adopted from Uses and Gratifications theory with the motivations identified from Self-Determination theory, we have to ask for each gratification (information, self-presentation, parasocial interaction, and entertainment), what type of intrinsic or extrinsic motivation (or amotivation) is realized in the concrete situation.

However, there is another form of motivation found on some information systems, namely motives driven by gamification [55]. The implementation of game mechanics and dynamics in non-game contexts is used to increase one's engagement, motivation and activity. Therefore, Web information systems and mobile applications already utilize it [56]. For Deterding [57], gamification means designing for motivation to adopt and to repeatedly use an information system. Typical gamification elements for producers and for participants on SLSSs are, for example, getting fans (becoming a fan), getting positive comments (giving comments), receiving gifts (making gifts), getting subscribers (becoming a subscriber) as well as getting shares and likes (giving shares and likes). For consumers (as well as for the other two groups) gamification elements as rankings, levels, coins or badges are possibly motivating.

Sometimes, media producers, participants or consumers may experience total absorption in an activity as well as the non-self-conscious enjoyment of it. Csikszentmihalyi [58] called such an optimal experience "flow." Flow can be reached if there is an optimal challenge. "Too much challenge relative to a person's skills leads to anxiety and disengagement, whereas too little leads to boredom and alienation" [49, p. 260]. Flow theory is compatible with Self-Determination theory, as Deci and Ryan state, "(w)hen people experience flow, their activity is said to be autotelic, which means that the purpose of the activity is the activity itself, and we often spoke of flow as the prototype of intrinsically motivated activity" [49, p. 260].

5 The Model

Paraphrasing Klapper [23], on social media, with user-generated content, we study what people do with social media *and* what social media do with the people. In "classical" communication science as of Lasswell or Katz we spoke of the "audience" of media (especially of TV), with the advent of the internet and especially of the social media the term changed to "users" [34, p. 505]. Nowadays, on social media, audience members are *users*. However, in SLSSs, they are very special users. As SLSSs combine (live) TV with social media the people working with SLSSs are both, TV audience and social media users. In this way the different research lines of communication science (studying the audience) on the one side and of HCI research (studying users) on the other side get together.

The special position of SLSSs in the field of all social media is mirrored by the kind of social interaction. While all parasocial interaction (on TV as well as on social media, but not on SLSSs) lacks proximity, and bodily contact, and (in many cases) reciprocity, SLSS-mediated interaction may be reciprocal, if the producer and the participant communicate live via the system. Of course, also on SLSSs there is no spatial proximity; however, there is a temporal proximity as all happens real-time (Table 1). So, SLSS-mediated interaction is closer to "normal" social interaction than to parasocial interaction.

Table 1. Forms of interactions

	Reciprocity	Spatial proximity	Temporal proximity	Bodily contact
Social interaction	Yes	Yes	Yes	Yes
Parasocial interaction	Sometimes	No	No	No
SLSS-mediated interaction	Yes	No	Yes	No

In Fig. 6 (flowchart) and 7 (database model), users search for intrinsically or extrinsically motivated gratifications through entertainment, information, SLSS-mediated social interaction, and self-presentation or through gamification elements (insofar provided by the service). In our flowchart, user X is a producer and user Y a participant. If another user Y' stops at the building block "information reception," she or he is a consumer.

In Fig. 6, you will find the building blocks of the Lasswell/Braddock model (from Fig. 2) on the way from user X (Who?), his/her motives (What purpose?) and the circumstances of the communication act (What circumstances?) via the production of a publication, i.e. a live video, with its particular content (What?), distributed on an SLSS (Which Channel?) to the information receivers Y or Y' (To Whom?) and their reactions (What Effect?). The Uses and Gratifications approach (from Fig. 3) starts with user Y resp. Y' (Audience member). The model stresses the importance of user characteristics as the circumstances, demographic data and the user's role/s in the entire

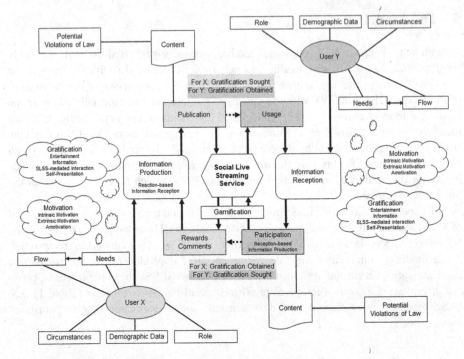

Fig. 6. Building blocks of information behavior on SLSSs as a flowchart

communication process (Social and psychological origins) as well as the user's motivation (Needs).

Former experiences with SLSSs (and other social media) lead to certain Expectation of Media and the use of SLSSs (Media exposure) or alternatively the use of Other Sources, leading to the satisfaction of the motivation (Need gratification) and to Other consequences (e.g., changing leisure behavior due to stark SLSS usage). Of course, our model also considers the relation between gratifications sought and gratifications obtained (from Fig. 4).

SLSSs offer spectators the possibilities to (only passively) view a video (as user Y' does) or to (actively) participate as a guest in a live stream (as in Fig. 1), to write chat messages and to reward the streamer (so does user Y). Producers (as our user X) interact with the viewers in real-time through their publications, i.e. their live streams. Additionally, they read the chat messages of the participants (now acting as consumers) and can instantly respond in their stream (now acting as participants). Gratification is sought by streaming (user X), by watching (user Y') as well as by commenting and donating (user Y); gratification is found by the satisfaction of one's motives. Therefore, user X will be satisfied when viewers react to the streams and reward him or her; user Y will be satisfied when the streamer or other viewers react to the comments resp. to the rewards; finally, user Y' will be satisfied when she or he receives the wanted live video.

Producers as well as participants distribute content. It is possible that this content is "contaminated" with juridical problems. If music is playing while broadcasting, this could be a copyright infringement. If the video shows other people without their written permission, say on a street, this is an object of personality rights violation.

Finally, users have characteristics, most important their roles (as producers, participants, or consumers), their gender, their nationality, and their age resp. their generational cohort. Additionally, we have to consider specific circumstances, in which the users behave and which influences the users' information behavior, e.g. their position as opinion leaders [59], influencers [60], micro-celebrities [61] or as stakeholders of companies, political parties or religious associations. On SLSSs, only producers are identifiable, while consumers and participants may stay anonymous. Also on other social media, users can decide whether they want to act identifiable or to remain anonymous. However, on some specific social media, e.g. Jodel[16] or the closed down service Yik Yak, users are always anonymous. As anonymity has impact on the information behavior of the users, the model has to pay attention to this differentiation.

The aim of the entity-relationship model [62] corresponding to our flowchart, is to describe the inter-related information of our specific domain of knowledge, the information behavior of SLSSs' users (Fig. 7). This way, we are able to generate a database which can hold lots of data for easy access and future analyses. The entity 'Consumer' is in relation to the entity 'Social Live Streaming Service,' since we want to analyze which user interacts with which kind of SLSSs. It would have been possible to attach the entity 'Social Live Streaming Service' to the 'Live-Stream' entity as an attribute, but we wish to gain insight into the gamification elements that influence the user and this relation would be lost, since a game mechanic is not attached to a stream per se, but to the SLSS. Analyzing the demographic data of a 'Consumer' (age, gender, country), and if the user was anonymously online, for which we chose Boolean values, is also our goal. Since flow is a state which is experienced or not we likewise chose Boolean values to answer if the user was immersed in the stream. We want to further inquire the motivational aspects and the different forms of gratifications a user searches for and in return receives, so we added the attributes 'Motivation,' 'Gratification sought' and 'Gratification obtained' which will later be filled with the applicable norm entries corresponding to the Self-determination and the Uses and Gratifications theory by the researchers.

Since the participant is a consumer who also writes comments, thus interacts with the stream, and a producer is a special form of participant who creates live-streams and can be an influencer, micro-celebrity or other kind of personality, we choose to implement 'is-a'-relationships to better distinguish between the three kinds of users of an SLSS. Further research could focus on the comments the participant writes during a live-stream, so we save the content of the comments and what kind of user writes them.

We are interested in different aspects of a live-stream; its duration, number of viewers, the number of likes, as well as the content, and therefore added attributes for them. Furthermore, since each live-stream can display several breaches of the law, we implemented them as an entity for easier analyses.

---

[16] https://jodel.com/de/.

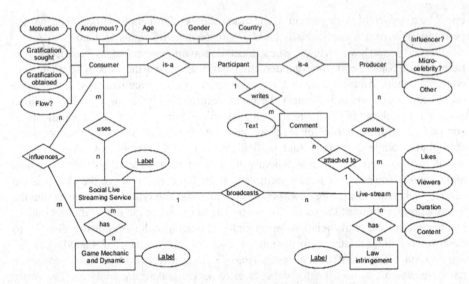

Fig. 7. Information behavior on SLSSs as entity-relationship model

6 Measuring Information Behavior on SLSSs

All building blocks depicted in Figs. 6 and 7 are measurable. However, how? How can we arrive at sound data on information behavior on SLSSs? As HCI and communication researchers, we are able to use four different sources for data gathering, namely (1st) log files of the information systems, (2nd) performing experiments with probands in controlled test situations, (3rd) asking the users (by quantitative surveys and qualitative interviews), and, finally, (4th) systematic observations of the streams.

As log files' data are not very meaningful (it is impossible to get data on users' motives and streams' content) we can only use this source for some basic data as, for instance, for describing few user characteristics (e.g., country of dial-up) and some technical interaction data (e.g., time spent on the SLSS) [63, 64].

If we are able to identify certain dimensions of information behavior it is possible to analyze those variables in a test situation. Wilk, Wulffert, and Effelsberg [65], for instance, performed experiments on the behavior of SLSS viewers concerning the effects of gamification elements. With experiments, it is possible to arrive at precise data on single variables; however, as the data were collected in a controlled situation, they are not necessarily the same outside the laboratory. To cover real-life information behavior, researchers have to go into the "wilderness," i.e. they have to study (real) users when they interact with (real) information services.

With Katz et al. [20, p. 511] we believe, that "people are sufficiently self-aware to be able to report their interests and motives in particular cases." Therefore SLSS researcher may conduct online surveys with SLSS users as participants and perform qualitative interviews with prototypical users. Indeed, many empirical investigations on SLSSs made use of surveys, for instance, concerning Twitch [66–69] or general SLSSs as YouNow [6, 11, 70], the former services Qik [71] and Meerkat [72], the Chinese

SLSSs Douyu TV and YY Live [73] as well as live streams via SNSs [74]. Additionally, there are surveys on the information behavior of special user groups as, e.g., teens [75].

As a further methodological approach, researchers will realize systematic observations of a sufficiently large amount of streams and evaluate the videos' content as well as the streamers' motivations (insofar they are observable) via content analysis [76, 77]. If there are open questions during the observations the researchers are able to ask both, the streamers as well as the viewers, during the live-sessions. We found some content analyses on SLSSs, e.g. analyzing user-generated content on YouNow [78] or comparing content on Periscope, Ustream and YouNow [79, 80]. Additionally, one may statistically analyze the word distribution of the chats [81].

The main perspective of our theoretical models is the *user* of an SLSS and her or his *information behavior* concerning the services and their environment. If we turn the angle to an evaluation of the *information service*, we are able to identity additional theoretical models which help us to structure research tasks on SLSSs. Until today many different evaluation models (among others, TAM, TAM 2, TAM 3, UTAUT and MATH) have been developed to measure the quality and acceptance of these services. However, those models consider only subareas of the whole concept that represents an information service. As a holistic and comprehensive approach, the Information Service Evaluation (ISE) model [82] studies five dimensions that influence adoption, use, impact and diffusion of the information service: information service quality, information user (here is the contact area with the information behavior models), information acceptance, information environment and time. All these aspects have a great impact on the final grading and of the success (or failure) of the service. Concerning SLSSs, ISE found application in an evaluation of the general SLSS YouNow [11, 70].

Table 2. Theoretical foundations of SLSS studies

	Examples
Lasswell formula	[83]
Uses and gratifications theory	[66–68]
Self-determination theory	[69]
Theory of flow	[74]
Information service evaluation model	[11, 70]

7 Conclusion

The aim of our article was the development of a heuristic theoretical model for the scientific description, analysis and explanation of users' information behavior on SLSSs in order to gain better understanding of the communication patterns in real-time social media. Our theoretical framework makes use of the classical Lasswell formula of communication, the Uses and Gratifications theory of media usage as well as the Self-Determination theory (including the theory of Flow). Additionally, we shortly mentioned the ISE model to consolidate studies on the information service. In the current literature on SLSSs, indeed all addressed theories and models could be identified (Table 2); however, in most cases only one of the theories.

The combined model of information behavior on SLSSs (as shown in Figs. 6 and 7), if necessary connected with the Information Service Evaluation model, has two main advantages:

- it addresses all building blocks of the entire communication process on SLSSs (leading scientists simply not to forget important research aspects),
- it establishes a common basis for comparable results from different research teams.

Albeit we constructed the model for understanding user behavior on SLSSs it is (with small changes) suitable for all kinds of social media. As other social media are mostly asynchronous, there is no direct backchannel from the audience to the producers. However, the building blocks of the research model will be the same for most of the known social media services.

Acknowledgement. We would like to thank Kaja J. Fietkiewicz and Isabelle Dorsch for valuable improvement suggestions on earlier versions of this paper.

References

1. Toffler, A.: The Third Wave. Morrow, New York (1980)
2. Linde, F., Stock, W.G.: Information Markets: A Strategic Guideline for the I-Commerce. De Gruyter Saur, Berlin, New York (2011)
3. Bruns, A.: Blogs, Wikipedia, Second Life, and Beyond: From Production to Produsage. Peter Lang, New York (2008)
4. Knautz, K., Baran, K.S. (eds.): Facets of Facebook: Use and Users. De Gruyter Saur, Berlin, Boston (2016)
5. Khoo, C.S.G.: Issues in Information Behavior on Social Media. Libres **24**(2), 75–96 (2014)
6. Scheibe, K., Fietkiewicz, K.J., Stock, W.G.: Information behavior on social live streaming services. JISTaP **4**(2), 6–20 (2016)
7. Bates, M.J.: Information behavior. In: Bates, M.J., Maack, M.N. (eds.) Encyclopedia of Library and Information Sciences, 3rd edn., pp. 2381–2391. CRC, New York (2009)
8. Wilson, T.D.: Human information behavior. Inf. Sci. **3**(2), 49–55 (2000)
9. Fisher, K.E., Erdelez, S., McKechnie, L.: Preface. In: Fisher, K.E., S., Erdelez, K.E., McKechnie, L. (eds.) Theories of Information Behavior, pp. xix–xxii. Information Today, Medford (2005)
10. Robson, A., Robinson, L.: Building on models of information behavior: linking information seeking and communication. J. Doc. **69**(2), 169–193 (2013)
11. Fietkiewicz, K.J., Scheibe, K.: Good morning … good afternoon, good evening and good night: adoption, usage and impact of the social live streaming platform YouNow. In: Proceedings of the 3rd International Conference on Library and Information Science, Sapporo, Japan, 23–25 August 2017, pp. 92–115. International Business Academics Consortium, Taipeh (2017)
12. Lasswell, H.D.: The structure and function of communication in society. In: Bryson, L. (ed.) The Communication of Ideas, pp. 37–51. Harper & Brothers, New York (1948)
13. Braddock, R.: An extension of the 'Lasswell Formula'. J. Commun. **8**(2), 88–93 (1958)

14. McQuail, D., Windahl, S.: Communication Models for the Study of Mass Communication, 2nd edn. Longman, London (1993)
15. Shannon, C.E.: A mathematical theory of communication. Bell Syst. Tech. J. **27**(379–423), 623–656 (1948)
16. Stock, W.G., Stock, M.: Handbook of Information Science. De Gruyter Saur, Berlin, Boston (2013)
17. Wenxiu, P.: Analysis of new media communication based on Lasswell's "5 W" model. J. Educ. Soc. Res. **5**(3), 245–250 (2015)
18. Jan, M.V.P.: Knowing what is said on enterprise social media: towards the development of an analytical communication framework. Revista Internacional de Relaciones Públicas **7**(13), 5–22 (2017)
19. Auer, M.R.: The policy sciences of social media. Pol. Stud. J. **39**(4), 709–736 (2011)
20. Katz, E., Blumler, J.G., Gurevitch, M.: Uses and gratifications research. Public Opin. Quart. **37**(4), 509–523 (1973)
21. Blumler, J.G., Katz, E.: The Uses of Mass Communications: Current Perspectives on Gratifications Research. Sage, Newbury Park (1973)
22. Papacharissi, Z., Rubin, A.M.: Predictors of internet use. J. Broadcast. Electr. Media **44**(2), 175–196 (2000)
23. Klapper, J.T.: Mass communication research: An old road resurveyed. Public Opin. Q. **27**(4), 515–527 (1963)
24. Rubin, A.M.: Audience activity and media use. Commun. Monographs **60**, 98–103 (1993)
25. Rubin, A.M.: Uses-and-gratifications perspective on media effects. In: Bryant, J., Oliver, M. B. (eds.) Media Effects. Advances in Theory and Research, pp. 165–182. Taylor & Francis, Hoboken (2008)
26. Rubin, A.M.: Uses and gratifications: an evolving perspective of media effects. In: Nabi, R. L., Oliver, M.B. (eds.) The SAGE Handbook of Media Processes and Effects, pp. 147–159. Sage, Los Angeles (2009)
27. Ruggiero, T.E.: Uses and gratifications theory in the 21st century. Mass Commun. Soc. **3**(1), 3–37 (2000)
28. Palmgreen, P., Wenner, L.A., Rayburn II, J.D.: Relations between gratifications sought and obtained: a study of television news. Commun. Res. **7**(2), 161–192 (1980)
29. Blumler, J.G., Katz, E.: The Uses of Mass Communications: Current Perspectives on Gratifications Research. Sage, Minneapolis (1975)
30. McQuail, D.: Mass Communication Theory: An Introduction. Sage, London (1983)
31. Argyle, M.: Social Interaction. Aldine Transaction, New Brunswig (1969)
32. Giles, D.C.: Parasocial interaction: a review of the literature and a model for future research. Mediapsychology **4**, 279–305 (2003)
33. Horton, D., Wohl, R.R.: Mass communication and para-social interaction. Psychiatry J. Study Interpers. Proc. **19**(3), 215–229 (1956)
34. Sundar, S.S., Limperos, A.M.: Uses and grats 2.0: new gratifications for new media. J. Broadcast. Electr. Media **57**(4), 504–525 (2013)
35. Joinson, A.N.: 'Looking at', 'looking up' or 'keeping up with' people? Motives and uses of Facebook. In: CHI 2008, Proceedings of the SIGCHI Conference on Human Factors in Computing Systems, pp. 1027–1036. ACM, New York (2008)
36. Stafford, T.F., Stafford, M.R., Schkade, L.L.: Determining uses and gratifications for the internet. Decis. Sci. **35**(2), 259–288 (2004)
37. Shao, G.: Understanding the appeal of user-generated media: a uses and gratification perspective. Internet Res. **19**(1), 7–25 (2009)

38. Park, N., Kee, K.F., Valenzuela, S.: Being immersed in social networking environments: Facebook groups, uses and gratifications, and social outcomes. Cyberpsychol. Behav. **12**(6), 729–733 (2009)
39. Raacke, J., Bonds-Raake, J.: MySpace and Facebook: applying the uses and gratifications theory to exploring friend-networking sites. Cyberpsychol. Behav. **11**(2), 169–174 (2008)
40. Grissa, K.: What "uses and gratifications" theory can tell us about using professional networking sites (E.G. LinkedIn, Viadeo, Xing, SkilledAfricans, Plaxo...). In: Jallouli, R., Zaïane, Osmar R., Bach Tobji, M.A., Srarfi Tabbane, R., Nijholt, A. (eds.) ICDEc 2017. LNBIP, vol. 290, pp. 15–28. Springer, Cham (2017). https://doi.org/10.1007/978-3-319-62737-3_2
41. Chen, G.M.: Tweet this: a uses and gratifications perspective on how active Twitter use gratifies a need to connect with others. Comput. Hum. Behav. **27**(2), 755–762 (2011)
42. Mo, R., Leung, L.: Exploring the roles of narcissism, uses of, and gratifications from microblogs on affinity-seeking and social capital. Asian J. Soc. Psychol. **18**(2), 152–162 (2015)
43. Sheldon, P., Bryant, K.: Instagram: Motives for its use and relationship to narcissism and contextual age. Comput. Hum. Behav. **58**, 89–97 (2016)
44. Hanson, G., Haridakis, P.: YouTube users watching and sharing the new: a uses and gratifications approach. J. Electr. Publ. **11**(3) (2008)
45. Karapanos, E., Teixeira, P., Gouveia, R.: Need fulfillment and experiences on social media: a case on Facebook and Whatsapp. Comput. Hum. Behav. **55**, 888–897 (2016)
46. Chai, J.X., Fan, K.K.: User satisfaction and user loyalty in mobile SNSs: WeChat in China. In: International Conference on Applied System Innovation, IEEE ICASI 2016, Article No. 7539936. IEEE, Washington, DC (2016)
47. Ryan, R.M., Deci, E.L.: Self-Determination Theory. Basic Psychological Needs in Motivation, Development, and Wellness. Guilford Press, New York, London (2017)
48. Ryan, R.M., Deci, E.L.: Self-determination theory and the facilitation of intrinsic motivation, social development, and well-being. Am. Psychol. **55**(1), 68–78 (2000)
49. Deci, E.L., Ryan, R.M.: The "what" and "why" of goal pursuits: human needs and the self-determination of behavior. Psychol. Inq. **11**(4), 227–268 (2000)
50. Deci, E.L., Ryan, R.M.: Self-determination theory: a macrotheory of human motivation, development, and health. Can. Psychol. **49**(3), 182–185 (2008)
51. Ang, C.-S., Abu Talib, M., Tan, K.-A., Tan, J.-P., Yaacob, S.N.: Understanding computer-mediated communication attributes and life satisfaction from the perspectives of uses and gratifications and self-determination. Comput. Hum. Behav. **49**, 20–29 (2015)
52. Ryan, R.M., Deci, E.L.: Intrinsic and extrinsic motivations: classic definitions and new directions. Contemp. Educ. Psychol. **25**, 54–67 (2000)
53. Vallerand, R.J.: Toward a hierarchical model of intrinsic and extrinsic motivation. Adv. Exp. Soc. Psychol. **29**, 271–360 (1997)
54. Weber, M.: Economy and Society. University of California Press, Berkeley, Los Angeles (1978)
55. Zichermann, G., Cunningham, C.: Gamification by Design: Implementing Game Mechanics in Web and Mobile Apps. O'Reilly, Sebastopol (2011)
56. Deterding, S., Dixon, D., Khaled, R., Nacke, L.: From game design elements to game-fulness: defining "gamification". In: Proceedings of the 15th International Academic MindTrek Conference: Envisioning Future Media Environments, pp. 9–15. ACM, New York (2011)
57. Deterding, S.: Gamification: designing for motivation. Interactions **19**(4), 14–17 (2012)
58. Csikszentmihalyi, M.: Flow: The Psychology of Optimal Experience. Harper & Row, New York (1990)

59. Mohr, I.: Going viral: an analysis of YouTube videos. J. Market. Dev. Competitiveness **8**(3), 43–48 (2014)
60. Uzunoğlu, E., Misci Kip, S.: Brand communication through digital influencers: leveraging blogger engagement. Int. J. Inf. Man **34**, 592–602 (2014)
61. Khamis, S., Ang, L., Welling, R.: Self branding, 'micro celebrity' and the rise of social media influencers. Celebrity Stud. **8**(2), 191–208 (2017)
62. Chen, P.: The entity-relationship model – toward a unified view of data. ACM Trans. Database Syst. **1**(1), 9–36 (1976)
63. Stohr, D., Li, T., Wilk, S., Santini, S., Effelsberg, W.: An analysis of the YouNow live streaming platform. In: 40th Local Computer Networks Conference Workshops, Clearwater Beach, FL, 26–29 October 2015, pp. 673–679. IEEE, Washington, DC (2015)
64. Favario, L., Siekkinen, M., Masala, E.: Mobile live streaming: insights from the Periscope service. In: 2016 IEEE 18th International Workshop on Multimedia Signal Processing, Article No. 7813395. IEEE, Washington, DC (2016)
65. Wilk, S., Wulffert, D., Effelsberg, W.: On influencing mobile live video broadcasting users. In: 2015 IEEE International Symposium on Multimedia, pp. 403–406. IEEE, Washington, DC (2016)
66. Sjöblom, M., Törhönen, M., Hamari, J., Macey, J.: Content structure is king: an empirical study on gratifications, game genres and content type on Twitch. Comput. Hum. Behav. **73**, 161–171 (2017)
67. Sjöblom, M., Hamari, J.: Why do people watch others play video games? An empirical study on the motivations of Twitch users. Comput. Hum. Behav. **75**, 985–996 (2017)
68. Gros, D., Wanner, B., Hackenholt, A., Zawadzki, P., Knautz, K.: World of streaming. Motivation and gratification on twitch. In: Meiselwitz, G. (ed.) SCSM 2017. LNCS, vol. 10282, pp. 44–57. Springer, Cham (2017). https://doi.org/10.1007/978-3-319-58559-8_5
69. Zhao, Q., Chen, C.-D., Cheng, H.-W., Wang, J.-L.: Determinants of live streamers' continuance broadcasting intentions on Twitch: a self-determination theory perspective. Telematics Inform. (2017). https://doi.org/10.1016/j.tele.2017.12.018
70. Friedländer, M.B.: And action! Live in front of the camera: an evaluation of the social live streaming service YouNow. Int. J. Inf. Commun. Technol. Hum. Dev. **9**(1), 15–33 (2017)
71. Dougherty, A.: Live-streaming mobile video: production as civic engagement. In: Proceedings of the 13th International Conference on Human Computer Interaction with Mobile Devices and Services, pp. 425–434. ACM, New York (2011)
72. Tang, J.C., Venolia, G., Inkpen, K.M.: Meerkat and Periscope: i stream, you stream, apps stream for live streams. In: Proceedings of the 2016 CHI Conference on Human Factors in Computing Systems, pp. 4770–4780. ACM, New York (2016)
73. Hu, M., Zhang, M., Wang, Y.: Why do audiences choose to keep watching on live video streaming platforms? An explanation of dual identification framework. Comput. Hum. Behav. **75**, 594–606 (2017)
74. Chen, C.-C., Lin, Y.-C.: What drives live-stream usage intention? The perspectives of flow, entertainment, social interaction, and endorsement. Telematics Inform. **35**(1), 293–303 (2018)
75. Lottridge, D., Bentley, F., Wheeler, M., Lee, J., Cheung, J., Ong, K., Rowley, C.: Third-wave livestreaming: Teens' long form selfie. In: Proceedings of the 19th International Conference on Human-Computer Interaction with Mobile Devices and Services MobileHCI 2017, Article No. 20. ACM, New York (2017)
76. Krippendorff, K.: Content Analysis: An Introduction to Its Methodology, 3rd edn. Sage, Thousand Oaks (2012)
77. Recktenwald, D.: Toward a transcription and analysis of live streaming on Twitch. J. Pragmat. **115**, 68–81 (2017)

78. Honka, A., Frommelius, N., Mehlem, A., Tolles, J.N., Fietkiewicz, K.J.: How safe is YouNow? An empirical study on possible law infringements in Germany and the United States. J. MacroTrends Soc. Sci. **1**(1), 1–17 (2015)

79. Friedländer, M.B.: Streamer motives and user-generated content on social live-streaming services. JISTaP **5**(1), 65–84 (2017)

80. Zimmer, F., Fietkiewicz, K.J., Stock, W.G.: Law infringements in social live streaming services. In: Tryfonas, T. (ed.) HAS 2017. LNCS, vol. 10292, pp. 567–585. Springer, Cham (2017). https://doi.org/10.1007/978-3-319-58460-7_40

81. Zhang, C., Liu, J., Ma, M., Sun, L., Li, B.: Seeker: topic-aware viewing pattern prediction in crowdsourced interactive live streaming. In: Proceedings of the 27th ACM Workshop on Network and Operating Systems Support for Digital Audio and Video, NOSSDAV 2017, pp. 25–30. ACM, New York (2017)

82. Schumann, L., Stock, W.G.: The Information Service Evaluation (ISE) model. Webology **11**(1) (2014). Article No. 115

83. Scheibe, K., Zimmer, F., Fietkiewicz, K.J.: Das Informationsverhalten von Streamern und Zuschauern bei Social Live-Streaming Diensten am Fallbeispiel YouNow [Information behavior of streamers and viewers on social live-streaming services: YouNow as a case study]. Information – Wissenschaft & Praxis **68**(5–6), 352–364 (2017)

Author Index

Printed in the United States
By Bookmasters